Series:

Genome Designing of Crops

Series editor: Chittaranjan Kole, Kolkata, West Bengal, India

1. Biofortification for Nutrient-Rich Crops
   Monika Garg, Saloni Sharma and *Apoorv Tiwari (Editors), 2025*
2. Omics and System Biology Approaches for Delivering Better Cereals
   Dinesh Kumar Saini and *Chittaranjan Kole (Editors), 2025*
3. The Pomegranate Genome
   Zhaohe Yuan and *Julian Bartual (Editors), 2025*

Genome Designing of Crops
Series editor: Chittaranjan Kole

The Pomegranate Genome

Editors
Zhaohe Yuan
Full Professor
College of Forestry, Nanjing Forestry University
Nanjing, Jiangsu, China

Julian Bartual
Director
Conselleria de Agricultura, Agua, Ganaderia y Pesca
Agricultural Experiment Station of Elche
Elche, Spain

CRC Press is an imprint of the
Taylor & Francis Group, an **informa** business
A SCIENCE PUBLISHERS BOOK

First edition published 2025
by CRC Press
2385 NW Executive Center Drive, Suite 320, Boca Raton FL 33431

and by CRC Press
4 Park Square, Milton Park, Abingdon, Oxon, OX14 4RN

CRC Press is an imprint of Taylor & Francis Group, LLC

Library of Congress Cataloging-in-Publication Data (applied for)

ISBN: 978-1-032-71966-5 (hbk)
ISBN: 978-1-032-75824-4 (pbk)
ISBN: 978-1-003-47577-4 (ebk)

DOI: 10.1201/9781003475774

Typeset in Times New Roman
by Prime Publishing Services

Series Preface

Food security has been the major challenge in crop improvement programs and remains the primary concern, specifically in the changing scenarios of increasing population. Plant breeders have started emphasizing nutrition security in the recent past considering the huge population suffering from malnutrition. However, trust and dependence on plant nutraceuticals are increasing more and more now leading to the addition of another challenge for the plant breeders. Fast depletion of fossil fuel has led to the exploration for alternative energy sources and thus identified crop and farm residues as a promising option. Finally, degradation of natural resources pertinent to agriculture and environment pollution, in general, have added a critical challenge to the plant breeders. The present mission of crop improvement should focus on addressing food, health, nutrition, energy, and environment (FHNEE) security and that invoke for revolutionary changes in the concept and strategies of crop improvement.

Fortunately, the science of crop improvement has made outstanding advancement during the last four decades. The techniques of molecular breeding developed in the 1990s have facilitated mapping, monitoring, and manipulating economic genes and thus developing crop varieties within shorter time and with relatively more precision. At the same time, the genetic engineering technique has contributed enormously by resulting in a large number of varieties improved by insertion of alien genes. Whole genome sequencing of rice, the leading staple food crop of the world, in the beginning of this century was immediately followed by sequencing genomes of an array of crops. Later on, functional genomics studies complemented the structural genomics outcomes and thereby facilitated various genomics-assisted breeding techniques making plant breeding more rapid and more precise. High-throughput genotyping and phenotyping in crops have thus led to the development of the advanced technique often termed as 'next-generation breeding'. The most recent technique of gene editing has enormously reinforced crop improvement leading to the targeted tailoring in crop plants. The modern plant breeding technique can safely and rightly be termed as 'genome designing' that could pave the way for genetic tailoring of crop genomes according to the breeding objectives and changing agroclimatic scenarios.

The book series on 'Genome Designing in Crops' includes book volumes containing deliberations on enumeration of crop genomes and their genetic improvement. I am thankful to the editors of these volumes and the authors of their chapters. I also thank all the staff of the publisher for their assistance in the entire process of editing this book series.

Kolkata **Chittaranjan Kole**

This Book Series is

Dedicated to

Prof. Sudhansu Bhusan Chattopadhyay

The Founder Vice Chancellor of Bidhan Chandra Krishi Viswavidyalaya (Agricultural University), Mohanpur, West Bengal, India

The best teacher, administrator and human being I have seen in my life.

Preface

In the realm of botanical exploration and genomic inquiry, it is our distinct privilege to present this seminal work, *The Pomegranate Genome*. As professors deeply engaged in the pursuit of knowledge at Nanjing Forestry University, Nanjing, China and Estaci6n Experimental Agraria de Elche. (Alicante), Spain, the journey into the intricate world of *Punica granatum* L. has been both intellectually enriching and academically rewarding. This comprehensive volume embarks on a meticulous examination of the pomegranate, a fruit with a storied past, an ancient cultivation history, and a promising future as an emerging and profitable crop.

Pomegranate, a member of the Lythraceae family native to central Asia, brings together historical significance and contemporary agricultural importance. Its appeal lies not only in the visually striking bright red appearance, but also being a rich source of bioactive compounds for human health found in both its peel and aril. Through the collective efforts of researchers and scholars, this book unfolds as a repository of knowledge, dissecting the various dimensions of the pomegranate genome.

The exploration commences with a thorough review on the botany and taxonomy of the pomegranate. This primary understanding serves as a springboard for an in-depth examination of the pomegranate germplasm's conservation and utilization strategies. The genomic landscape studied in subsequent chapters shows, the intricacies of the pomegranate genome, unveiling a tapestry of molecular information that defines this unique fruit.

Genomic analyses, showcased in "Multiomics in Pomegranate", underscore the complexity and interconnectedness of various molecular components. The examination extends to the chloroplast genome of pomegranate, providing insights into the genetic elements that contribute to its distinctive features. Molecular genomics takes center stage in the exploration of floral organ differentiation and seed coat development, unraveling the genetic determinants that orchestrate these critical aspects of the pomegranate's life cycle.

The pomegranate is not merely a visual delight; it encapsulates the molecular basis for fruit coloration, as explored in detail in subsequent chapters. Fruit development is a multifaceted process, and this volume dissects the mechanisms behind crucial phenomena such as fruit cracking and salt tolerance. The latter, in particular, aligns with contemporary agricultural challenges, offering valuable insights into the adaptability of the pomegranate.

The core of this book is based in the genomic exploration of various gene families, each of wich contributes a unique chapter to the pomegranate's biological narrative. From the investigation of MAPK and MAPKK gene families to the in-depth study of MIKC-Type MADS-Box Genes and the comprehensive identification and expression analysis of the TALE gene family, each segment unveils the genomic intricacies that govern specific aspects of pomegranate biology.

This collective endeavor represents the culmination of rigorous research, collaboration, and dedication of experts across the field of genomics. We sincerely hope that *The Pomegranate Genome* will serve as a beacon, and guide students, researchers, and enthusiasts into the nuanced world of pomegranate genomics. We would be delighted if this comprehensive work could be used as a reference book, we traverse the pages of this comprehensive work, may it ignite curiosity, inspire further inquiry, and pave the way for future advances in the knowledge of the genetic tapestry of this extraordinary fruit of this extraordinary fruit's genetic tapestry.

Zhaohe Yuan
Julián Bartual

Contents

List of Contributors

Candelario Mondragón Jacobo
Facultad de Ciencias Naturales, Universidad Autónoma de Querétaro, Santiago de Querétaro, México

Cuiyu Liu
Research Institute of Subtropical Forestry, Chinese Academy of Forestry, Hangzhou, China

Elena Zuriaga Garcia
Instituto Valenciano de Investigaciones Agrarias, Conselleria de Agricultura, Ganadería y Pesca, Moncada, Valencia, Spain

Gaihua Qin
Anhui Academy of Agricultural Sciences, Hefei, China

Jiangli Shi
College of Horticulture, Henan Agricultural University, Zhengzhou, China

Jiyu Li
Anhui Academy of Agricultural Sciences, Hefei, China

Julián Bartual Martos
Estación Experimental Agraria de Elche, Conselleria de Agricultura, Ganadería y Pesca, Elche, Alicante, Spain

María Luisa Badenes
Instituto Valenciano de Investigaciones Agrarias, Conselleria de Agricultura, Ganadería y Pesca, Moncada, Valencia, Spain

Ming Yan
College of Forestry, Nanjing Forestry University, Nanjing, China

Ran Wan
College of Horticulture, Henan Agricultural University, Zhengzhou, China

Sha Wang
College of Forestry, Nanjing Forestry University, Nanjing, China

Wei Zhang
Anhui Academy of Agricultural Sciences, Hefei, China

Xinhui Zhang
College of Forestry, Nanjing Forestry University, Nanjing, China

Xueqing Zhao
College of Forestry, Nanjing Forestry University, Nanjing, China

Yuan Ren
College of Forestry, Nanjing Forestry University, Nanjing, China

Yujie Zhao
College of Horticulture, Henan Agricultural University, Zhengzhou, China

Yuying Wang
College of Forestry, Nanjing Forestry University, Nanjing, China

Zhaohe Yuan
College of Forestry, Nanjing Forestry University, Nanjing, China

List of Abbreviations

2,4-D	2,4-dichlorophenoxyacetic acid
ABA	Abscisic acid
ABCG	ATP-binding cassette-G subfamily
ABF	ABA-responsive element binding factor
ABI	ABA-insensitive
ABRE	*cis*-acting element involved in the abscisic acid responsiveness
ACA	Ca^{2+}-ATPases
ADP	Adenosine-diphosphate
AFLP	Amplified fragment length polymorphism
AGL	AGAMOUS-Like
AKT	Affinity potassium transporter
AMP	Adenosine-monophosphate
ANR	Anthocyanidin reductase
ANS	Anthocyanin synthase
ANT	AINTEGUMENTA
AOMT	Anthocyanin O-methyltransferases
AP1	APETALA1
AP2/ERF	Apetala2/ethylene response factor
APG IV	Angiosperm Phylogeny Group IV
APX	Ascorbate peroxidase
AQP	Aquaporin
ARE	*cis*-acting regulatory element essential for the anaerobic induction
ARF	Auxin Response Factor
ATH1	*A. thaliana* homeobox 1
ATHB8	*A. thaliana* homeobox 8
ATP	Adenosine-triphosphate
atpB	Adenosine Triphosphate synthase subunit B
AuxRR-core	*cis*-acting regulatory element involved in auxin responsiveness
B	Boron
BDG	BODYGUARD
BELL	BEL1-like homeobox
bHLH	Basic helix-loop-helix
bp	Base pair

BR	Brassinosteroids
Br^-	Bromide ion
BS	B-SISTER
bZIP	Basic Leucine Zipper
Ca	Calcium
CAD	Cinnamyl alcohol dehydrogenase
CaM/CML	Calcium-binding proteins
CAT-box	*cis*-acting regulatory element related to meristem expression
CAX	Cation/H^+ antiporter
CBS	Cystathionine beta synthase
CCGTCC-box	*cis*-acting regulatory element related to meristem specific activation
CCX	Cation/calcium exchanger
cDNA	Complementary DNA
CDPK	Calcium-dependent protein kinase
CDS	Coding Sequence
CesA	Cellulose synthase
CGTCA-motif	methyl jasmonate response element
CHI	Chalcone isomerase
CHS	Chalcone synthase
CIPK	CBL-interacting protein kinase
Cl^-	Chloride ion
CLC	Chloride channel
CM	Cuticular membrane
CNGC	Cyclic nucleotide-gated channel
Cp	Chloroplast
CRC	CRABS CLAW
CSCs	Cellulose synthase complexes
Cy	Cyanidin
DAI	Days after initiation
DAP	Days after pollination
DEG	Differentially expressed gene
DELs	Differentially expressed lncRNAs
DFR	Dihydroflavonol 4-reductase
DNA	Deoxyribonucleic Acid
Dp	Delphinidin
DREB/CBF	Dehydration-responsive element-binding protein/C-repeat binding factor
E/Glu	Glutamate
EBGs	Early biosynthetic genes

EC	Electrical conductivity
ER	Endoplasmic reticulum
ERE	ethylene-responsive element
F3′5′H	Flavonoid 3′5′-hydroxylase
F3′H	Flavonoid 3′-hydroxylase
F3H	Flavanone 3-hydroxylase
F5H	Ferulate-5-hydroxylase
FB	Fruit non-cracking under bagging
FC	Fruit cracking
FIL	FILAMENTOUS FLOWER
FLC	FLOWERING LOCUS C
FNC	Fruit non-cracking
FUL	FRUITFULL
GA_3	Gibberellin
GAE	Gallic Acid Equivalents
GARE-motif	gibberellin-responsive element
GC	Guanine Cytosine
GC-motif	enhancer-like element involved in anoxic specific inducibility
GC-MS	Gas chromatography-mass spectrometry
GCN4_motif	*cis*-regulatory element involved in endosperm expression
GCV	Genotypic coefficient of variation
GLR	Glutamate receptor
GO	Gene ontology
GPAT6	Glycerol-3-Phosphate Acyltransferase
GRAS	GAI, RGA, SCR
GRAVY	Grand average of the hydrophobicity
GSDS	Gene Structure Display Server
GST	Glutathione s-transferase
HAT 1	Homeobox from *Arabidopsis thaliana*
HAT 2	Homeobox from *A. thaliana* 2
HB	Homeobox
HCO_3^-	Bicarbonate ion
HD-ZIP	Homeodomain-leucine zipper
HKT	High-affinity potassium transporter
HLL	HUELLENLOS
HMM	Hidden Markov Models
HOX	Homeobox domain
HPLC	High performance liquid chromatography
HSF	Heat shock transcription factor
HSP	Heat shock protein

I^-	Iodide ion
INO	INNER NO OUTER
IRs	Inverse Repeat Regions
KEGG	Kyoto encyclopedia of genes and genomes
KNAT2	KNOTTED-like from A. thaliana 2
KNOX	KNOTTED-like homeobox
LAR	Leucoanthocyanidin reductase
LBD	Lateral organ boundaries domain
LBGs	Late biosynthetic genes
LD	Luminidependens
LDOX	Leucoanthocyanidin oxidase
LEA	Late embryogenesis abundant protein
LSC	Large Single Copy
LTR	*cis*-acting element involved in low-temperature responsiveness
MAPK	Mitogen-activated protein kinase
MAPKK	Mitogen-activated protein kinase kinase
MAS	Marker-assisted selection
MBS	MYB binding site involved in drought-inducibility
MBSI	MYB binding site involved in flavonoid biosynthetic genes regulation
MBW	MYB-bHLH-WD40
MCScanX	Multiple Collinearity Scan toolkit
miRNAs	microRNAs
ML	Maximum Likelihood
Mv	Malvidin
Mw	Molecular Weight
MYB	V-myb avian myeloblastosis viral oncogene homolog
NAC	No apical meristem, Arabidopsis thaliana AF1/2, cup-shaped cotyledon
NaCl	Sodium chloride
NDX	Nodulin homeobox genes
NGS	Next generation sequencing
NHX	Na^+/H^+ antiporter
NO_3^-	Nitrate ion
NRT	Nitrate transporter
OFP	OVATE family protein
P	*Punica*
P/Pro	Proline
P1-P8	Phase 1-Phase 8
P-box	gibberellin-responsive element

PCR-RFLP	Polymerase Chain Reaction - Restriction Fragment Length Polymorphism
PCV	Phenotypic coefficient of variation
Pg	Pelargonidin
PgCRC	Pomegranate CRC gene
PgINO	Pomegranate INO gene
PgMADSs	Pomegranate MADS-box family genes
PGRs	Plant Growth Regulators
PgYABBY	Pomegranate YABBY family genes
PHD	Plant homeodomain
pI	Theoretical Isoelectric Point
PI	PISTILLATA
Plant TFDB	Plant Transcription Factor Database
PLINC	Plant zinc finger
Pn	Peonidin
PNW	Pacific Northwest
POD	Peroxidase
PP2C	Protein phosphatase 2C
Pt	Petunidin
pv	Pathovars
PYR/PYL	Pyrabatin resistance/ Pyrabatin resistance 1-like
qRT-PCR	Quantitative real-time polymerase chain reaction
QTL	Quantitative trait loci
RAPDS	Random Amplified Polymorphic DNA
rbcL	Ribulose-1,5-bisphosphate carboxylase/oxygenase large subunit
RIL	Recombinant inbred line
RNA	Ribonucleic acid
ROS	Reactive oxygen species
rps16	Ribosomal Protein S16 gene
rRNA	Ribosomal Ribonucleic Acid
RY-element	*cis*-acting element involved in seed-specific regulation
S/Ser	Serine
SAM	Shoot apical meristem
SEP	SEPALLATA
SHP	SHATTERPROOF
SNPs	Single Nucleotide Polymorphism
SnRK	Sucrose non-fermenting-related protein kinase
SO_4^{2-}	Sulfate ion
SOC1	SUPPRESSOR OF OVEREXPRESSION OF CONSTANS1
SOD	Superoxide dismutase

SPL/NZZ	SPOROCYTELESS/NOZZLE
SSC	Small Single Copy
SSRs	Simple Sequence Repeats (microsatellites)
STK	*SEEDSTICK*
STM	Shoot meristemless
SUMO	Small ubiquitin-like modifier
SVP	SHORT VEGETATIVE PHASE
TAC	Total Anthocyanins Content
TALE	Three amino-acid loop extension
TATC-box	*cis*-acting element involved in gibberellin-responsiveness
TCA-element	*cis*-acting element involved in salicylic acid responsiveness
TC-rich	repeats *cis*-acting element involved in defense and stress responsiveness
TF	Transcription factor
TGACG-motif	*cis*-acting regulatory element involved in the MeJA-responsiveness
TGA-element	auxin-responsive element
TMHs	Transmembrane helices
TPM	Transcripts Per Kilobase per Million Mapped reads
tRNA	Transfer Ribonucleic Acid
trnF	Transfer RNA Phenylalanine
trnL	Transfer RNA Leucine
trnQ	Transfer RNA Glutamine
UFGT	UDP-glucose: flavonoid 3-O-glucosyltransferase
UGTs	UDP glycosyltransferases
UPOV	International Union for the Protection of New Varieties of Plants
VHA	H^+-ATPase
WGD	Whole-genome duplication
WOX	Wuschel homeobox
WUN-motif	wound-responsive element
WUS	WUSCHEL
YAB1/2/3/4/5	YABBY1/2/3/4/5
ZM-HOX	*Zea mays* homeobox

1

Botany and Taxonomy of Pomegranate

Jiangli Shi[1*]

This chapter immerses readers in pomegranate's extensive history and global impact, dating back to 4000 and 3000 BCE in Central Asia and Persia. It unveils the narrative of pomegranate's cultivation, exploring its symbolic significance and diverse applications across industries. The chapter scrutinizes the contemporary perception of pomegranate as a "miracle fruit" or "super fruit", highlighting its popularity for nutritional and medicinal attributes. Delving into pomegranate's taxonomy, recent evidence places it in the Lythraceae family, emphasizing vast genetic diversity among cultivars. Ongoing breeding programs aim to enhance traits like larger fruits, softer seeds, and increased yield. The exploration of pomegranate botany meticulously details growth patterns, varieties, and organoleptic characteristics. Cultivar categorization, based on taste profiles, maturity indices, and chemical compositions, is thoroughly investigated. Navigating intricate details of pomegranate anatomy, from roots to flowers, the chapter encompasses factors influencing fruit quality, including aril and husk color. Culminating in a detailed analysis of pomegranate fruit morphology, the chapter scrutinizes size, weight, peel thickness, aril composition, and seed hardness. Emphasizing correlations with crucial quality indicators like antioxidant activity, anthocyanin content, and ascorbic acid levels, this overview establishes the groundwork for a deeper exploration of pomegranate's facets in subsequent chapters.

1. Introduction

Pomegranate (*Punica granatum* L.) has been familiar with human since ancient times (4000 and 3000 BCE), native to Central Asia and Persia (modern-day

[1] College of Horticulture, Henan Agricultural University, 450002, Zhengzhou, China.
[*] Corresponding author: sjli30@henau.edu.cn

Iran). It is an ancient medicinal fruit crop grown worldwide (Holland et al., 2009; Karimi et al., 2017). Later, the cultivation area spread over ancient Egypt, India, Asia Minor, North Africa, and the Mediterranean coast. Now, pomegranate is grown commercially in China, India, Iran, Pakistan, Turkey, United States, Spain, and also in the subtropical areas of South America (Akyıldız et al., 2020; Karimi et al., 2017). Pomegranate blooms as a symbol of life, permanence, well-being, femaleness, fertility, knowledge, immortality, and holiness (Mahdihassan, 1984), its blossom and flower, the shape of the pome, with a pointed or crowned tip, and its shining red color have been emblems of power since ancient times; its innumerable ruby-red seeds hint at fecundity; and the regular geometry of the seeds is a replica of the divine order (Nigro and Spagnoli, 2018).

Nowadays, pomegranate is regarded as a "miracle fruit" or "super fruit", popular with consumers due to higher nutritional values and medicinal purposes (Adhami et al., 2009; Chaves et al., 2020; Hegazi et al., 2021; Karimi et al., 2017; Teixeira da Silva et al., 2013). In fact, pomegranate has been regarded as the most significant medication even in the writings of Hippocrates, Pliny, Soranus, and Dioscorides. The increasing researches on pomegranate as a medicinal and nutritional food source and frequent consumption of fruits promoted the rapid development of the pomegranate industry. Pomegranate is consumed in different forms like fresh juice, fresh fruit, concentrated juice, and products such as teas, jellies, wine, jam, paste and coloring beverages, seed oils pharmaceutical, skin care products, and medicinal products in industry (Holland et al., 2009). Many researches showed that pomegranate fruit, flowers, leaves, bark, and root, even husk extracts, contained bioactive phytochemicals exhibiting potent physiological effects. Among them, pomegranate juice has become more popular because of its diversely documented health benefits such as anticancer action (Bishayeea et al., 2016; Chaves et al., 2020; Khwairakpam et al., 2018), antioxidant properties (Melgarejo-Sánchez et al., 2015; Tehranifar et al., 2010; Verotta et al., 2018), antimicrobial attributes (Salgado et al., 2009; Dahham et al., 2010) and anti-inflammatory action (Verotta et al., 2018).

Mediterranean-type conditions are optimal for pomegranate, arid climates with high exposure to sunlight, annual total precipitation of 170–560 mm, mild winters with minimal temperature not lower than −18°C, and summers without precipitation, long, hot, dry during the last stages of the fruit development (Teixeira da Silva et al., 2013).

2. Taxonomy

Pomegranate (*Punica granatum* L.) was regarded as belonging to the Punicaceae family for a long time, which had a single genus, Punica, including two species, *P. granatum* and *P. protopunica*. *Punica granatum* is derived from "Pomuni granatum", Pomum (apple), granatus (grainy), which translates to "seeded apple" (Teixeira da Silva et al., 2013). However, recent morphological (Graham and Graham, 2014) and molecular (Berger et al., 2016) evidence, as well as the new classification in the APG IV system (Byng et al., 2016), has suggested that pomegranate belongs to Lythraceae family (Yuan et al., 2018; Yan et al., 2019).

Pomegranate consists of more than 800 genotypes which are collected and retained in the Yazd and Saveh germplasm facilities. China Pomegranate Germplasm Resource Garden (Shandong, China) has collected and preserved 291 pomegranate genotypes, and combines pomegranate germplasm resources with pomegranate culture, typical training, and sightseeing. More than 500 cultivars of pomegranate have been named all over the world, and only 50 are grown commercially (Teixeira da Silva et al., 2013). Pomegranate was characterized by high genetic diversity of morphological and biochemical quality traits. A large number of studies have shown significant differences in fruit characteristics such as quality, antioxidant activity, polyphenol, anthocyanin, organic acid, sugar contents among cultivars or clones (Alcaraz-Mármol et al., 2017; Chen et al., 2022; Hasnaoui et al., 2011; Melgarejo-Sánchez et al., 2015; Tzulker et al., 2007). Nowadays, the most interesting characteristics of the breeding programs focus on bigger fruits, larger arils, soft seeds, higher juice yield, red colored rind and arils, and higher yield.

3. Botany

Pomegranate grows as a shrub or small tree that naturally tends to develop multiple trunks with a bushy appearance. It usually grows up to 6–10 m and lives for a long time. Pomegranate trees are mostly deciduous; however, some evergreen cultivars are also reported particularly in India (Holland et al., 2009).

Generally, there are two pomegranate varieties: edible and ornamental types. Edible varieties are divided into three groups according to their organoleptic characteristics and chemical compositions: sour (SV, >2.725 g/100 g), sour-sweet (SSWV, 0.317–2.725 g/100 g), and sweet (SWV, <0.317 g/100 g), based on total organic acids content of 40 Spanish pomegranate cultivars (Melgarejo et al., 2000). Maturity index (MI) was calculated by the ratio to soluble solids content and titratable acidity, as one of the most reliable indicators of pomegranate fruit maturity; it depends on the cultivar and climatic conditions (Shwartz et al., 2009; Fawole and Opara, 2013). Martínez et al. (2006) classified the pomegranate fruits based on their maturity index (MI, total soluble solids/titratable acidity) as sweet (67.1–111), sour-sweet (40.2–66.4), and sour (7.85–21.8). Chen et al. (2022) established MI classification criteria for Chinese pomegranate cultivars, the sweet cultivars for 31–66, sour-sweet cultivars for 16–28, and sour cultivars for 8–13. Alcaraz-Mármol et al. (2017) found that citric acid predominated over malic acid in sour cultivars while the concentration of malic acid was higher in sweet cultivars. Glucose and fructose were the most abundant sugars found in pomegranate fruits. The fructose concentration was greater than glucose (Alcaraz-Mármol et al., 2017; Fawole and Opara, 2013; Mena et al., 2011). However, the cultivar and/or the agroclimatic effect were evident in other studies in which the glucose content was higher than that of fructose (Hasnaoui et al., 2011). Surely, such a chemical composition of the pomegranate depends mainly on the ripening stage and type of cultivar. Generally, pomegranates cultivars are also classified into two groups: soft-seed and hard-seed according to the seed firmness, which is also an important economic trait. There exist other standards of classification. Seeds in soft-seed pomegranate cultivars are easily swallowed and edible, attracting more consumers. However, the disadvantage is spitting seeds for hard-seed pomegranate cultivars, which seriously affects the taste and consumer appreciation. As we know, lignin is a principal structural component of cell walls in higher plants, and the pomegranate seed formation is closely related to lignin biosynthesis and metabolism.

3.1 Root

Pomegranate roots are yellow-brown, and divided into extension roots and absorbing roots. Extension roots grow vertically and horizontally in soil, with the vertical distribution of 20–70 cm and the horizontal distribution extending outward for 100–300 cm of the tree crown under cultivation management conditions, which depend on the soil and environmental conditions and cultural practices (Hiwale et al., 2011). A large number of fibrous roots grow on the main roots, which is easy to produce root tillers.

3.2 Stem

The newly growing stem is smooth, some showing red-brown. Later it becomes gray, with some thorns on branches. One-year branches grow above 100 cm long, and their hardness is often related to cultivar traits. Generally, the branches of soft-seeded pomegranate cultivars are more pliable than those of hard-seeded ones. The bark of the old tree tends to split and detach from the trunk, exhibiting the light-brown wood.

3.3 Bud

Pomegranate buds are located on the tips of the new branches and next to the axils of the leaves, with a short stalk. Rajaei and Yazdanpanah (2015) reported two types of ecodormant scaly buds in cv. Rabbab-e-Neyriz–narrow buds with sharp scales, and large buds with oval-shaped scales.

3.4 Leaf

Leaves have an oblanceolate shape with an obtuse (rounded) apex and an acuminate (tapered) base. Mature leaves are 2–8 cm long simple, bright green, exstipulate (stipules usually absent), short-petioled, glabrous (hairless), and glandular (Holland et al., 2009; Teixeira da Silva et al., 2013).

3.5 Flower

Flowering occurs about one month after buds break on newly developed branches. The petals are obovate, very delicate, and slightly wrinkled. Flowers of pomegranates are bisexual, with well-formed female (stigma, style, ovary) and male (filaments and anthers). Both self- and cross-pollination occurs in the pomegranate. In the northern hemisphere, flowering occurs from April to May and can continue until the end of summer; however, late flowering results in poorly matured fruits as the cold season approaches. There are two pomegranate varieties according to the petal types, double-petals and single-petal. Thick calyx is triangular or vase-shaped, and the upper end of the calyx is often 5–7 lobed. Flower colors are rich, such as red, yellow, white, and pink, but most of which are not 'fertile', or produced fruits are not edible with lower quality.

According to the shapes of calyx, two types of flowers are divided, vase-shaped flowers (inflated calyx) and bell-shaped flowers (sharp calyx) (Figure 1). Vase-shaped flowers have a well-developed ovary and an urceolate calyx. Their stigma is higher than the anthers, thus allowing self-pollination as well as pollination by insects, resulting in a set fruit. Cross-pollination in pomegranate is performed mainly by insects, including beetles, black ants, honey bees, and so on. Bell-shaped flowers that are poorly developed or have no pistil, often drop or fail to set as a fruit.

3.6 Fruit

The fruit ripens 5–8 months after fruit set, depending on the variety. The pomegranate fruit is composed of husk (also called rind, peel), aril, seed, along with an endocarp and a mesocarp (Figure 1C). The multi-ovule chambers (locules) are separated by membranous walls (septum) and a fleshy mesocarp. The chambers are filled with numerous seeds, surrounded by a juicy aril that develops entirely from the outer epidermal cells of the seed.

Pomegranate parts fit for human consumption are aril, which is about 52% of the total weight of fruit and is mainly composed of 78% juice and 22% seeds (Karimi et al., 2017). Juice is a good source of potassium, phosphorous, calcium, iron, manganese, zinc, and copper. Nowadays, aril color shows creamy white (e.g., Sanbai and Baihuayushizi), pink (e.g., Tunisia, Yudazi and Mollar), red (e.g., Red angle), and dark purple (e.g., Zimei) (Figure 1D).

The color of the fruit husk has always been an important appearance quality attribute, which influences consumer behavior, especially, the intensive red color. Husk colors vary largely, creamy white, green yellowish, light red, dark red, etc. Figure 1E shows the phenotypes of 20 pomegranate cultivars from China by the photos taken naturally in the laboratory (Chen et al., 2022). Among them, 'Daqingpitian' exhibited a greenish-yellowish color, "Red angle" showed dark red, and the others appeared from light to intensive red. As we know, pomegranate plants are very sensitive to climatic conditions. It is noted that 'Tunisia' husks from four pomegranate production areas exhibit slight differences in red color due to climate factors and cultivation managements.

Morphological diversity of different pomegranate accessions has been evaluated in different pomegranate producer region. Fruit size is a varietal characteristic that may fluctuate depending on climatic and agricultural conditions

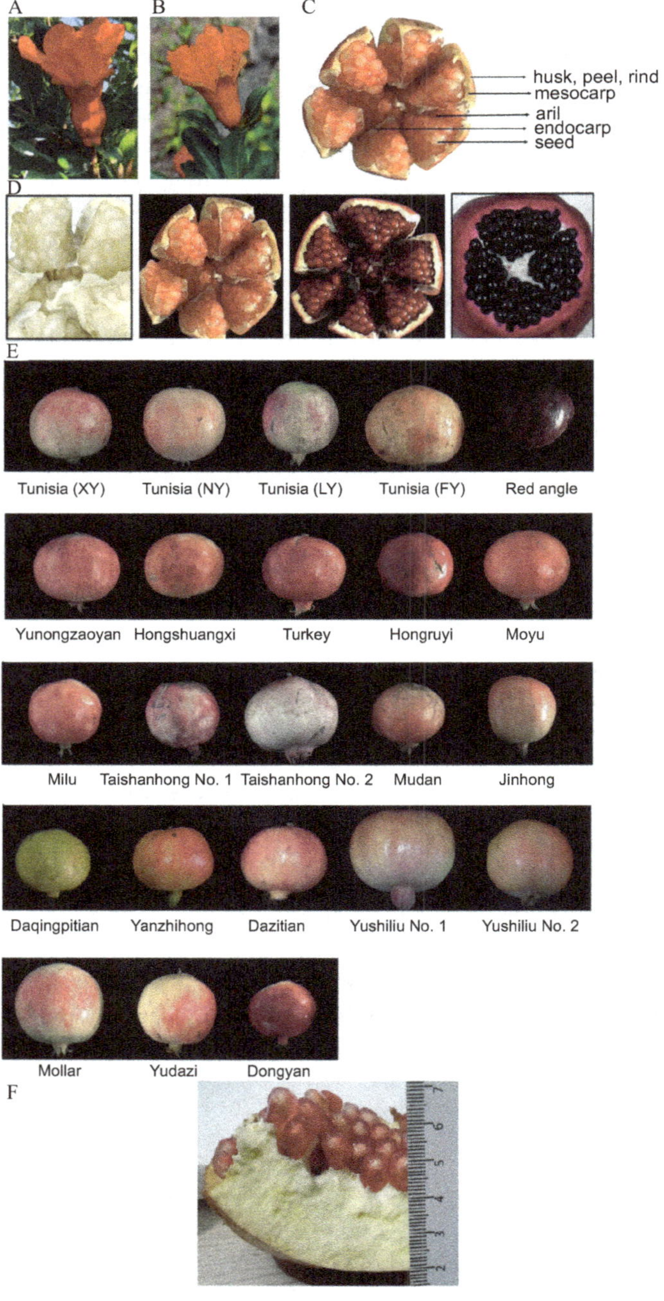

Fig. 1 Morphological investigation of pomegranate. A. Vase-shaped flowers (inflated calyx). B. Bell-shaped flowers (sharp calyx). C. Anatomical figure of pomegranate fruit. D. Aril color of pomegranate cultivars. E. The phenotypes of 20 pomegranate cultivars from China. F. Peel thickness of pomegranate fruit.

except for genotypes (Ferrara et al., 2011). The mean fruit weight is 444.26 g among 20 cultivars planted in China, ranging from 289.25 g–1072.35 g (Chen et al., 2022), and 363 g among 30 Tunisia ones, ranging 196–674 g (Hasnaoui et al., 2011). Also, single fruit weight was reported that from 300.10–854.63 g of six clones grown in Spain (Melgarejo-Sánchez et al., 2015); 186–551 g of 29 ones in Israel (Dafny-Yalin et al., 2010); 106.60–496.91 g of 120 ones in Iran (Tehranifar et al., 2010; Khadivi et al., 2018); and 168.9–574.9 g of eight clones in Italy (Ferrara et al., 2011). Among them, Mollar is cultivated widely in Spain, about 350.04 g of fruit weight (Hernández et al., 2014); 256.0 g in Morocco (Martínez et al., 2006); and 449.96 g in China (Chen et al., 2022).

Peel thickness is about 2.4–5.0 mm for most of the pomegranate cultivars. According to our investigation, we found the maximum 22.2 mm of peel thickness (Figure 1F).

Hundred-aril weight is the key trait for evaluating edible portion. Generally, it ranged between 24.26 g (Mudan, China) and 70.3 g (Chelfi, Tunisia) (Hasnaoui et al., 2011; Chen et al., 2022). Seeds constitute about 30% of the fruit weight which is mainly made of sugars, vitamins, polysaccharides, polyphenols, minerals (Melgarejo et al., 2000) and a small amount of oil that contains polyunsaturated (n-3) fatty acids (Singh et al., 2002).

Based on seed hardness, different groups are applied in pomegranate cultivars. Lu et al. (2006) divided them into three groups: soft seed (<3.67 kg/cm^2), semi-soft seed (3.67–4.2 kg/cm^2), and hard seed (>4. 2 kg/cm^2). Zarei et al. (2013) divided them into four groups: soft seed (8.16–15.31 kg/cm^2), semi-soft seed 20.41–22.45 kg/cm^2, semi-hard seed (30.61–42.86 kg/cm^2, and hard seed (45.92–64.29 kg/cm^2).

A positive correlation has been reported between seed hardness and TA, and color-related characteristics including fruit and aril color, as well as anthocyanin content (Khadivi-Khub et al., 2015; Zarei, 2017). a* value of husk had significantly positive correlation with anthocyanin content in arils (Chen et al., 2022). Antioxidant activity has also shown positive correlations with the amount of ascorbic acid and gallic acid in arils (Sarkhosh et al., 2008).

Fruit shape index ranged between 0.80–0.97 among 20 pomegranate cultivars planted in China, closing to roundness (Chen et al., 2022), and between 0.88–1.61 among 20 Iranian pomegranate cultivars (Tehranifar et al., 2010).

References

Adhami V.M., Khan N. and Mukhtar, H. 2009. Cancer chemoprevention by pomegranate: Laboratory and clinical evidence. *Nutrition and Cancer*, *61*: 811–15.

Akyıldız, A., Karaca, E., Ağçam, E., Dündar, B. and Çınkır, Nİ. 2020. Changes in quality attributes during production steps and frozen-storage of pomegranate juice concentrate. *Journal of Food Composition and Analysis*, *92*: 103548.

Alcaraz-Mármol, F., Nuncio-Jáuregui, N., García-Sánchez, F., Martínez-Nicolás, J.J. and Hernández, F. 2017. Characterization of twenty pomegranate (*Punica granatum* L.) cultivars grown in Spain: Aptitudes for fresh consumption and processing. *Sci. Hortic.*, *219*: 152–60.

Berger, B.A., Kriebel, R., Spalink, D. and Sytsma, K.J. 2016. Divergence times, historical biogeography, and shifts in speciation rates of Myrtales. *Mol. Phylogen. Evol.*, *95*: 116–36.

Bishayee, A., Mandal, A., Bhattacharyya, P. and Bhatia, D. 2016. Pomegranate exerts chemoprevention of experimentally induced mammary tumorigenesis by suppression of cell proliferation and induction of apoptosis. *Nutr. Cancer*, *68*: 120–30.

Byng, J.W., Chase, M.W, Christenhusz, M.J.M., Fay, M.F., Judd, W.S., Mabberley, D.J., Sennikov, A.N., et al. 2016. An update of the angiosperm phylogeny group classification for the orders and families of flowering plants: APG IV. *Bot. J. Linn. Soc.*, *181*: 1–20.

Chaves, F.M., Pavan, I.C.B., da Silvam L.G.S., de Freitas, L.B., Rostagno, M.A., Antunes, A.E.C., Bezerra, R.M.N. and Simabuco, F.M. 2020. Pomegranate juice and peel extracts are able to inhibit proliferation, migration, and colony formation of prostate cancer cell lines and modulate the Akt/mTOR/S6K signaling pathway. *Plant Foods Hum. Nutr.*, *75*: 54–62.

Chen, Y., Gao, H., Wang, S., Liu, X., Hu, Q., Jian, Z., Wan, R., Song, J. and Shi, J. 2022. Comprehensive evaluation of 20 pomegranate (*Punica granatum* L.) cultivars in China. *Journal of Integrative Agriculture*, *21*(2): 434–45. doi: 10.1016/S2095-3119(20)63389-5.

Dafny-Yalin, M., Glazer, I., Bar-Ilan, I., Kerem, Z., Holland, D. and Amir, R. 2010. Color, sugars, and organic acids composition in aril juices and peel homogenates prepared from different pomegranate accessions. *Journal of Agricultural and Food Chemistry*, *58*: 4342–52.

Dahham, S.S., Ali, M.N., Tabassum, H. and Khan, M. 2010. Studies on antibacterial and antifungal activity of pomegranate (*Punica granatum* L.). American-Eurasian *Journal of Agricultural and Environmental Sciences*, *9*: 273–81.

Fawole, O.A. and Opara, U.L. 2013. Changes in physical properties: Chemical and elemental composition and antioxidant capacity of pomegranate (cv. 'Ruby') fruit at five maturity stages. *Scientia Horticulturae*, *150*: 37–46.

Ferrara, G., Cavoski, I., Pacifico, A., Tedone, L. and Mondelli, D. 2011. Morpho-pomological and chemical characterization of pomegranate (*Punica granatum* L.) genotypes in Apulia region, southeastern Italy. *Scientia Horticulturae*, *130*: 599–606.

Graham, S.A. and Graham, A. 2014. Ovary, fruit, and seed morphology of the Lythraceae. *Int. J. Plant Sci.*, *175*: 202–40.

Hasnaoui, N., Jbir, R., Mars, M., Trifi, M., KamalEldin, A., Melgarejo, P. and Hernandez, F. 2011. Organic acids, sugars, and anthocyanins contents in juices of Tunisian pomegranate fruits. *Int. J. Food Prop.*, *14*: 741–57.

Hegazi, N.M., El-Shamy, S., Fahmy, H. and Farag, M.A. 2021. Pomegranate juice as a super-food: A comprehensive review of its extraction, analysis, and quality assessment approaches. *Journal of Food Composition and Analysis*, *97*: 103773.

Hernández, F., Legua, P., Martínez, R., Melgarejo, P. and Martínez, J.J. 2014. Fruit quality characterization of seven pomegranate accessions (*Punica granatum* L.) grown in southeast of Spain. *Scientia Horticulturae*, *175*: 174–80.

Hiwale, S.S., More, T.A. and Bagle, B.G. 2011. Root distribution pattern in pomegranate 'Ganesh' (*Punica granatum* L). *Acta Hortulturae*, *890*: 323–26.

Holland, D., Hatib, K. and Bar-Ya'akov, I. 2009. Pomegranate: Botany, horticulture, breeding. *Horticultural Reviews*, *35*: 127–91.

Karimi, M., Sadeghi, R. and Kokini, J. 2017. Pomegranate as a promising opportunity in medicine and nanotechnology. *Trends Food Sci. Technol.*, *69*: 59–73.

Khadivi, A., Ayenehkar, D., Kazemi, M. and Khaleghi A. 2018. Phenotypic and pomological characterization of a pomegranate (*Punica granatum* L.) germplasm collection and identification of the promising selections. *Scientia Horticulturae*, *238*: 234–45.

Khadivi-Khub, A., Kameli, M., Moshfeghi, N. and Ebrahimi, A. 2015. Phenotypic characterization and relatedness among some Iranian pomegranate (*Punica granatum* L.) accessions. *Trees*, *29*(3): 893–901. doi:10.1007/s00468-015-1172-9.

Khwairakpam, A.D., Bordoloi, D., Thakur, K.K., Monisha, J., Arfuso, F., Sethi, G., Mishra, S., Kumar, A.P. and Kunnumakkara, A.B. 2018. Possible use of *Punica granatum* (Pomegranate) in cancer therapy. *Pharmacol Res.*, *133*: 53–64.

Lu, L.J., Gong, X.M. and Zhu, L.W. 2006. Study on seed hardness of pomegranate cultivars in China. *Journal of Anhui Agricultural University*, *33*(3): 356–59. (In Chinese)

Mahdihassan, S. 1984. Outline of the beginnings of alchemy and its antecedents. *American Journal of Chinese Medicine*, *12*: 32–42.

Martínez, J.J., Melgarejo, P., Hernández, F., Salazar, D.M. and Martínez, R. 2006. Seed characterization of five new pomegranate (*Punica granatum* L.) varieties. *Scientia Horticulturae*, *110*: 241–46.

Melgarejo, P., Salazar, D.M. and Artes, F. 2000. Organic acids and sugars composition of harvested pomegranate fruits. *European Food Research and Technology*, *211*: 185–90.

Melgarejo-Sánchez, P., Martínez, J.J., Legua, P., Martínez, R., Hernández, F. and Melgarejo, P. 2015. Quality, antioxidant activity, and total phenols of six Spanish pomegranates clones. *Sci. Hortic.*, *182*: 65–72.

Mena, P., García-Viguera, C., Navarro-Rico, J., Moreno, D.A., Bartual, J., Saura, D. and Martí, N. 2011. Phytochemical characterization for industrial use of pomegranate (*Punica granatum* L.) cultivars grown in Spain. *J. Sci. Food Agric.*, *91*: 1893–1906.

Nigro, L. and Spagnoli, F. 2018. Pomegranate (*Punica granatum* L.) from Motya and its deepest oriental roots. *Vicino Oriente XXII*, 49–90.

Rajaei, H. and Yazdanpanah, P. 2015. Buds and leaves in pomegranate (*Punica granatum* L.): Phenology in relation to structure and development. *Flora-Morphology, Distribution, Functional Ecology of Plants*, *214*: 61–69. doi: 10.1016/j.flora.2015.05.002.

Salgado, L., Melgarejo, P., Meseguer, I. and Sánchez, M. 2009. Antimicrobial activity of crude extracts from pomegranate (*Punica granatum* L.). *Acta Horticulturae*, *818*: 257–64.

Sarkhosh, A., Zamani, Z., Fatahi, R., Ebadi, A. and Ranjbar, H. 2008. Evaluation of Iranian soft-seed pomegranate accessions by using simple and multivariate analyses. *Tree and Forestry Science and Biotechnology*, *2*(1): 18–25.

Shwartz, E., Glazer, I., Bar-Ya'akov, I., Matityahu, I., Bar-Ilan, I., Holland, D. and Amir, R. 2009. Changes in chemical constituents during the maturation and ripening of two commercially important pomegranate accessions. *Food Chemistry*, *115*: 965–73.

Singh, R.P., Murthy, K.C. and Jayaprakasha, G.K. 2002. Studies on the antioxidant activity of pomegranate (*Punica granatum*) peel and seed extracts using *in vitro* models. *Journal of Agricultural and Food Chemistry*, *50*(1): 81–86.

Tehranifar, A., Zarei, M., Nemati, Z., Esfandiyari, B. and Vazifeshenas, M.R. 2010. Investigation of physico-chemical properties and antioxidant activity of twenty Iranian pomegranate (*Punica granatum* L.) cultivars. *Sci. Hortic.*, *126*: 180–85.

Teixeira da Silva, J.A., Rana, T.S., Narzary, D., Verma, N., Meshram, D.T. and Ranade, S.A. 2013. Pomegranate biology and biotechnology: A review. *Scientia Horticulturae*, *160*: 85–107.

Tzulker, R., Glazer, I., Bar-Ilan, I., Holland, D., Aviram, M. and Amir, R. 2007. Antioxidant activity, polyphenol content, and related compounds in different fruit juices and homogenates prepared from 29 different pomegranate accessions. *J. Agric. Food Chem.*, *55*: 9559–70.

Verotta, L., Panzella, L., Antenucci, S., Calvenzani, V., Tomay, F., Petroni, K., Caneva, E. and Napolitano, A. 2018. Fermented pomegranate wastes as sustainable source of ellagic acid: Antioxidant properties, anti-inflammatory action, and controlled release under simulated digestion conditions. *Food Chem.*, *246*: 129–36.

Yan, M., Zhao, X., Zhou, J., Huo, Y., Ding, Y., et al., 2019. The complete chloroplast genomes of *Punica granatum* and a comparison with other species in Lythraceae. *International Journal of Molecular Sciences*, *20*(12): 2886.

Yuan, Z., Fang, Y., Zhang, T., Fei, Z., Han, F., Liu, C., Liu, M., et al., 2018. The pomegranate (*Punica granatum* L.) genome provides insights into fruit quality and ovule developmental biology. *Plant Biotechnology Journal*, *16*: 1363–74.

Zarei, A. 2017. Biochemical and pomological characterization of pomegranate accessions in Fars Province of Iran. *SABRAO Journal of Breeding and Genetics*, *49*(2): 155–67.

Zarei, A., Zamani, Z., Fatahi, R., Mousavi, A.S. and Seyed, A. 2013. A mechanical method of determining seed-hardness in pomegranate. *Journal of Crop Improvement*, *27*(4): 444–59.

2

Pomegranate Germplasm

Conservation and Utilization

Julián Bartual,[1*] *Candelario Mondragón*,[2] *Elena Zuriaga*[3] and *María Luisa Badenes*[3]

Pomegranate germplasm banks are an essential source of living information and consequently they must be properly managed and maintained to protect these genetic resources. The long history of domestication of the pomegranate tree gave rise to the development of numerous traditional cultivars in the different areas where it spread. In this sense, more than 3,000 accessions are maintained in genebanks and official collections established in the world, including mostly local cultivars, but also internationally significant varieties, ancient and semi-domesticated accessions. Among the main policies behind germplasm conservation are the long-term conservation and providing information with validated phenotypic and genetic descriptions for use by plant breeders or researchers. There is a growing concern about the adaptation to changing climatic conditions of the pomegranate cultivars, which includes new challenges in the face of physiological disorders, pests and fungal diseases, and also the adaptation of fruit characteristics to new tastes or consumption habits. This chapter summarizes the distribution of genebank collections, *in situ* and *ex situ* conservation methods, specific traits and their source, and the use of germplasm in breeding programs.

[1] Estación Experimental Agraria de Elche. Ctra. CV-855 Km.1, 03290 Elx (Alicante), Spain.

[2] Universidad Autónoma de Querétaro. Cerro de las Campanas, s/n, 76010. Santiago de Querétaro, México.

[3] Instituto Valenciano de Investigaciones Agrarias. Ctra. CV-315, Km 10,7 46113 - Moncada (València), Spain.

* Corresponding author: bartual_jul@gva.es

1. Introduction

Pomegranate was one of the first fruit crops to be domesticated, as it has been cultivated in the Middle East for more than 5,000 years (Janick, 2005). As suggested by de Candolle, Iran and surrounding regions are the origin of pomegranates (Holland et al., 2009). From there, it was introduced into North Africa and Europe through the Mediterranean basin, and to the rest of Asia. Later, Spanish sailors and Jesuit missionaries introduced pomegranates into Mexico and California in the 1500s and it arrived in Florida 200 years later (Holland and Bar-Ya'akov, 2018). Nowadays, the pomegranate tree grows in dry, subtropical, and tropical regions around the world (Chandra et al., 2010). Wild pomegranates are still growing today in Central Asia, from Iran and Turkmenistan to northern India (Chandra et al., 2010). Pomegranate is considered a minor fruit and it is far from the top of the list of consumed fruits, such as apple, banana, grapes, and citrus; however, it is one of the most interesting fruits in terms of traditional and local importance and increasing consumption as functional food. Global area planted for pomegranates is estimated at 700,000 hectares (Bartual et al., 2021), producing 7.8 million tons of fruit in 2020; however, pomegranate statistics are not precise for the international trade since it is included in a harmonized code with other fresh fruits.

The long history of pomegranate domestication resulted in development of local cultivars in the different areas where it spread (Holland and Bar-Ya'akov, 2018). Diversity in phenological and pomological traits has been found in collections from Turkey, India, Pakistan, and Europe (Caliskan and Bayazıt, 2013; Durgaç et al., 2008; Ferrara et al., 2011; Martínez et al., 2012; Varasteh et al., 2009). Moreover, pomegranate fruits also have a high diversity in their chemical profile (Drogoudi et al., 2005; Hasnaoui et al., 2011), including relevant content in minerals, vitamins, sugars, organic acids, and polyphenols. These compounds provide a lot of healthy properties to the juice, in which the high content in antioxidants is remarkable (Sadeghi et al., 2009; Mena et al., 2011; Melgarejo-Sánchez et al., 2015).

Nowadays, the interest on the pomegranate fruit has reached a plateau, as many other fruits have been promoted as sources of antioxidants like starfruit (*Averrhoa carambola*), guava (*Psidium guajava*), passion fruit (*Passiflora edulis*), acai berry (*Euterpe oleracea*), dragon fruit (*Hylocereus undatus*), or uchuva (*Physalis peruviana*). Many others included in the ever-changing group of exotic fruits are taking up the attention of the consumers. Common features of most exotic fruits are the tropical and subtropical origin, their low per capita consumption, and the short period of availability in the market.

Maintaining the interest of the consumers in the product, either as a fresh fruit or raw material for the food and pharmaceutical industries, may boost planted acreage. Therefore, the need of new orchards will require more plant material which in turn may stimulate the nursery business. However, problems of crop management such as pests and diseases, and adaptation to variable environmental conditions driven by climate change must continue to be faced. In this context, the demand for new varieties is clearly justified and genetic improvement is the best strategy.

Varietal replacement is another factor to be considered when breeding new varieties, especially in underdeveloped countries, where the farmers cannot afford the high costs of frequent renewal of productive orchards unless there is a clear market pressure. The long productive life of pomegranate is also a factor that hampers the acceptance of new cultivars. High-scale growers or companies, with quick access to improved varieties and international markets, usually are the best customers for new varieties.

Genetic variability is critical for crop improvement, and the centers of origin and domestication are its natural reservoirs. Germplasm conservation should include wild relatives, semi-domesticated and cultivated accessions, with special emphasis on local varieties. However, to improve germplasm utilization, it must be cataloged, evaluated, and used, an endeavor requiring ample resources and clear long-term policies.

In this scenario, the following issues must be addressed: actual availability of genetic resources, extent and nature of variability, the challenges of the long-term conservation, alternatives to rational use of genetic resources and its immediate connection to the development of new varieties.

2. *Punica Species:* Distribution of Diversity

The *Punica* genus, included in the order Myrtales, has been placed under the family Myrtales, Punicaceae and, finally, Lythraceae. This genus contains just two species: *P. granatum* L., commonly known as pomegranate, and *P. protopunica* Balf. f., endemic to Socotra Islands (Yemen). Additionally, some taxonomists consider the dwarf *P. granatum* var. *nana* as a different species, named *P. nana* L. (Rana et al., 2010). Ovary color distinguishes two subspecies of *P. granatum*, *chlorocarpa*, and *porphycarpa.* Due to its common propagation method, mainly by cuttings, core *P. granatum* cultivars found today reflect local priorities (Holland and Bar-Ya'akov, 2018).

Several genetic diversity analyses have been carried out using different kinds of molecular markers, but in general each one mainly includes germplasm from a specific country and some accessions from another place. This is very useful for the management and use of these collections, but it does not allow us to know in a global way the structure of all the available pomegranate germplasm. Recent examples of them using SSRs are the ones conducted using germplasm from India (Shahsavari et al., 2021; Patil et al., 2020), Italy (Giancaspro et al., 2023), Pakistan (Aziz et al., 2020), Spain (Zuriaga et al., 2022) or the Slovenian and Croatian areas of Istria (Višnjevec et al., 2017). Other kinds of molecular markers have been also used previously, such as AFLPs with germplasm from the Çoruh Valley in Turkey (Ercisli et al., 2011), RAPDS with Indian (Sarkhosh et al., 2006; Ranade et al., 2009) or Tunisian germplasm (Hasnaoui et al., 2010), or PCR-RFLP with Spanish accessions (Melgarejo et al., 2009). These studies corroborate the high variability present in the pomegranate germplasm, as well as the existence of certain groups of closely related accessions. The reduction of the costs of massive sequencing has allowed us to reach a greater level of detail, studying diversity at the genomic scale. As far as we know, the first attempt in this sense was carried out by Ophir and collaborators (2014). In this work, pomegranate transcriptomes were sequenced from two accessions (Nana and Black), using RNA from different tissues such as leaves, roots, flowers, and fruits. After *de novo* assembly and functional annotation, 480 SNPs were selected to screen 105 pomegranates from the Agricultural Research Organization (ARO) located at the Newe Ya'ar Research Center in northern Israel. The results revealed two major groups. The first one included ornamental, inedible accessions, and materials from India, China and Iran. The second one included accessions from the Mediterranean region, Central Asia, and California, and among them the "Wonderful-like" accessions appear clearly grouped. The pink skin and aril color and soft seeds cultivars group belong to a Turkmen and Spanish subgroup; and all evergreen accessions are placed in the eastern group. Despite the efforts made to date, a more exhaustive analysis including samples covering the entire distribution range of the species in a more balanced way could provide more information on the relationships between the different groups of accessions and the origin of the species.

Complete genome sequences have been also analyzed providing information about the punicalagin biosynthesis (Qin et al., 2017), fruit quality, and ovule developmental biology (Yuan et al., 2018), and the divergence between soft- and hard-seeded cultivars (Luo et al., 2020). In the last one, the soft-seed Tunisia cultivar was sequenced to obtain a genome assembly, and other 26 varieties with different seed hardness were sequenced. The hard-seeded varieties appeared grouped and separated from the rest of semi soft- and soft-seeded. More recently, chloroplast genome sequencing and comparative analysis of 16 accessions have been reporting, including 11 commercial cultivars, two ornamental, and three wild types (Singh et al., 2021). Also, the mitochondrial genome was sequenced and assembled to further understanding of organization, variation, and evolution of mitogenomes of this tree species (Feng et al., 2023). All these data are highly valuable to increase the knowledge and tools for pomegranate breeding.

3. Commercial Cultivars

The commercial growth reported in this century in countries with a history of pomegranate cultivation was achieved using local selections obtained by farmers through empirical selection or old improved varieties. This is the case of 'Kandahari' and 'Bedana' in Afghanistan; 'Mollar de Elche' and 'Valenciana' in Spain; 'Bhagwa', 'Dolkha', 'Ganesh' in India or 'Taishanhong' in South Central China (Bartual et al, 2012; Ran et al., 2015; Hosamani and Alur, 2015). Pomegranate cultivation with 'Wonderful' expanded in California to satisfy the national demand powered by the reports on the benefits of functional properties on human health.

The biodiversity present in the pomegranate germplasm makes possible to extend its cultivation with marketable cultivars to different areas, from the semi-arid limits of the temperate zones, and into Mediterranean climatic geographical region in the Southern Hemisphere. Following this trend, new orchards were planted in Central Chile, Southern Peru, and Argentina, mainly looking for exports to Europe.

Commercial orchards of pomegranate trees are grown (Table 1) on different scales in Asia (Azerbaijan, Jordan, Iran, Iraq, Saudi Arabia, Afghanistan, Pakistan, India, China, Bangladesh, Myanmar, Vietnam, Thailand, Kazakhstan, Turkmenistan, Tajikistan, Kirgizstan, Armenia, and Georgia) and the Mediterranean Basin (North

Table 1 Global pomegranate production (metric tons), area (ha) and export (metric tons), percentage of own production for export, and share in global export by leading countries (2020).

Country	*Total production (tons)*	*Area (ha)*	*Export (tons)*	*Export (%)*	*Share in global export (%)*
India	3,186,000	283,000	68,000	2.1	11.17
China	1,220,000	133,000	12,000	1.0	1.97
Iran	1,086,000	90,000	12,000	1.1	1.97
Turkey	600,021	28,463	155,000	25.8	25.45
Egypt	380,000	33,096	127,000	33.4	20.86
USA	218,052	8,606	28,000	12.8	4.60
Afghanistan	194,386	18,018	39,000	20.1	6.40
Azerbaijan	181,060	22,948	22,000	12.2	3.61
Morocco	115,000	12,300			
Tunisia	98,045	13,257	10,000	10.2	1.64
Algelia	84,870	9,190			
Iraq	82,370	5,970			
Spain	73,341	5,666	53,000	72.3	8.70
Syria	69,820	5,450			
Saudi Arabia	65,310	1,230			
Italy	15,700	1,303			
Peru	48,320	2,327	40,000	82.8	6.57
Greece	41,888	4,189	16.000	38.2	2.63
Pakistan	37,300	7,300			
Israel	31,250	2,500	9,500	30.4	1.56
South Africa	12,984	1,024	6,369	49.1	1.05
Chile	9,129	537	5,325	58.3	0.87
Argentina	9,000	500	2,200	24.4	0.36
Mexico	7,144	1,068	3,000	42.0	0.49
Australia	4,000	400	526	13.2	0.09
Total	**7,870,990**	**691,392**	**608,920**		**10.000**

Africa, Egypt, Israel, Syria, Lebanon, Turkey, Greece, Cyprus, Italy, France, Spain, Portugal, Albania). Pomegranates are also grown in North and South America (mainly in the USA, Chile, Argentina, Brazil, and Peru), South Africa, and Australia (Holland et al., 2009; Bartual et al., 2021).

Growers look for new cultivars that fit specific market demands or produce higher yields. The global number of pomegranate cultivars in germplasm collections is above 3,000; however, there is higher the degree of relatedness among them due to the widespread practice of vegetative propagation of the cultivars well accepted in national and international markets. Farmers and consumers classified and named pomegranates mainly after the peculiarities of its flowers and fruits, such as the shape and color of the flowers, fruits or arils, taste, the hardness of seeds, or origin. As a result, there is considerable redundancy among accessions within and among collections and it's common to find homonymies (genotypes with the same name but different characteristics) or synonymies (identical genotypes with different names).

Currently, more than 20 pomegranate commercial cultivars are available for cultivation in **India**, among these, 'Bhagawa' is present in more than 80% of the planted area. This is followed by 'Ganesh', 'Arakta', 'Mridula' and 'Ruby'. 'Bhagawa' was probably derived from hybridization between 'Ganesh' with 'Gul-e-Shah', a Russian variety. Reportedly, each state has its own favorite varieties: 'G-137' (Maharashtra), 'Jyoti' (Karnataka), 'Dholka' (Gujarat), 'Jalore Seedless' (Rajasthan), 'Kandhari' (Himachal Pradesh), and 'Yercaud-I' (Tamil Nadu). The commercial varieties grown in India tend to have medium to large fruits, yellow with red tinge to dark red peel, sweet flavor with low acidity, light pink to deep red arils, and very soft to soft seeds (Shilpa et al., 2021).

Due to the long history of **Iranian** pomegranate cultivation where the pomegranate is native, a large variability in quantitative morphological traits in fruits including shape, color, and juice have been reported. In this country, superior and productive genotypes have been traditionally planted together. Pomegranate cultivars 'Malase-Tourshe-Saveh', 'Rabab-e-Neyriz', 'Shise-Kepe-Ferdous', 'Malase-e-Yazdi', and 'Bejestani' are among the well-known in Iran (Zarei et al., 2020). Although the red and sour-sweet cultivars are predominant in Iran, 'Shirin-e-Shahvar' and 'Bihaste Ravar' are popular cultivars with white yellow peel and sweet taste. On the other hand, 'Shishe-Kab' and 'Yousef Khani' have been reported as moderately tolerant to salinity but sensitive to aril browning (Karimi and Hasanpour, 2014).

Pomegranate spread to **China** more than 2,000 years ago, from the western regions. Several production areas and different cultivars have been produced in China during domestication (Zhao et al., 2013). Phenotypic diversity indicates rich diversity within the Chinese pomegranate germplasm resources (Peng et al., 2020) that was also observed studying the genetic diversity (Zhao et al., 2013). 'Huaibeiruanzi3', 'Lintongsanbaitian' and 'Baiyushizi' have yellow-white peel and white arils. Pink peel 'Dabenzi' was the first pomegranate to have its genome sequenced and published as a reference (Qin et al., 2017).

Among the best-known cultivars that could be highlighted, 'Huaibeiqingpiruanzi' has a very large fruit, reaching 600 g weight. Also, the well common commercial red cultivars known as 'Taishanhong' and pink reddish 'Jianshuihongzhenzhu' are cultivated (Cao and Hou, 2013). 'Yicheng Kanghan' showed the best cold resistance (–14.23°C) and 'Tunisiruanzi' (very popular, introduced from Tunisia) soft-seed pomegranate showed the worst cold resistance (Luo et al., 2018).

In **Afghanistan**, the famous 'Kandahari' ('Pand post') and 'Nazak post' accessions have thick and thin rind, respectively. The cream-yellow skin, sweet, and very soft-seeded 'Bedana' is a notable variety. The black peel fruits of 'Tor anar' are appreciated for ornamental purposes (Chater et al., 2020).

In **Central Asia** (Kazakhstan, Kyrgyzstan, Tajikistan, Turkmenistan, Uzbekistan) and Caucasus (Armenia, Azerbaijan, Georgia), there are valuable genetic resources. In Azerbaijan, 'Bala Mursal' is the most popular cultivar. 'Girmiz Gabig' and 'Shirin Girmizi' are two selections with very weak aril septation, while 'Valas' and 'Azernaijan Guleyshasi' are used for juice (Akparov et al., 2015).

In the countries of the Mediterranean area, there are many ecological regions, and so there are many pomegranate cultivars adapted to different regions attending the consumer preferences. **In Turkey**, 'Hicaznar' is the main commercial and exported cultivar, characterized by its dark red peel and dark red juice. 'Lefan' and 'Katirbashi' present a yellow-pinkish peel, and they are widely grown in the eastern Mediterranean and southeastern Anatolian regions. Other local cultivars of lesser importance are: 'Fellahyemez', 'Eksilik', 'Ernar', and 'Asinar'.

In **Greece**, 'Ermioni' is the most important population-cultivar. In **Italy,** four types of 'Dente di cavallo' have been described (regular, late, hard-seeded, and soft-seeded), the very sweet 'Primosole' and the pinkish yellow 'Grossa di Faenza'. In **Portugal**, the only cultivar known is 'Assaria'.

Among the **Egyptian** pomegranates, the most common cultivars are 'Manfalouty' which has a pinkish red fruit, and the very early and sour flavored 'Assuity' with a long history of cultivation, both growing in different environmental conditions. 'Sefri' is a well-known cultivar in Morocco characterized by its yellow skin. In Tunisia, pomegranate cultivation is concentrated in semiarid areas where drought stress is frequent. Yellow-pink fruits and reddish pink arils are the main traits of 'Gabsi', regarded as the typical Tunisian cultivar originated in the Gabes region. Several ecotypes of 'Gabsi' and 'Tounsi' were reported by Elkar et al. (2011).

In the **United States of America**, the most important cultivar is 'Wonderful', which is mainly cultivated in California. It is considered the industry standard (Stover and Mercure, 2007). 'Wonderful' is a unique case of an old cultivar, dating back to 1896, which is well accepted by the European and North American markets (Ashton, 2006). It has demonstrated a remarkable adaptation, since it is ubiquitously present in almost all producing regions of the world including parts of semiarid tropical India, the dry southern Spain, the hot desert conditions of Israel and Australia, the drylands of North Central Chile, and Southern Peru. It has even been planted in the subtropical highlands of Central Mexico and the semiarid north with great success. It is also a variety present in most of the germplasm collections.

'Early Wonderful' and 'Early Foothill', reportedly bud mutations of Wonderful, 'Granada', and some other cultivars introduced to the USDA germplasm in the 1990s, have been tried in Georgia, Texas, and Florida at a small scale.

The natural genetic variability has been increased apart from cultivars considered as standard varieties given its wide distribution beyond its country of origin such as Wonderful (USA), Bhagwa (India), Mollar de Elche (Spain), Hicaznar (Turkey), and Acco (Israel) with some promising new cultivars with high commercial potential released mostly after 2000 (**Table 2**).

Table 2 Improved varieties of pomegranate developed by crossbreeding programs from Research Centers in India, Spain, Israel, Turkey, and China (released after year 2000).

Origin	*Cultivar*	*Outstanding Traits*	*References*
India	'Solapur Lal'[1], 'Kandhari Seedless', and 'Cazri Vishal'	[1]Very high iron, zinc, Vitamin-C, and anthocyanin contents.	(Shilpa et al., 2021)
Spain	'Iliana', 'Tast'm', and 'Rugalate'	Reddish-pink peel and red arils.	(Villamon et al., 2017)
Israel	'Emek', 'Shani-Yonay', and 'Neta-NY'	Predominantly early ripening, red peel, and red arils.	(Holland D. and Bar-Ya'akov I., 2018)
Turkey	'Batem Esinnar', 'Efenar 35', 'Tezeren 35', and 'Canernar'	Red dark peel, sweet, and soft-seeded.	(Yilmaz, 2007) (Aksoy and Dalkilic, 2019)
China	'Yushiliu 4', 'Baiyushizi', 'Hongyushizi'[2]	[2]Resistance to Lythia and Cercospora diseases.	(Zhao et al., 2004, 2006) (Zhu et al., 2005)

3.1 Clonal Selection

It has been proposed that some of the commercial and traditional pomegranate cultivars are in fact 'cultivar-populations', or a group of very similar clonal genotypes, except for a few phenotypical differences, that could be the case of 'Mollar de Elche' 'Wonderful', 'Ermioni', 'Dente di Cavallo' or 'Sefri' (Bartual et al., 2015; Holland D. and Bar-Ya'akov I., 2018; Chater et al., 2020). Taking advantage of this heterogeneity the clones 'Iviagrana49' and 'Iviagrana55' were selected in Spain, both derived from "Mollar de Elche". In India, the oldest cultivar 'Ganesh' is a clonal selection from 'Alandi', and 'G-137' is a clonal selection from 'Ganesh'. Besides this, other selections or hybrids for anardana purpose are also released. Clonal selections in Turkey are 'Izmir-23', 'Izmir-26' (pink), and 'Izmir-1264' and 'Izmir 1523'(red), similar to 'Hizcanar' (Yilmaz, 2007).

4. Genetic Resources

4.1 Availability of Pomegranate Genetic Resources

Considering the recent scientific reports, the genetic resources of pomegranate are vast. The descriptions published during the previous two decades show wide phenotypical variability and ample distribution. In all cases the preferred method of conservation is *in vivo*, assembling field collections of full-size trees. According to their genetic origin, they include wild ancestors, semi-domesticated landraces, local and regional varieties, and new cultivars, but in some cases, they also maintain foreign varieties as well as ornamental specimens. The collections are usually planted without a clear idea for how long they will be maintained and who can have access to vegetative or seed material. Data collected is not standardized, therefore varies among collections, but it is usually focused mainly on fruit traits, but also on tree features.

The number of accessions is significantly high in those countries located close to the center of origin. Historically, the most frequently cited collection is the one assembled at Turkmenistan, which reportedly included 1,100 accessions (Levin, 2006). It was located at the Turkmenistan Research Station of Garrygala, but according to its curator it is no longer accessible. Iran has reported up to 770 entries maintained at the Horticultural Research Center of the University of Tehran (Zamani et al., 2013). A couple of germplasm collections containing 200 and 300 accessions was also reported by Levin (2006) for Azerbaijan and Tajikistan. In Turkey, three major collections have been reported, all belonging to governmental public institutions; the largest one conserving 180 accessions in charge of the Alata Horticultural Research Institute; another one kept by the Plant Genetic Resources Department with 158 accessions; and the smallest one maintained by the Cukurova University with 33 entries.

Considering only the aforementioned collections the number of accessions reaches 2,370 entries. Some other collections around the same geographical area have been reported in Russia (800 accessions), Ukraine (370 accessions), and Turkey (180 accessions) (Onur and Kaska, 1985; Yezhov et al., 2005). Given the geographical proximity of these countries as well as the long history of trading and exchange, it is expected to find a high degree of redundancy within and among collections. However, it is well-known that political and economic unrest in the region directly influences the human and economic resources devoted to the long-term maintenance of these collections, threatening its very existence and, more importantly, the effective access for breeders from other countries that are interested in specific traits.

Small field collections have been reported in Egypt, Jordan, Albania, Greece, Portugal, Italy, Morocco, Tunisia, South Africa, and Hungary. Usually, they can be regarded as exploratory trials for future development of the crop.

Countries far away from the main center of variability and early dispersal, such as Argentina, Chile (Frank, 2012), Mexico (Mondragon, 2012; Mondragon et al., 2018) and Australia, have reported genotypes resembling those from European ancestry and interestingly some generated *in situ*, explained by the allogamous reproductive behavior of the plant and the easiness of clonal propagation.

4.2 Comprehensive Management of Genetic Resources of Pomegranate

The experience in other crops has shown that the existence of a wide genetic variability does not guarantee that a particular crop will become an economic alternative for farmers in a particular country. Significant planted acreage, organized farmers, a well-defined consumer base or target markets either national or international, and scientific expertise, are some of the ingredients for a successful integration of conservation and utilization of genetic resources program.

This approach of germplasm management is followed by China, India, Spain, and Israel. These countries participate in the conservation and evaluation of genetic resources, with direct links to breeding, food science research, education, and production. Their programs are focused on local and foreign genotypes of immediate commercial value or specific accessions that can be readily included in breeding programs.

China has a long history of cultivation and utilization of pomegranates. Dating back to the I or II century B.C., a continuous flow of seeds, trees, and products along the Silk Road allowed for the introduction of interesting genotypes. Adapted trees also reproduced sexually, which in time allowed the development of local genotypes. Among the Chinese farmers, the tree and its products are well known and extensively used. Its cultivation received an important boost after the 1990s, reaching 90,000 ha in 2004. Nowadays, the pomegranate crop is widely distributed, taking advantage of the enormous agro-ecological variability found in the country; from temperate to subtropical and tropical areas, starting low altitude of 50 up to 1,800 m above sea level. The wide distribution of the plants prompted the creation of production clusters, mainly in Shandong Province, Anhui Province, Henan Province, Shanxi Province, Hebei Province, Yunan Province, Sichuan Province, and Xinjiang Autonomous Region. Local cultivars and some new superior selected genotypes are prevailing in these areas (Yuan and Zhao, 2019).

Since the 1980s, the Chinese authorities noted the importance of the pomegranate genetic resources, research and utilization promoting the systematic research of the genetic resources. These activities were strengthened by the inception of the National Pomegranate Repository in the Yicheng district of the Shangdong Province, which is reportedly holding 289 entries, mostly of Chinese origin (Yuan and Zhao, 2019). Besides this location, there is another field collection at the Yunnan province holding 85 accessions (Yang et al., 2015). It is also known that there are many smaller local collections in the seven regions noted by its pomegranate production; however, they are not reported in the English literature. All are supported by provincial funding and connected to universities and farmers' organizations.

Research on the evolution, botany, physiology, and agronomy of the crop is very active. Pomegranate is included in the research topics of the Chinese Society of Horticultural Science and other regional academic and scientific organizations. The recent interest in the biotechnology, the genomics of pomegranate, and the chemistry of metabolites will certainly improve the benefits of the crop.

Regarding the development of new variability, there has been a remarkable progress. Cao et al. (2013) reported the discovery and description of 238 pomegranate local cultivars obtained from farmers' fields. They also mentioned the existence of 50 cultivars obtained by hybridization, radiation, and bud mutation, with outstanding fruit quality. Niu et al. (2018) reported the development of three soft-seeded cultivars using these techniques. The genetic pool is also continuously enriched introducing foreign germplasm.

India also has a comprehensive strategy of pomegranate germplasm conservation and utilization. This crop is included in the activities of 14 research centers, each holding at least 60 accessions each. They are scattered from the south central semiarid, east arid, up to the northern temperate regions bordering Pakistan and China. The major collections are found at the National Research Center on Pomegranate which belongs to the Indian Council of Agricultural Research (ICAR) in Sholapur, Maharashtra, with 187 accessions; a collection containing 90 accessions located at Uttarakhand, the collection of the Central Institute for Arid Horticulture in Rajasthan reportedly maintains 190 entries. Not less important is the collection held at New Delhi, holding 170 entries, that belongs to the National Bureau of Plant Genetic Resources (Chandra et al., 2010). Most of these collections are coordinated with state universities, allowing for genetic and agronomic studies. Given the importance of the pomegranate crop in the Indian agriculture, the collections are intensively used for generating improved varieties, which are readily propagated and promoted among farmers. It remains to be known the genetic redundancy present within and among these collections.

There are two important places containing pomegranate germplasm in Spain. First one is the collection of the Escuela Superior de Orihuela, which belongs to the Universidad Miguel Hernandez, located in Alicante Spain, which contains 59 accessions obtained from the different growing regions of Spain. This collection started in 1992 and a description of the trees is regularly updated. The collection is also used for fruit quality studies and industrial derivatives (Martinez et al., 2012). The second Spanish collection is located at Agricultural Experiment Station (EEA)/Instituto Valenciano de Investigaciones Agrarias (IVIA) in Elche, Alicante (Bartual et al., 2012). It is an active collection and currently contains 225 accessions originated from 22 different countries, including samples of old varieties and segregant populations (Figure 1). Its main aim is the development of new varieties of high quality and productivity for southern Spain.

The collection from Israel is also a working collection, utilized intensively in the development of new varieties mainly through hybridization and selection. It is located at the Newe Ya'har Research Center of the Agricultural Research Organization, and contains more than 150 accessions, commercial varieties, hybrids and segregant populations (Holland et al., 2014).

In North America, the National Clonal Repository of the Agricultural Research Service of the United States Department of Agriculture (ARS-USDA) based in Davis, California has assembled an interesting collection of genotypes obtained from various origins around the world. According to its policies, the institution received

Fig. 1 Segregating progeny of the crosses between Spanish and foreign cultivars.

the plant material as donation to the USDA with the compromise to maintain them for term usage and to share with any interested party, private or official, national or foreign organization, keeping a high standard of sanitation. Unfortunately, despite its effectiveness, this model has not been reproduced in other countries. The USDA also maintains a simple database that allows for quick selection of the available entries, at the same time allowing the user to prepare a request small samples of planting material, regularly dedicated to experimental purposes. A search conducted at the end of 2021 showed 179 accessions available (https://npgsweb.ars-grin/gov/gringlobal/search).

A potential handicap for the utilization of valuable genes of these pools are the official restrictions imposed to import and export of genetic resources, which in most countries is a sensitive issue. Additionally, in the case of pomegranate, phytosanitary requirements for exchange of vegetative and seed materials are in place in all countries.

4.3 Conserving Genetic Resources for the Future

The perennial nature of the pomegranate tree imposes similar challenges as many other fruit crops, since the preferred means of conservation is live full-size trees, a choice associated to high maintenance costs, sometimes higher than a regular commercial orchard and without the return of fruit sales. Field collections are easier to handle if they belong or are somehow associated to universities or agricultural colleges, since orchard management and regular data collection can be justified as teaching activities.

Lack or insufficient funding, disinterest from the official entities for conservation and disregard towards the obvious benefits of maintaining the gene pool for future generations, are common in underdeveloped countries, where the public interest is on most immediate needs.

4.4 The Dilemma and Opportunities and of Wild Stocks of Pomegranate

Wild forms of pomegranate have been reported in those countries included in the center of origin, such as Iran, Afghanistan, Turkmenistan, Pakistan, and northern India. They are also represented in formal field collections in the Iranian National Pomegranate Collection located at Yazd and Saveh and the Faculty of Agriculture of the University of Tehran and other collections. However, the information available does not allow to discriminate if they are unique or simple replicates. The extension of the genetic variability present in nature, as well as the assembled *ex situ* collections, need to be evaluated. Completion of accurate descriptors may allow the breeders to acknowledge which traits are useful to integrate them into the improved cultivars of the future.

Usually, the conservation of wild stocks is a costly endeavor, and not very amenable for donors or governmental programs, especially if there is not accurate information of the extent of the variability remaining in the wild stocks, its immediate or future value and also on the extremely long-term nature of those projects. Local governments usually lack the resources needed for maintenance. The establishment of nature reserves or national parks has been proven useful in other plants and animal species. Until now there are no reports of any initiative focused on the conservation of pomegranate wild genetic resources.

5. Conservation

The pomegranate germplasm collections are used as a significant source of materials for characterization and research. However, *ex situ* field collections have to face maintenance and organization problems. Core collections obtained from structured samples of bigger collections are a useful tool to improve germplasm management, although it has several serious problems, including exposure to natural disasters, attacks by pests and pathogens, and high maintenance costs. Determination of levels of genetic diversity is also essential requirement for conservation of genetic resources. *In situ* long-term conservation of the wild relatives of the cultivated pomegranate could be the most adequate strategy. However, as in many other fruit crops, many of those areas are frequently allotted as communal lands utilized for animal grazing, extraction of firewood and other products of the forest, activities that endanger the establishment of seedlings, and in turn the repopulation and maintenance of the natural stock. The questions still remain, who and how will be conserving the pomegranate wild germplasm and for how long.

An example of the great value of the wild genotypes is the use of the wild Indian ‘Daru’ pomegranate in crosses with commercial varieties with the aim to obtain resistance to bacterial blight (*Xanthomonas axonopolis* pv. punicae), a devastating pomegranate disease affecting Indian orchards which is difficult to contain with pesticides. This program is undertaken at the Indian Institute of Horticultural Research and the progenies are at the stage of field testing (Kumar, 2016) with very promising results.

6. Characterization

A common problem found when comparing entries from different countries is the lack of standard descriptors. Concerted efforts to understand and elucidate the actual variability can be sped up assembling databases of basic botanic and agronomic traits. Some platforms like GRIN Global (http://www.grin-global.org/), already in use at the USDA pomegranate and other fruit crops repositories, could improve the organization and access to plant data. It is freely available to interested parties, but its inception and operation usually depend on the economic importance of the crop for a particular government. Unfortunately, pomegranate is not included in the list of valuable or strategic crops of many countries.

Technical examinations of new candidate varieties of *Punica granatum* L in the European Union are carried out according to the Technical Protocol for tests on distinctness, uniformity, and stability (UPOV, 2014). In the CPVO-TP/284/1, 40 characteristics of the tree, shoots, leaves, flowers, and fruits of pomegranate are analyzed.

High pomological diversity of pomegranate genotypes implies a wide range of different fruit characteristics, nevertheless new approaches using molecular techniques have been established for pomegranates and they are very esteemed to complement phenotypical descriptions. Morphological and phytochemical descriptors have been essentially based on fruits traits, seeds, and juice composition to characterize pomegranate genotypes. However, some descriptors corresponded to date of ripening, growth of the tree (e.g., vigor, habit), thorniness, type and color of shoots, leaf and flower shape, have revealed a high variability among accessions. For instance, the plant height varies from 50 cm to more than 4.5 m.

The properties of pomegranate fruits depend on the genetics of the cultivars, but observed differences can also be explained by plant nutrition, ripening season or growing locations. Offering cultivars with unique traits, could increase the diversity of pomegranate in the marketplace.

It is essential to get a better understanding of the genetic control of key traits, as well as the range of variation for significant traits for the purposes of breeding. In this regard, a compilation of several studies from different germplasm collections (Iran, China, India, Turkey, Israel, and Spain) of some important fruit and juice characteristics are shown (Table 3). The percentage of juice per fruit (in weight) ranged from 30.1–55.4g. Two interesting traits for the acceptability of pomegranate fresh fruit for consumers are aril size and seed hardness. It was found that the weight of seeds in commercial cultivars ranged from less than 22–84.4g. We outline that having more than 25g is desirable for commercial purposes.

Table 3 Percentage of juice, weight of seeds, hardness and levels of soluble solids content (SSC), acidity (TA), sugar content, total poliphenols and anthocyanins in aril juices of different varieties from collections grown in different countries.

Trait	*Iran*	*China*	*India*	*Turkey*	*Spain*	*Israel*
Ratio juice / fruit (w/w %)	40.5–54.4	35.7–55.6	30.1–48.5	37.1–47.4	35.2–48.1	37–46
Weight 100 seeds (g)	35.5–44.6	25.8–84.4	35.2–44.5	24.0–62.7	22.0–47.5	22.4–45.1
Seed hardness (N)	8.1–64.2	2.86–7.11	3.41–7.05	nd	9.6–59.5	nd
TA (expr. citric ac.%)	0.33–4.11	0.28–3.59	0.12–4.20	0.26–4.60	0.18–2.03	0.22–3.10
SSC (°Brix)	11.4–18.3	12.4 –19.0	15.5–17.6	13.9–16.5	13.5–17.6	13.7–17.8
Sugar content (%)	13.2–21.7	8.7–15.9	12.6–15.0	13.2–21.7	11.0–15.8	4.8–6.6
Total poliphenols (mg GAE/L)	2380–9303	932–1096	876–1536	1245–2076	1500–4500	934–2057
Total anthocyanins (mg/L)	15.0–252.4	27.9–82.2	26.3–78.1	6.1–219.0	6.0–105.2	20–235

Note: nd = no data. Yang and Yang, 2015; Kalaycıoglu, Z., and Erim, F. 2017; Poyrazohua et al., 2002; Dafny-Yalin et al., 2010; Tzulker et al., 2007; Szychowski et al. 2015.

Seed hardness is a major sensory attribute for fruits considered for fresh consumption. If the seeds are too hard, and thus too difficult to masticate, consumer satisfaction can be considerably reduced. Different methodologies have been used to measure seed hardness and only some results can be compared among the studies that used the same texture profile analysis (TPA) or puncture test (PT) (Alcaraz-Marmol et al., 2015; Szychowski et al., 2015). Zarei et al., (2013) categorized hardness of the pomegranate genotypes into four distinct groups in Newtons' force for rupturing them with a flat cylinder as soft-seeded (80–150 N), semi-soft seeded (200–220 N), semi-hard seeded (300–420 N), and hard seeded (450–630 N). All genotypes had lignin components in their seeds, being higher in the hard-seeded genotypes (Zarei et al., 2016). Other hardness systems are considered, as the percentage of fiber in their seed coat.

Juice acidity is the main factor that determines taste in arils, and consumer acceptance varies according to the country of origin. For instance, sweet varieties are preferred in many Mediterranean Basin countries, India, and China. On the contrary, sweet-sour to sour cultivars are preferred in Azerbaijan, Afghanistan, or Iran. Sour cultivars could be described as having over 1.85% (titratable acidity

(TA) expressed as citric acid equivalents at maturity). In most breeding programs, sweet-sour (0.6–1.0%) cultivars are preferred, following the trend in the North European, Asian, the USA, and Australian markets. Measurements of cultivars studied, revealed significant differences between sweet and sour taste fruits, as total acidity values ranged from 0.12 (China) to 4.60 (Turkey). It has been noted that the aril juice has shown a relatively narrow range of total soluble solids (TSS) from 11.4–19.0 (°Brix). The level of the sugars in pomegranate juice is highly correlated with the level of soluble solids content (SSC). The sugar measurements ranged between 8.7 and the highest 21.7%.

Primary and secondary metabolites show extensive variability since the fruit are used for different purposes. The highest values of phenolics corresponded to the Iranian and Spanish references. Total polyphenols varied more than 10-fold from 876–9,303 (mg GAE/L). These varieties are valuable sources for extracts and further isolation.

The contents and ratio of anthocyanins are different in the skin of pomegranate fruits. Anthocyanin composition in pomegranate peel varies depending on the cultivars, fruit developmental stages, and coloration time. The external fruit color ranges from yellow, green, or pink overlaid with pink to deep red or deep purple. Cyanidin 3,5-diglucoside, pelargonidin 3,5-diglucoside, pelargonidin 3-glucoside, cyanidin 3-glucoside, cyanidin 3-rutinoside, delphinidin 3-glucoside, delphinidin 3,5-diglucoside, cyanidin hexoxide, and cyanidin pentoxide are the main anthocyanins identified and quantified in the peel (Fischer et al., 2011). To determine the phenotypic and color variability among pomegranate varieties a representative sample belonged to the EEA/IVIA Spanish pomegranate collection were analyzed. Among the six primary anthocyanin molecules, total anthocyanin levels and composition were measured for both peel and aril extracts. Delphinidin 3-glucoside content on peel was 20-fold higher in the black accession '3/1-13' to red 'Rugalate' (Table 4), like the results of Zhao et al. (2013) reported in black accessions, large amounts of cyanidin mono-glycoside and delphinidin mono-glycoside (over 100 mg/100 g) were found in the skin. The color of the juice can

Table 4 Anthocyanin content on pomegranate peel accessions (mg/Kg).

Accession	*Dp 3,5-*	*Cy 3,5-*	*Pg 3,5-*	*Dp 3-*	*Cy 3-*	*Pg 3-*	*Total anthocyanins*
Rugalate (Red)	0.22±0.09	4.54±1.45	0.60±0.21	0.20±0.13	4.50±1.04	0.43±0.13	10.49±2.93
3/1-137 (Black)	1.23±0.39	2.97±0.63	2.40±0.07	4.68±0.08	4.28±0.77	0.37±0.08	13.93±2.60
Mollar (Cream)	nd	0.02±0.01	nd	nd	0.04±0.02	nd	0.07±0.03

***Note*:** Mean ± standard deviation (n=4). nd = not-detected concentration using the method employed by the laboratory. Pelargonidin 3-glucoside (Pg 3-), Pelargonidin 3,5-diglucoside (Pg 3,5-), Cyanidin 3-glucoside (cy 3-), Cyanidin 3,5-diglucoside (Cy 3,5-), Delphinidin 3-glucoside (Dp 3-), Delphinidin 3,5-diglucoside (Dp 3,5-).

vary from white to deep red (Holland et al., 2009). The highest TAC was found in Iran cultivars, as total anthocyanins of juice reach 252.4 (mg/L), and it was about 16 times higher than the lower value.

Trained sensory panels are used to evaluate the genetic diversity in flavor attributes and preferences for fresh pomegranate arils. Sensorial characteristics are assessed by a 9-point Hedonic scale for the following fruit quality traits: taste (sweet, sour, and bitter), odor (red wine, and pomegranate fruity notes); and mouth-feel (astringency, juiciness, and seed hardness). Bitter taste in juice is due to the astringent properties of ellagitannins (Chater et al., 2018). Flavor pomegranate fruit preferences are relatively high sweetness, moderate to low sourness, high red wine and pomegranate fruity flavor notes, low bitterness and astringency, and medium to soft seeds (Mayuoni-Kirshenbaum et al., 2013). In addition, a lexicon for pomegranate juices was established that includes attributes describing a range of flavors such as brown spice, fermented, molasses, vinegar, wine-like, woody, apple, berry, cranberry, cherry, and grape (Koppel and Chambers IV, 2010).

7. *In vitro* Conservation

The use of *in vitro* techniques for germplasm conservation of different species has been reviewed by Panis et al. (2020) and Rajasekharan and Sahijram (2015), including micropropagation, somatic embryogenesis, cell cultures, embryo rescue, and *in vitro* cold storage. Wherever possible they can complement the efforts of conservation based on field collections, allowing the clonal reproduction of clean propagules, disease- and virus-free, of selected genotypes with unique allelic combinations. Cryopreservation, on the other hand, may be the technique that could reduce the need for valuable land required to maintain field collections; however, it is not widely available. Cryogenic preservation, commonly used for banana and apple (Panis et al., 2020), is still in the experimental stages for pomegranate. A study to observe the effect of the cryopreservation (at –196°C) on the pollen germination rate of pomegranate was reported by Sahu et al. (2014).

Regarding pomegranate micropropagation, reviewed by Kumar et al. (2017), da Silva et al. (2013), and Chandra and Babu (2010), as in other species, is highly influenced by genotype, explant type, season, culture media, and plant growth regulators (PGRs). The high phenolics' content in pomegranate explants, as in many other fruit tree species, is a major problem for tissue culture since it causes medium browning. To control this problem, Murkute et al. (2003) suggested the subculture of explants 3 times at 24h intervals. Similarly, Singh and Patel (2016) suggested the use of subculturing followed by the addition of activated charcoal into the medium. Different sources of pomegranate explants have been studied in different approaches such as direct and indirect organogenesis and somatic embryogenesis. Deepika and Kanwar (2010) used cotyledon, hypocotyl, leaf, and internode sections from one-month old *in vitro* seedlings to induce 'calli' formation. A percentage higher than 75% was obtained in all cases, but it was greater using cotyledons (85.5%). The highest regeneration rate was also obtained with cotyledons as explant source. Somatic embryogenesis was observed from seedlings explants (Jaika and Mehra, 1986) and also from petals (Nataraja and Neelambika, 1996). One of the main problems is the low rate of complete plants obtained. Recently, Desai et al. (2018) developed an efficient micropropagation protocol through axillary shoot proliferation for pomegranate variety 'Bhagwa', the most popular in India. These authors suggest the sequential application of various antifungal and antibacterial agents, as sterilization method of the explants, and five translocations of the explants in the same media bottle every 24h, as an effective method to control exudation. The shoot induction response was better with Woody Plant Medium (WPM) supplemented with 6-benzylaminopurine, and cobalt chloride proved to control defoliation. WPM supplemented with activated charcoal was suggested as the best treatment for root induction. Finally, cocopeat was the best growing medium for acclimatization of plantlets.

8. Utilization

Conservation and utilization of genetic resources depends largely on the availability of information regarding phenotypic and genotypic characteristics of the species of interest. In this sense, pomegranate genetic resources are kept in collections that include wild, local populations, cultivars, and few rootstocks. Pomegranate has been grown for thousands of years and is one of the oldest plants cultivated in Asia and Mediterranean countries. Levin (2006) indicated that pomegranate domestication started in the Middle East about 5,000 years ago. The best wild species were selected by the ancient growers and cultivated in family gardens. The pomegranate has been traditionally regarded as a fresh fruit, but its juice is also the raw material for a wide array of processed products concocted at a home level such as beverages, teas, liquors, jellies and jams, and at the same time industrial applications in the food, cosmetics and pharmaceutical industries are common.

(a) **Fresh Fruit:** At present, major production is consumed as fresh fruit, but in Iran, China, and other Asian countries, it is very common that people use them for juicing at home.

(b) **Juices and Processed Products:** The pomegranate juice and its concentrate segment accounts for the second highest share in terms of revenue in the near future, as pomegranate juice concentration is used in cooking (like pomegranate wine, vinegar, and sour sauce). The promotion during the first decade of XX century of the functional properties of the juice, based mainly on its antioxidant effects (Lansky and Newman, 2007), boosted the global demand of fresh fruit for direct consumption as well as the development of numerous processed products, mainly processed juices and mixed drinks at industrial scale. As a result, it boosted the interest on its cultivation in new areas, such as the USA, Peru, Chile, South Africa, Australia, among others, that are away from the traditional production areas from the Near East and the Mediterranean basin.

(c) **Ornamental Types (Ganesh ornamental, etc.):** Produce attractive (orange, white, red), simple and double flowers; most of them without bearing any fruit. Characters of tested ornamental pomegranate cultivars of thrifty five ornamental pomegranate cultivars had abundant genetic diversity in China (Zhang et al., 2008). Single and double flower form; White, agate, pink, red color of petals; Cyan-green, white, pink, red, purplish red pericarp color. There are many ornamental cultivars with different flower colors. For instance, "Golnar Farsi Sarvestan" is one of the Iranian ornamental cultivars grown. *P. granatum* var. *nana* is a natural dwarf variant that grows only up to 60–75 cm. A dwarf cultivar that produces small fruits is "Yellow Nana", a new ornamental pomegranate by crossing "Kabul Yellow" × 'Nana' with hard seeds.

(d) **Suitable Varieties for *Anardana* Purpose:** ('Daru', "Goma khatta", 'Amlidana', "Solapur anardana", "Phule Anardana") with high titratable acidity until 4.8%.

(e) **Health:** Some medicinal value has been traditionally attributed to the dry flowers, bark, fruit rinds and seeds, with recipes still prevalent in the herbalism of India, China, and other countries. According to Holland (2009) and Viuda-Martos et al., (2010), there are evidences that pomegranate has been cultivated and used by people for many health purposes. Flowers, bark, and leaf tissues contain bioactive phytochemicals suitable for wide-scale medical consumption. During more than 2,000 years, it has been traditionally used as an anthelmintic and vermifuge, to eliminate parasites, to treat mouth ulcers, diarrhea, acidosis, dysentery, hemorrhage, microbial infections, and respiratory pathologies, and as an antipyretic.

To compare the nutrition values of different cultivars, many studies reported pomegranate as source of various compounds of high chemical value. Phenolics, flavonoids, tannins, alkaloids, terpenoids, phytosterols, lignin, glycosides, saponins, fatty acids, amino acids, or carbohydrates (Dafny et al., 2010; Bar-Ya'akov et al., 2019; Viuda-Martos et al., 2010). 'Poost Siah', meaning black skin, is an interesting medicinal and ornamental cultivar grown in different regions in Iran.

(f) **Rootstocks:** Liu et al. (2019) reported a grafting system of pomegranate *in vitro*. These authors established that the optimal combination of pomegranate tube grafting with the survival rate of grafting was reached using the stem tip with four blades as scion and the seedlings without cotyledon as rootstock.

9. Hybridization

Domesticated pomegranates usually have two types of flowers: a fully functional hermaphrodite flower, able to set fruit, and a functionally male flower with an undeveloped gynoecium and complete androecium (Wetzstein et al., 2011). Both types usually bear abundant pollen. Profuse and long-lasting flowering season, up to a month depending on genotype and local weather conditions, allows for repeated attempts at crossing.

9.1 Hybridization Technique

Hybridizing pomegranates is a fairly easy process, given that the plant bears large campanulate flowers with well exposed anthers and stigmata (Figure 2A–C).

It is recommended to collect pollen from functional flowers the day before its opening or before sunrise of the opening day, since pollen shedding usually starts mid-morning. Due to the abundance of pollen, processing 4–6 flowers yields enough pollen for many crossings. Anthers should be carefully removed from the flowers and placed on a piece of paper or small plastic tray, removing any residues of styles and petals. Then, allow the sample to dry for 24–48 h at room temperature. If local weather conditions are humid, then the pollen sample can be dried under a dehydrator to enhance anther dehiscence. Next day the pollen can be collected and stored in glass vials or glassine envelopes. Pomegranate pollen is naturally a dry powder, and it can be stored in a refrigerator at 4–6°C for a year without significantly compromising its viability.

When selecting flowers for crossings, it is advisable to select the largest in the whorl, which is usually the earliest, and completely remove the rest to avoid competition among them. The flowers selected for crossings can be emasculated by removing only the anthers or by completely shaving the filaments and anthers with a scalpel or tweezers. Both methods are easy to perform. Once emasculated, they can be isolated with a small glassine envelope and a paper clip. Figure 2 shows the complete process of hybridization, from emasculation, pollen collection, manual pollination, and isolation. In pomegranate, there is no need to repeat the pollination the next day, even though the stigma may still be receptive. For crosses conducted during the normal flowering season, fruit set is not a problem nor compatibility among cultivated diploid cultivars. In our experience, it is easy to obtain more than 100 seeds per fruit, therefore completing three or four flowers per cross should yield ample supply of seeds.

9.2 Seed Germination and Seedling Management

Once the fruits from the crosses are harvested, the seeds can be extracted and cleaned using a strainer to remove the juice and aril residues, rinsing them thoroughly with clean water. Then, they can be sorted, selecting only the largest and fully developed seeds. To reduce the risk of loss of crosses in the field, the fruits can be harvested before they reach the commercial maturity. This moment is reached as soon as the arils start to accumulate pigments and at this stage the seeds are fully developed.

It is advisable to disinfect the seeds prior to planting submerging them in a solution of 2–3% domestic bleach (Clorox® or similar cleaner based on sodium hypochlorite) for 5 minutes. This is enough to prevent seed rot during germination. The germination of pomegranate seed can be accomplished using a regular mix of peat moss and perlite as a substrate (ratio 70/30). This mix provides good drainage as well as enough moisture retention. Clean coco coir powder can

Fig. 2 Hybridization technique: (A) Hermaphrodite functional flower suitable for emasculation. (B and C) Petal and anther removal. (D) Emasculated flower. (E) Pollination. (F and G) Labeling and isolation.

also be used as a substitute of peat moss, using the same rate. Under greenhouse conditions seed germination takes place between 6–10 days, batches of fresh seeds usually attain germination rates above 80%.

According to Levin (2006) the seed germination percentage varies from 7–98%, while the time needed for germination varies from 10 to more than 100 days depending upon the variety, seed hardness, and length of storage period. Longer periods for germination have been attributed to the presence of water-soluble inhibitors in the seed coat, thickness, and hardness of the seeds (Jalikop, 2010), residual phenols from the sarcotesta and physiological dormancy. Karimi et al. (2011) reported that physiological dormancy can be overcome by alternating treatment of cold and warm mist. Depending upon the geographical origin some cultivars may need stratification, such was the case of the cultivar 'Malas-e-Yazdi', which reported the highest germination after 40 days of stratification. Some wild pomegranates collected from temperate regions also have been reported to require up to 60 days of stratification at 5°C.

The clean seed can be planted immediately after disinfection. We usually plant them in polystyrene trays with large cells (60 count), where they grow until they reach 15–25 cm of height. Then they are transfer to plastic pots (2–3L) where they can be maintained up to a year, depending on the planting season. Later, they can be planted in the field as soon as they reach 40–50 cm tall or as a one-year-old plants. Young pomegranate plants can also be maintained for longer periods in the nursery. For this purpose, they are transplanted to large plastic or textile bags (8–12 L). Under these conditions they can be maintained for up to 3 years, which could be convenient in some instances.

9.3 Juvenilility

The length of time required to reach the reproductive stage from plants propagated by seed depends upon the species and the variety. In general, it shows a large variation in pomegranate. The ornamental dwarf *P. granatum* var. *nana* is able to bear flowers and set fruits during the same year of planting, very similar to plants regenerated from leaf segments of the same variety according to Terakami et al. (2007). These authors developed a protocol for genetic transformation based on *Agrobacterium*, and obtained fully mature transformed plants in 12 months. Based on this, these authors have proposed dwarf pomegranate as a model plant for genetic studies of other fruit crops. However, for those cultivars of standard size, the length of the juvenile phase varies from 2–4 years, if they are transplanted in the field as a one-year-old plants. Shorter juvenile periods, 2 years, have been reported for Israeli cultivars and segregants, but the first fruits are usually smaller than those produced by older trees propagated by cuttings (Holland et al., 2009).

10. Future Thrust

Climate change is modifying the scenario for the agriculture sector with increasing weather variability, higher temperatures, changing hydrology, and extreme events predicted. An integrated pomegranate genetic resource and its diversity management are essential for breeding programs toward new selection objectives, such as those needed to anticipate climate change.

Pomegranate breeding programs focus on improving external and internal fruit quality, prolonging the harvest season, good storage performance and developing resistance and/or tolerance to biotic and abiotic stress. Other important characteristics demanded by consumers and growers are:

(a) Attractive color of fruit skin, arils, and juice.
(b) Fruit weight (around 300–350 g).
(c) High tree yield.
(d) Tolerance to water deficiency, frost, or salinity.
(e) Tolerance to pests and diseases (Alternaria sp, Bacterial blight, ...).
(f) Clean fruits free of physiological disorders (Fruit cracking, sunburn, aril browning, ...).
(g) Good quality after postharvest and shelf-life.
(h) High content of nutritional and bioactive compounds.

New research and experimentation must be done regarding adaptation of traditional cultivars and new genetic material to new pomegranate orchard tendency, which is radically different from those of the past, thanks to the use of trellis or the use of covered and protected net systems. Currently, we can set up medium and even high-density orchards, which bear fruit of better quality (less sunburn damages) but with higher orchard establishment and initial management costs per ha. Average yield ranges from 8 to 25 tons per hectare, but it is expected that in intensive production yield could reach up to 35–40 tons per hectare.

According to our experience, for this purpose, some local (autochthonous, folk, primitive) cultivars of Asian, European, or American pomegranates have better potential compared to some commercial cultivars due to their higher resistance to pest and disease.

References

Akparov, Z.I., Bayramova, D.B., Mustafayeva, Z.P. and Mammadov, A.T. 2015. Pomegranate genetic diversity in Azeirbaijan. *Acta Hort., 1089*: 253–61.

Aksoy and Dalkilic, Z. 2019. Determination of Blooming, Pollen, and Fruit Set Characteristics in *Punica granatum*. *Notulae Botanicae Horti Agrobotanici Cluj-Napoca, 47*(4): 1258–63. https://doi.org/10.15835/nbha47411216.

Alcaraz-Mármol, F., Calín-Sánchez, Á., Nuncio-Jáuregui, N., Carbonell-Barrachina, A.A., Hernández, F. and Martínez, J.R. 2015. Classification of pomegranate cultivars according to their seed hardness and wood perception. *J. Texture Stud., 46*: 467–74.

Ashton, R. 2006. Meet the pomegranate. *In*: *The Incredible Pomegranate Plant and Fruit*, 5–8. Third Millenium Pub. Tempe Az. USA.

Aziz, S., Firdous, S., Rahman, H., Awan, S.I., Michael, V. and, Meru G. 2020. Genetic diversity among wild pomegranate (*Punica granatum*) in Azad Jammu and Kashmir region of Pakistan. *Electronic Journal of Biotechnology, 46*: 50–54. doi:10.1016/j.ejbt.2020.06.002.

Bartual, J., Zuriaga, E. and Badenes, M.L. 2021. International scenario of pomegranate production and utilization: Challenges and opportunities. *In*: *Ancient Fruit in Modern Horticulture*, 1–5. Ed. ICAR-NRCP and SARP, Solapur, Maharashtra, India.

Bartual, J., Valdes, G., Andreu, J., Lozoya, A., García-González, J. and Badenes, M. 2012. Pomegranate improvement through clonal selection and hybridization in Elche. *Options Mediterraneennes, 103*: 71–77.

Bartual, J., Palou, L. and Pérez-Gago, M.B. 2015. Characterization of fruit traits from 'Mollar de Elche' pomegranate progenies. *Acta Hortic.*, *1106*: 25–30. doi:10.17660/ActaHortic.2015.1106.5.

Bar-Ya'akov, I., Tian, L., Amir, R. and Holland, D. 2019. Primary Metabolites, Anthocyanins, and Hydrolyzable Tannins in the Pomegranate Fruit. *Frontiers in Plant Science*, *10*. doi:10.3389/fpls.2019.00620.

Caliskan, O. and Bayazıt, S. 2013. Morpho-pomological and Chemical Diversity of Pomegranate Accessions Grown in Eastern Mediterranean Region of Turkey. *Journal of Agricultural Science and Technology*, *15*: 1449–60.

Cao, S.Y. and Hou, L.F. 2013. *China Fruit Monograh-Punica Granatum*. China Forestry Publishing House, Beijing, China.

Chandra, R. and Babu, K.D. 2010. Propagation of pomegranate: A review. *Fruit, Vegetable, and Cereal Science and Biotechnology*, *4* (special issue 2): 51–55.

Chandra, R., Jadhav, V.T. and Sharma, J. 2010. Global scenario of pomegranate (Punica granatum L.) culture with special reference to India. *Fruit, Veg., Cereal Sci. Biotechnol.*, *4*: 7–18.

Chater, J.M., Merhaut, D.J., Jia, Z., Arpaia, M.L., Mauk, P.A. and Preece, J.E. 2018. Effects of Site and Cultivar on Consumer Acceptance of Pomegranate. *Journal of Food Science*, *83*: 1389–95. https://doi.org/10.1111/1750-3841.14101.

Chater, J.M., Yavari, A., Sarkhosh, A., Merhaut, D., Preece, J., Cossio, F, Qin, G., Liu, C., Li, J., Parashuram, S., Babu, D., Sharma, J., Yılmaz, C., Bartual, J., Mustafayeva, Z., Saeedi, M., Awad, N., Moersfelder, J. and Zhou, L. 2020. World Pomegranate Cultivars. *In*: Sarkhosh, Yavari and Zmani (Eds.), *The Pomegranate: Botany, Production and Uses*. https://doi.o/g/10.1079/9781789240764.0157.

da Silva, J.A.T., Rana, T.K., Narzary, D., Verma, N., Meshram, D.T.and , Ranade, S.A. 2013. Pomegranate biology and biotechnology: A review. *Scientia Horticulturae*, *160*: 85–107. doi:10.1016/j.scienta.2013.05.017.

Dafny-Yalin, M., Glazer, I., Bar-Ilan, I., Kerem, Z., Holland, D. and Amir, R. 2010. Color, sugars, and organic acids composition in aril juices and peel homogenates prepared from different pomegranate accessions. *J. Agric. Food Chem.* *58*: 4342–52. doi:10.1021/jf904337t.

Deepika, R. and Kanwar K. 2010. *In vitro* regeneration of *Punica granatum* L. plants from different juvenile explants. *J. Fruit Ornam. Plant Res.*, *18*: 5–22.

Desai, P., Patil, G., Dholiya, B., Desai, S., Patel, F.and , Narayanan S. 2018. Development of an efficient micropropagation protocol through axillary shoot proliferation for pomegranate variety 'Bhagwa'. *Annals of Agrarian Science*, *16*: 444–50. doi:10.1016/j.aasci.2018.06.002.

Drogoudi, P.D., Tsipouridis, C. and Michailidis, Z. 2005. Physical and Chemical Characteristics of Pomegranates. *Hort. Sci.*, *40*: 1200–1203. doi:10.21273/HORTSCI.40.5.1200.

Durgaç, C., Özgen, M., Özhan, E., Kaçar, Y.A., Kıygal, Çelebi, S., Gündüz, K. and Serçe, S. 2008. Molecular and pomological diversity among pomegranate (*Punica granatum* L.) cultivars in Eastern Mediterranean region of Turkey. *African Journal of Biotechnology*, *7*: 1294–1301.

Elkar, C., Ferchichi, A., Attia, F. and Bouajila, J. 2011. Pomegranate (*Punica granatum*) Juices: Chemical Composition, Micronutrient Cations, and Antioxidant Capacity. *Journal of Food Science*. *76*: C795–800. 10.1111/j.1750-3841.2011.02211.x.

Ercisli, S., Kafkas, E., Orhan, E., Salih Kafkas, Yildiz Dogan and Ahmet Esitken. 2011. Genetic characterization of pomegranate genotypes (*Punica granatum* L) by AFLP markers. *Biol. Res.*, *44*: 345–50. doi:10.4067/S0716-97602011000400005.

Feng, L., Wang, Z., Wang, C., Yang, X., An, M. and Yin Y. 2023. Multichromosomal mitochondrial genome of *Punica granatum*: Comparative evolutionary analysis and gene transformation from chloroplast genomes. *BMC Plant Biol.*, *23*(1): 512. doi: 10.1186/s12870-023-04538-8.

Ferrara, G., Cavoski, I., Pacifico, A., Tedone, L. and Mondelli, D. 2011. Morpho-pomological and chemical characterization of pomegranate (*Punica granatum* L.) genotypes in Apulia region, Southeastern Italy. *Scientia Horticulturae*, *130*: 599–606. doi:10.1016/j.scienta.2011.08.016.

Fischer, U.A., Carle, R. and Kammerer, D.R. 2011. Identification and quantification of phenolic compounds from pomegranate (*Punica granatum* L.) peel, mesocarp, aril, and differently produced juices by HPLC-DAD–ESI/MSn. *Food Chemistry*, *127*: 807–21.

Franck, N. 2012. The cultivation of Pomegramate cv. Wonderful in Chile. *In*: Melgarejo, P. Valero, D. (Eds.). *II International Symposium on the Pomegranate*, 97–99. Zaragoza: CIHEAM / Universidad Miguel Hernández, (Options Méditerranéennes: Série A. Séminaires Méditerranéens; n. 103).

Giancaspro, A., Giove, S.L., Marcotuli, I., Ferrara, G. and Gadaleta, A. 2023. Datasets for genetic diversity assessment in a collection of wild and cultivated pomegranates (*Punica granatum* L.) by microsatellite markers. *Data Brief.*, *49*: 109346. doi:10.1016/j.dib.2023.109346.

Hasnaoui, N., Jbir, R., Mars, M., Trifi, M., Kamal-Eldin, A., Melgarejo, P. and Hernandez, F. 2011. Organic Acids, Sugars, and Anthocyanins Contents in Juices of Tunisian Pomegranate Fruits. *Inter. J. Food Proper.*, *14*: 741–57. doi:10.1080/10942910903383438.

Hasnaoui, N., Mars, M., Chibani, J. and Trifi, M. 2010. Molecular Polymorphisms in Tunisian Pomegranate (*Punica granatum* L.) as Revealed by RAPD Fingerprints. *Diversity*, *2*(1): 107–114. doi:10.3390/d2010107.

Holland, D., Hatib, K. and Bar-Ya'akov, I. 2009. Pomegranate: Botany, Horticulture, Breeding. *Hortic. Rev.*, *35*: 127–91. doi:10.1002/9780470593776.ch2.

Holland, D. and Bar-Ya'akov, I. 2018. Pomegranate (*Punica Granatum* L.) Breeding. *In*: Al-Khayri J., Jain S. and Johnson D. (Eds.), *Advances in Plant Breeding Strategies: Fruits*. Springer, Cham. https://doi.org/10.1007/978-3-319-91944-7_15.

Holland, D., Bar-Ja'kob, I. and Hatib, K. 2014. 'Emek' a red and very early-ripening new pomegranate cultivar. *HortScience*, *49*: 968–70.

Hosamani, S.B. and Alur, A.S. 2015. Pomegranate: An Indian scenario. *Acta Hort.*, *1089*: 333–43.

Jaika, K. and Mehra, P.N. 1986. Morphogenesis in *Punica granatum* (pomegranate). *Can. J. Bot.*, *64*: 1644–53.

Jalikop, S.H. 2010. Pomegranate breeding. En: *Fruit, Vegetable, and Cereal Science and Biotechnology*. Global Science Books.

Janick, J. 2005. The Origins of Fruits, Fruit Growing, and Fruit Breeding. *In*: J. Janick (Ed.), *Plant Breeding Reviews*. doi:10.1002/9780470650301.ch8.

Kalaycıoglu, Z. and Erim, F. 2017. Total phenolic contents, antioxidant activities, and bioactive ingredients of juices from pomegranate cultivars worldwide. *Food Chem.*, *221*: 496–507. doi:10.1016/j.foodchem.2016.10.084.

Karimi, H.R. and Hasanpour, Z. 2014. Effects of salinity and water stress on growth and macro nutrients concentration of pomegranate (*Punica granatum* L.). *Journal of Plant Nutrition*, *37*12): 1937–51. doi:10.1080/01904167.2014.920363.

Karimi, H.R., Esmaelizadeh, M. and Abolipur M. 2011. Preliminary study of stratification treatment on germination of pomegranate (*Punica granatum* L.). *Iranian Horticultural Science Congress*. Isfahan. Iran. September.

Koppel, K. and Chambers IV, E. 2010. Development and application of a lexicon to describe the flavor of pomegranate juice. *Journal of Sensory Studies*, *25*: 819–37. doi:10.1111/j.1745-459X.2010.00307.x.

Kumar, R., Jakhar, M.L., Verma, R. and Jat, H.R. 2017. Pomegranate Micropropagation: A Review. *Int. J. Pure App. Biosci.*, *5*: 1138–49. doi: 10.18782/2320-7051.5124.

Kumar, B.S. 2016. Disease-resistant wild varieties have been hybridized with commercial varieties. *The Hindu*, May 7.

Lansky, E.P. and Newman, R.A. 2007. *Punica granatum* (pomegranate) and its potential for prevention and treatment of inflammation and cancer. *J. Ethnopharmacol.*, Jan. 19; *109*(2): 177–206. doi: 10.1016/j.jep.2006.09.006.

Levin, G.M. 2006. *Pomegranate Roads: A Soviet Botanist's Exile from Eden*. Floreant Press. Forestville, CA. USA.

Li, Q., Tan, W., Zhao, L., Luo, H., Zhou, Z., Zhang,Y., Bi, R. and Zhao, L. 2023. A Comprehensive Evaluation of 45 Pomegranate (Punica Granatum L.) Cultivars Based on Principal Component Analysis and Cluster Analysis. *International Journal of Fruit Science*, *23*(1): 135–50. doi:10.10 80/15538362.2023.2223312.

Liu, Z., Zhao, Y., Hu, Q., Tan, B., Chen, Y., Jiaan, Z., Shi, J. and Wan, R. 2019. Study on micro-grafting *in vitro* of pomegranate. *Journal of Fruit Science, 36* (4): 521–28 (in Chinese with English abstract) doi:10.13925/j.cnki.gsxb.20180422.

Luo, H., Zhao, L., Chen, Y., Wang, Y. Wang, Y. and Hao, Z. 2018. *Evaluation on Cold Resistance of 24 Pomegranate Cultivars*. doi:10.14083/j.issn.1001-4942.2018.01.009. (in Chinese with English abstract).

Luo, X., Li, H., Wu, Z., Yao, W., Zhao, P., Cao, D., Yu, H., Li, K., Poudel, K., Zhao, D., Zhang, F., Xia, X., Chen, L., Wang, Q., Jing, D. and Cao, S. 2020. The pomegranate (*Punica granatum* L.) draft genome dissects genetic divergence between soft- and hard-seeded cultivars. *Plant Biotechnol J., 18*(4): 955–68. doi:10.1111/pbi.13260.

Martínez, J.J., Melgarejo, P., Legua, P., Martínez, R. and Hernández, F. 2012. Diversity of pomegranate (*Punica granatum* L.) germplasm in Spain. *Options Méditerranéennes: Série A. Séminaires Méditerranéens,. 103*: 53–56.

Mayuoni-Kirshenbaum, L., Bar-Ya'akov, I., Hatib, K., Holland, D. and Porat, R. 2013. Genetic diversity and sensory preference in pomegranate fruits. *Fruits, 68*: 517–24.

Melgarejo, P., Martínez, J.J., Hernández, F., Martínez, R., Legua, P., Oncina, R. and Martínez-Murcia, A. 2009. Cultivar identification using 18S–28S rDNA intergenic spacer-RFLP in pomegranate (*Punica granatum* L.). *Scientia Horticulturae, 120*: 500–503. doi:10.1016/j.scienta.2008.12.013.

Melgarejo-Sánchez, P., Martínez, J.J., Legua, P., Martínez, R., Hernández, F. and Melgarejo, P. 2015. Quality, antioxidant activity, and total phenols of six Spanish pomegranates clones. *Scientia Horticulturae, 182*: 65–72. doi: 10.1016/j.scienta.2014.11.020.

Mena, P., García-Viguera, C., Navarro-Rico, J., Moreno, D.A., Bartual, J., Saura, D. and Martí, N. 2011. Phytochemical characterization for industrial use of pomegranate (*Punica granatum* L.) cultivars grown in Spain. *J. Sci. Food Agric., 91*: 1893–1906. doi:10.1002/jsfa.4411.

Mondragon, J.C. 2018. Agronomical, physico-chemical, and functional characterization of Mexican pomegranates as compared to Wonderful pomegranate. *Acta Hort., 1089*: 299–306.

Mondragon, J.C. 2012. Breeding Mexican pomegranates to improve productivity and quality and increase versatility of uses. *Options Mediterraneennes, 103*: 61–66.

Murkute, A.A., Patil, S. and Mayakumari. 2003. Exudation and browning in tissue culture of pomegranate. *Agricultural Science Digest, 23*: 29–31.

Nataraja, K. and Neelambika, G.K. 1996. Somatic embryogenesis and plantlet from petal cultures of pomegranate, *Punica granatum* L. *Indian J. Exp. Biol., 34*: 719–21.

Niu, J., Hong-hua, T., Shang-yi, C., Haoxian, L., Pinli, Y., Diguang, Z. and Guhong, Z. 2018. Progress in soft seed pomegranate in China. *Acta Hort., 1089*: 379–83.

Onur, C. and Kaska, N. 1985. Selection of pomegranate (*Punica granatum* L.) from The Mediterranean region of Turkey. *Tr. J. Agric. Forest, 9*: 25–33.

Ophir, R., Sherman, A., Rubinstein, M., Eshed, R., Sharabi Schwager, M., Harel-Beja, R., Bar-Ya'akov, I. and Holland, D. 2014. Single-nucleotide polymorphism markers from *de novo* assembly of the pomegranate transcriptome reveal germplasm genetic diversity. *PLoS One, 9*(2): e88998. doi:10.1371/journal.pone.0088998.

Panis, B., Nagel, M., Van den Houwe, I. 2020. Challenges and Prospects for the Conservation of Crop Genetic Resources in Field Genebanks, in *In Vitro* Collections and/or in Liquid Nitrogen. *Plants (Basel), 9*: 1634. doi:10.3390/plants9121634.

Patil, P.G., Singh, N.V., Parashuram, S., Bohra, A., Mundewadikar, D.M., Sangnure, V.R., Babu, K.D. and Sharma, J. 2020. Genome wide identification, characterization, and validation of novel miRNA-based SSR markers in pomegranate (*Punica granatum* L.). *Physiol. Mol. Biol. Plants, 26*: 683–96. doi:10.1007/s12298-020-00790-6.

Peng, Y., Wang, G., Cao, F. and Fu, F. 2020. Collection and evaluation of thirty-seven pomegranate germplasm resources. *Appl. Biol. Chem., 63*: 15. https://doi.org/10.1186/s13765-020-00497-y

Poyrazoğlu, E., Gökmen, V. and Artık, N. 2002. Organic Acids and Phenolic Compounds in Pomegranates (*Punica granatum* L.) Grown in Turkey. *Journal of Food Composition and Analysis, 15*(5): 567–75. https://doi.org/10.1006/jfca.2002.1071.

Qin, G., Xu, C., Ming, R., Tang, H., Guyot, R., Kramer, E.M., Hu, Y., Yi, X., Qi, Y., Xu, X., Gao, Z., Pan, H., Jian, J., Tian, Y., Yue, Z. and Xu, Y. 2017. The pomegranate (*Punica granatum* L.) genome and the genomics of punicalagin biosynthesis. *Plant J., 91*(6): 1108–28. doi:10.1111/tpj.13625.

Rajasekharan, P.E. and Sahijram. 2015. *In vitro* conservation of plant germplasm. *In*: Bahadur, B., Venkat Rajam, M., Sahijram, L. and Krishnamurthy, K.V. (Eds.), *Plant Biology and Biotechnology*: Volume II: *Plant Genomics and Biotechnology*, 417–43. doi: 10.1007/978-81-322-2283-5_22.

Ran, K., Xialoi, S. and Wang, S. 2015. 'Taishanhong' pomegranate and its cultivation techniques with high yield and good quality. *Acta Hort., 1089*: 263–68.

Rana, T.S., Narzary, D. and Ranade, S.A. 2010. Systematics and taxonomic disposition of the genus *Punica* L. *Fruit, Vegetable, and Cereal Science and Biotechnology, 4* (special issue 2): 51–55.

Ranade, S.A., Rana, T.S. and Narzary, D. 2009. SPAR profiles and genetic diversity among pomegranate (*Punica granatum* L.) genotypes. *Physiol. Mol. Biol. Plants, 15*: 61–70. doi:10.1007/s12298-009-0006-x.

Richardson, E.A., Seeley, S.D., Walker, R.D., Anderson, J. and Aschcroft, G. 1975. Phenoclimatography of spring peach bud development. *HortScience, 10*: 236–37.

Sadeghi, N., Jannat, B., Oveisi, M.R., Hajimahmoodi, M. and Photovat, M. 2009. Antioxidant Activity of Iranian Pomegranate (*Punica granatum* L.) Seed Extracts. *JAST., 11*: 633–38.

Sahu, P., Murthy, S., Rajeshekharan, Angami, T., Devi, L., Hidangmayum, Babu, K. and Pal, R.K. 2014. Feasibility of pollen cryopreservation in pomegranate (*Punica granatum* L.). *In*: *National Seminar-cum-Exhibition on Pomegranate for Nutrition, Livelihood Security and Entrepreneurship Development*. ICAR-NRCP, Solapur.

Sarkhosh, A., Zamani, Z., Fatahi, R. and Ebadi, A. 2006. RAPD markers reveal polymorphism among some Iranian pomegranate (*Punica granatum* L.) genotypes. *Scientia Horticulturae, 111*(1): 24–29. doi:10.1016%2Fj.scienta.2006.07.033.

Shahsavari, S., Noormohammadi, Z., Sheidai, M., Farahani, F. and Vazifeshenas, M.R. 2021. Genetic structure, clonality, and diversity in commercial pomegranate (*Punica granatum* L.) cultivars. *Genet. Resour. Crop Evol., 68*: 2943–57. doi:10.1007/s10722-021-01167-8.

Shilpa, P., Roopa, P., Singh, N.V., Patil, P.G., Babu, K.D. and Marathe, R.A. 2021. *New Commercial Varieties of Pomegranate in India*, 24. Edn. Agro India.

Singh, N.V., Patil, P.G., Sowjanya, R.P., Parashuram, S., Natarajan, P., Babu, K.D., Pal, R.K., Sharma, J. and Reddy, U.K. 2021. Chloroplast Genome Sequencing, Comparative Analysis, and Discovery of Unique Cytoplasmic Variants in Pomegranate (*Punica granatum* L.). *Front. Genet., 12*: 704075. doi:10.3389/fgene.2021.704075.

Singh, P. and Patel, R.M. 2016. Factors affecting *in vitro* degree of browning and culture establishment of pomegranate. *African Journal of Plant Science, 10*(2): 43–49. doi:10.5897/AJPS2013. 1119.

Stover, E.D. and Mercure, E.W. 2007. The pomegranate: A new look at the fruit of paradise. *HortScience, 42*(5): 1088–92.

Szychowski, P., Frutos, M.J., Burló, F., Pérez-López, A.J., Carbonell-Barrachina, A.A. and Hernández, F. 2015. Instrumental and sensory texture attributes of pomegranate arils and seeds as affected by cultivar. *Food Science and Technology, 60*: 656–63. https://doi.org/10.1016/j.lwt.2014.10.053.

Terakami, S., Matsuta, N., Yamamoto, et al. 2007. Agrobacterium mediated transformation of the dwarf pomegranate (*Punica granatum* L. Var. nana). *Plant Cell Rep., 26*: 1243–51.

Tzulker, R., Glazer, I., Bar-Ilan, I., Holland, D., Aviram, M. and Amir, R. 2007. Antioxidant activity, polyphenol content, and related compounds in different fruit juices and homogenates prepared from 29 different pomegranate accessions. *J. Agric. Food Chem., 55*: 9559–70. doi:10.1021/jf071413n.

UPOV. *Protocol for Tests on Distincness, Uniformity and Stability in Pomegranate*. https://cpvo.europa.eu/sites/default/files/documents/punica_granatum.pdf (Accessed 12/12/2021).

Varasteh, F., Arzani, K., Zamani, Z. and Mohseni, A. 2009. Evaluation of the most important fruit characteristics of some commercial pomegranate (*Punica granatum* L.) cultivars grown in Iran. *Acta Hortic., 818*: 103–108. doi:10.17660/ActaHortic.2009.818.13.

Višnjevec, A.M., Ota, A., Skrt, M., Butinar, B., Možina, S.S., Cimerman, N.G., Nečemer, M., Arbeiter, A.B., Hladnik, M., Krapac, M., Ban, D., Bučar-Miklavčič, M., Ulrih, N.P. and Bandelj, D.

2017. Genetic, Biochemical, Nutritional, and Antimicrobial Characteristics of Pomegranate (*Punica granatum* L.) Grown in Istria. *Food Technol. Biotechnol.*, *55*: 151–63. doi:10.17113/ftb.55.02.17.4786.

Viuda-Martos, M., Fernández-López, J. and Pérez-Álvarez, J.A. 2010. Pomegranate and its many functional components as related to human health: A review. *Compr. Rev. Food Sci. Food Saf.*, *9*: 635–54. doi:10.1111/j.1541-4337.2010.00131.x.

Villamón, D., Palou, L., Bartual, J., Taberner, V., de la Fuente, B. and Pérez-Gago, M.B. 2019. Fruit quality attributes of a new Spanish pomegranate cultivar at harvest and during cold storage. *Acta Hortic.*, *1254*: 275–82. doi10.17660/ActaHortic.2019.1254.41.

Wetzstein, H.Y., Nadav, R.E., Wilkins, A. and Pinheiro, M. 2011. A Morphological and Histological Characterization of Bisexual and Male Flower Types in Pomegranate. *J. Amer. Soc. Hort. Sci.*, *136*(2): 83–92.

Yang, W. and Yang, F. 2015. Evaluation and Comparison of Fruit Quality of Pomegranate Cultivars in China. *Storage and Process*, *15* (2) : 62–67, 72.

Yezhov, V., Smykov, A.V. and. Smykov, V.K. 1995. Genetic resources of temperate and subtropical fruit and nut species at the Nikita Botanical Gardens. *HortScience*, *40*: 5–9.

Yılmaz, C. 2007. *Pomegranate*. Hasad Publisher, Istanbul.

Yuan, Z., Fang, Y., Zhang, T., Fei, Z., Han, F., Liu, C., Liu, M., et al. 2018. The pomegranate (*Punica granatum* L.) genome provides insights into fruit quality and ovule developmental biology. *Plant Biotechnol J.*, *16*(7): 1363–74. doi: 10.1111/pbi.12875.

Yuan, Z.H. and Zhao, X. 2019. Pomegranate genetic resources and their utilization in China. *Acta Hortic.*, 1254: 49–56. DOI: 10.17660/ActaHortic.2019.1254.8

Zamani Z., Adabi M. and Khadivi-Khub A. 2013. Comparative analysis of genetic structure and variability of wild and cultivated pomegranates as revealed by morphological variables and molecular markers. *Plant Sys. Evol.*, *299*: 1067–80.

Zarei, A., Zamani, Z. and Sarkhosh, A. 2020. Biodiversity, Germplasm Resources, and Breeding Methods. *In*: Sarkhosh, Yavari, and Zamani (Eds), *The Pomegranate: Botany, Production, and Uses*. https://doi.o/g/10.1079/9781789240764.0157.

Zarei, A., Zamani, Z., Fatahi, R., Mousavi, A., Salami, S.A., Avila, C. and Cánovas, F.M, 2016. Differential expression of cell-wall related genes in the seeds of soft and hard-seeded pomegranate genotypes. *Scientia Horticulturae*, *205*: 7–16. https://doi.org/10.1016/j.scienta.2016.03.043.

Zarei, A., Zamani, Z., Fatahi, R., Mousavi, A. and Salami, S.A. 2013. A Mechanical Method of Determining Seed-Hardness in Pomegranate. *Journal of Crop Improvement*, *27*(4): 444–59. doi:10.1080/15427528.2013.790867.

Zhang S.P., Wang, L.J. and Cao, S.Y. 2008. Analysis of genetic diversity of 23 pomegranate genotypes by SRAP. *Journal of Fruit Science*, *25*(5): 655–60 (in Chinese with English abstract).

Zhao, Y.L., Feng, Y.Z., Li, Z.H. and Cao, Q. 2006. Breeding of the new pomegranate cultivar 'Yushiliu 4'. *China Fruits*, *2*: 8–10.

Zhao, L., Li, M., Cai G., Pan T. and Shan, C. 2013. Assessment of the genetic diversity and genetic relationships of pomegranate (*Punica granatum* L.) in China using RAMP markers. *Scientia Horticulturae*, *151*: 63–67. doi:10.1016/j.scienta.2012.12.015.

Zhao, X., Yuan, Z., Yanming, F. and Yin, Y. and Feng, L. 2013. Characterization and evaluation of major anthocyanins in pomegranate (*Punica granatum* L.) peel of different cultivars and their development phases. *Eur. Food Res. Technol.*, *236*: 109–17. doi: 10.1007/s00217-012-1869-6.

Zhu, L.W., Jia, B., Zhang, Y.M., et al., 2004. 'Baiyushizi', a high quality pomegranate cultivar. *South China Fruits*, *33*(5): 69–70.

Zhu, L.W., Zhang, S.M., Gong, X.M., et al., 2005. A new soft-seeded pomegranate variety 'Hongyushizi'. *Acta Hortic Sin.*, *32*: 965.

Zuriaga, E., Pintová, J., Bartual, J. and Badenes, M.L. 2022. Characterization of the Spanish Pomegranate Germplasm Collection Maintained at the Agricultural Experiment Station of Elche to Identify Promising Breeding Materials. *Plants*. *11*(9): 1257. doi: 10.3390/plants11091257.

3

The Genome of Pomegranate

Wei Zhang,[1] *Jiyu Li* and *Gaihua Qin*[1*]

The pomegranate is a perennial fruit tree, bearing nutritious fruits. Genomics holds promise for discovering the genetic characteristics and evolution of pomegranates. Whole-genome sequencing and assembly are important tools for exploring genomics. This information can help to understand the genetic basis contributing to nutritional content biosynthesis, disease resistance, and adaptability, provide insights into the genomic basis of health-related compounds biosynthesis to maximize their potential health benefits, and also provide insights into the evolutionary history of pomegranates. High-throughput sequencing techniques, such as next-generation sequencing (NGS), have enabled the rapid and cost-effective sequencing of large genomes. Genome assembly combines the sequenced DNA fragments to reconstruct the complete genome. Assembling the pomegranate genome provides a reference for further genetic and genomic studies. Besides the nuclear genome, the genomes of mitochondrial, and chloroplast genomes are needed, the set of genomes allows for a more comprehensive understanding of the genetic makeup and evolutionary history of the species.

1. Introduction

The pomegranate tree, scientifically known as *Punica granatum* L. and belonging to the Lythraceae family, is indeed a perennial fruit tree that bears nutritious fruits and has been used as a medicinal resource for people (Li, 1998; Kim et al., 2002; Sánchez-Lamar et al., 2008; Yuan et al., 2008). The fruit of the pomegranate tree is highly valued for its distinctive flavor and nutrient content, and it has been

[1] Anhui Academy of Agricultural Sciences, Hefei 230001, China.
* Corresponding author: qghahstu@163.com

traditionally utilized for various medicinal purposes in different cultures. The rich nutritional profile of pomegranate fruit, along with its potential health benefits, has contributed to its significance as both a food source and a natural remedy. They are rich in nutrients such as vitamin C, vitamin K, and potassium, and are also a great source of antioxidants, particularly punicalagins and anthocyanins. These antioxidants have been linked to various health benefits, including reducing the risk of heart disease, lowering blood pressure, and having anti-inflammatory effects (Halvorsen et al., 2002; Johanningsmeir and Harris, 2011). In addition to its nutritional value, pomegranate has a long history of use in traditional medicine for various purposes, such as treating diarrhea, dysentery, and parasitic infections. The wide range of potential health benefits makes pomegranate a valuable addition to a healthy diet. Whole-genome sequencing and assembly are important tools for exploring pomegranates' genetic characteristics and evolution. By sequencing the entire genome, researchers can identify the genes responsible for traits such as polycarpic phenotype, ellagitannin synthesis, and anthocyanin accumulation. This information can be used to develop new varieties with improved fruit quality, disease resistance, and other desirable traits. Additionally, whole-genome sequencing can provide insights into pomegranates' evolutionary history. By comparing the genomes of different varieties and species, researchers can determine how these plants have evolved and how they are related to other plant species. This information can help us better understand the origins and diversification of pomegranates and their place in the broader tree of life.

In 2017, the first genome of pomegranate, specifically the cultivar 'Dabenzi', was published using next-generation sequencing (NGS) techniques. This milestone provided valuable insights into the genetic makeup and characteristics of pomegranates. The assembly of the 'Dabenzi' genome allowed researchers to identify and annotate genes, study gene expression patterns, and investigate the functional elements within the genome. It also facilitated the discovery of genetic variations and markers that can be used in breeding programs and genetic studies. The publication of the 'Dabenzi' genome was a significant advancement in pomegranate research, laying the foundation for further studies on this fruit tree species. Since then, additional genome sequencing efforts have been undertaken to explore the genetic diversity and evolutionary history of pomegranates, leading to a deeper understanding of this fascinating plant (Qin et al., 2017). The genome was assembled by SOAPdenovo with 328.38 Mb size. The genome of another pomegranate cultivar, 'Taishanhong', was sequenced using next-generation sequencing (NGS) techniques. The assembly of the 'Taishanhong' genome was achieved using the ALLPATHS-LG strategy and had a size of approximately 274Mb (Yuan et al., 2008). Genome sequencing efforts, such as the assembly of the 'Taishanhong' genome, contribute to our understanding of the genetic diversity within pomegranates. By comparing different pomegranate genomes, researchers can identify variations in gene sequences, structural variations, and other genomic features that may be associated with specific traits or characteristics. The availability of multiple pomegranate genomes provides a valuable resource for comparative genomics studies, breeding programs, and further exploration of the biological mechanisms underlying important traits in pomegranates. Third-generation single-molecule sequencing (TGS) technologies, such as the PacBio platform, have emerged as powerful tools for *de novo* genome assembly due to their ability to generate long reads. In 2020, the genome of a soft-seeded pomegranate cultivar, 'Tunisia', was published using the PacBio platform and assembled with the Canu software. The use of TGS technologies for genome assembly has significantly improved our ability to study complex genomes, such as those found in pomegranates, and has led to a deeper understanding of their biology and potential applications in agriculture (Luo et al., 2020).

The high-quality pomegranate genome assemblies provide a valuable resource for comparative and evolutionary genomics studies, as well as for investigating the genetic mechanisms underlying important biological traits. By comparing the genomes of different pomegranate cultivars, researchers can identify variations in gene sequences, structural variations, and other genomic features that may be associated with specific traits or characteristics. This information can be used to develop new varieties with desirable traits, improve cultivation practices, and enhance disease management strategies. Furthermore, the availability of high-quality genome assemblies allows for more precise identification and functional annotation of genes involved in important biological processes, such as fruit development, ripening, and quality. This information can be used to develop molecular markers and breeding strategies aimed at improving these traits.

2. The History of Genome Sequencing Techniques and Strategies

The rapid evolution of genome sequencing techniques and strategies has made it more accessible and affordable, allowing researchers and scientists to explore and unlock the secrets hidden within the DNA of various organisms. Sanger and Coulson (1975) developed the "plus and minus" method that was used to determine two sequences in bacteriophage φX174 DNA using the synthetic decanucleotide A-G-A-A-A-T-A-A-A-A and a restriction enzyme digestion product as primers (Sanger and Coulson, 1975); in addition, the era of genome sequencing has brought about an explosion of data, driving the growth of bioinformatics and computational biology. This interdisciplinary field plays a critical role in analyzing and interpreting the vast amount of genomic information, leading to a deeper understanding of biological systems and the development of novel technologies. Standen (1979) proposed the idea of "shotgun sequencing", a strategy that used bacterial vectors to clone the sequence of DNA, the random reads were sequenced and assembled according to their overlaps (R. Standen, 1979). Following the strategy, Sanger assembled the genome of phageλin 1982, even though it was only 48,502 bp long (Sanger et al., 1982). In 2005, there were new methods that came to be known as next-generation sequencing, as they were designed to employ massively parallel strategies to produce large amounts of sequence from multiple samples at very high-throughput and at a high degree of sequence coverage to allow for the loss of accuracy of individual reads when compared to Sanger sequencing (K. Kulski, 2016). The second-generation sequencing technology had many characteristics, shotgun sequencing of random fragmented genomic (fg) DNA or cDNA reverse transcribed from RNA is performed without the need for cloning via a foreign host cell: instead, linker and/or adapter sequences are ligated to the fgDNA or cDNA for construction of template libraries (K. Kulski, 2016).

NGS has an enormous contribution to viral, bacterial, and archaeal genomics. It has become easier, faster, and cheaper to assemble the whole genome, as the improved technology of bioinformatics and the NGS data were used to *de novo* assemble the whole genome. The raw sequencing signals were produced by a sequencing machine or system, then were converted into nucleotide bases of short read data and they were stored in FASTQ format (K. Kulski, 2016). It is necessary to quantify the raw reads, the FastQC software was used to check the quality of the raw data (S. Andrews, 2010). Then the Trimmomatic was employed to trim the raw data if the raw data had a high error rate (Bolger et al., 2014). At this step, it will remove reads with low quality, sequence errors, and sequences such as primers, vectors, adapters, tags, and tails that were introduced experimentally during the preparation of the sequencing libraries (K. Kulski, 2016; Hong et al., 2013). The clean data were fed into genome *de novo* assemblers, such as ALLPATHS-LG, SPAdes, SOAPdenovo, and Minia. These genome assembly tools would generate the scaffold file that was used to assemble chromosome-level by different strategies, such as according to chromosomal homology of related species, genetic map, bio-nano, and Hi-C technology.

Third-generation single-molecule sequencing technologies have indeed made significant advancements in reducing the cost of sequencing and improving the length of reads compared to traditional next-generation sequencing (NGS) methods (K. Kulski, 2016; M. Melzker, 2010; Schadt et al., 2010; Eid et al., 2009). Currently, there are three main platforms for third-generation sequencing: Single-Molecule Real-Time (SMRT) sequencing by Pacific Biosciences (PacBio), Nanopore sequencing by Oxford Nanopore Technologies (ONT), and a newer technology called Synthetic Long-Read Sequencing (K. Kulski 2016). Pacific Biosciences (PacBio) has recently developed a new sequencing technology called HiFi (High-Fidelity) sequencing, which is a variant of Single-Molecule Real-Time (SMRT) sequencing. HiFi sequencing combines the benefits of long read lengths with high accuracy, making it ideal for various applications, including genome assembly, structural variant detection, haplotype phasing, and transcriptomics (Hon et al., 2020). Third-generation sequencing technologies such as PacBio SMRT sequencing and Nanopore sequencing can produce reads that are significantly longer than those generated by traditional Next-Generation Sequencing (NGS) technologies. The length of reads produced by third-generation sequencing ranges from several kilobases to hundreds of kilobases. This is a major advantage over NGS, which typically generates reads that are only a few hundred base pairs in length. The generation of long continuous consensus sequences benefits greatly from the use of long reads, as they provide a more comprehensive view of the genomic landscape. In contrast, NGS short reads have difficulty in detecting overlaps due to their limited length (C-S Chin et al., 2016).

To assemble the ultra-long and noisy reads, many kinds of strategies were used, such as FALCON (Alkan et al., 2011) and Canu (Koren et al., 2017). It was outstanding in terms of genome assembly quality for both of them; however, they consume a great amount of computer resources. Currently, how to optimize the computational efforts for developing a *de novo* genome assembler is a research hotspot, such as Wtdbg2 (Ruan and Li, 2020), Flye (Kolmogorov et al., 2019), NextDenovo (https://github.com/Nextomics/NextDenovo), MECAT/ NECAT (Xiao et al., 2017; Chen et al., 2021), Raven (Vaser and Sikic, 2020), SMARTdenovo (Liu et al., 2020), minimum (H. Li, 2016). The PacBio HiFi sequencing method yields highly accurate long-read sequencing datasets with read lengths averaging 10–25 kb and accuracies greater than 99.5% (Hon et al., 2020; Wenger et al., 2019). In 2020, Chicano was published for *de novo* assembling HiFi reads (Nurk et al., 2020). In 2021, Chiliasm was published, it was a fast haplotype-resolved *de novo* assembler for PacBio HiFi reads (Cheng et al., 2021). Genome assembly is a complex process, and different software and strategies can produce varying results in terms of assembly quality. It is often beneficial to apply multiple assembly approaches and compare the results to select the optimal solution (Del Angel et al., 2018).

3. The NGS Technology Enables Pomegranate Genome Assembly

The development of genome sequencing techniques and strategies has indeed made it possible to assemble the genome of pomegranate. The use of NGS technology has played a crucial role in enabling the assembly of the pomegranate genome. In 2017, the first pomegranate genome was successfully assembled using NGS technology, marking a significant milestone in pomegranate genomics research (Qin et al., 2017). The sequencing of the pomegranate cultivar 'Dabenzi' using the Illumina HiSeq 2000 platform with 305.52x depth and paired-end reads of libraries ranging from 170bp–40kb is impressive. This sequencing effort should have provided a high-quality assembly of the pomegranate genome, given the depth of coverage and range of library sizes used. The evaluated genome size of 357Mb based on the k-mer method is consistent with previous estimates of pomegranate genome size, which ranges from 300–600 Mb depending on the variety (Table 1) (Qin et al., 2017). The *de novo* assembly of the pomegranate genome using SOAPdenovo2 and obtaining a final genome size of 328.38Mb is a significant achievement. The N50 contig length of 66.97 kb indicates that half of the genome is contained within contigs of at least this length. Achieving an N50 scaffold length of 1.89 Mb implies a reasonable level of continuity in the assembly at the scaffold level. A genetic map of 'Dabenzi' was constructed by sequencing 195 F_1 individuals of 'Dabenzi' and 'Baiyushizi' subspecies cross for constructing pseudo-chromosomes, nine linkage groups were established according to the haploid chromosome of pomegranate, which was by the chromosomes numbers in root tip cell of pomegranate 'Dabenzi' ($2n = 2x = 18$) (Table 2). The identification of 155.3 Mb of repetitive sequences that account for 46.1% of the pomegranate genome assembly is significant and consistent with previous estimates of repeat content in plant genomes (Qin et al., 2017). By estimating the insertion time of identified LTR retrotransposons based on the divergence of LTR sequences, the relatively lower divergence of Copia than Gypsy elements was found, suggesting a recent or ongoing wave of retrotrans position for Copia elements. Considering that Copia elements had been inserted in both gene-rich and gene-poor regions, their activities represent a potential source of genetic and epigenetic variation (Qin et al., 2017).

Table 1 Statistics for libraries with different insert sizes libraries used in genome assembly.

Pair-end Libraries	*Insert Size (Mean/SD)*	*Average Reads Length (bp)*	*Total Data (Gb)*	*Sequence Depth (×)[#]*
Filtered Reads	170bp	100_100	8.71	24.4
	500bp	100_100	32.42	90.81
	2kp	49_49	3.15	8.82
	5Kb	49_49	3.74	10.48
	10Kb	49_49	1.79	5.01
	20Kb	49_49	1.64	4.59
	40Kb	49_49	0.78	2.18
Total	/	/	52.22	146.27

Note: [#] The sequenced depth is based on genome size 357Mb. DNA libraries with different insert sizes were constructed and sequenced. In total, 109.07 GB of raw data were generated and the sequencing depth is about 305.52×. After data filtering, more than 146.27× clean data were used in the genome assembly.

In the gene annotation process, firstly, repeats were annotated by *de novo* and homology-based approaches located in the pomegranate genome. Then, the RNA-seq data of six different organs or tissues combined with *de novo* and homology-based prediction were used to train the prediction program 29,229 protein-coding gene models were annotated. There were 24,734 genes located on the chromosomes, 12,036 of them located on the chromosomes with reversed direction. All of the gene models were divided into 13,640 gene families based on sequence similarity. There are 6,230 orphan genes in pomegranate, which could not cluster with any genes of the other 12 species (Table 2). These orphan genes were enriched in terms including catalytic activity, metabolic process, cellular process, and cell and cell part GO terms and pathways for biosynthesis of secondary metabolites and phenylpropanoid according to KEGG pathway enrichment.

Table 2 Gene families clustered by OrthoMCL in 12 species.

Species	*Genes Number*	*Genes in Families*	*Un-clustered Genes*	*Family Number*	*Unique Families*	*Average Genes per Family*
P.granatum	29,229	22,999	6,230	13,640	806	1.69
A.thaliana	26,637	22,888	3,749	13,431	703	1.7
C.papaya	27,793	19,497	8,296	13,789	544	1.41
E.grandis	36,379	28,460	7,919	14,603	866	1.95
M.domestica	61,721	44,250	17,471	17,774	3640	2.49
O.sativa	38,964	26,486	12,478	13,404	2107	1.98
P.trichocarpa	40,963	33,033	7,930	15,608	892	2.12
T.cacao	28,624	23,772	4,852	15,357	587	1.55
V.vinifera	25,329	18,995	6,334	13,545	611	1.4
Z.jujuba	32,808	27,041	5,767	13,790	983	1.96
S.lycopersicum	33,585	25,357	8,228	14,464	1004	1.75
P.persica	27,792	24,163	3,629	15,369	383	1.57

Note: Un-clustered genes refer to special genes of current species; Unique families refer to special gene families of current species.

In 2018, another genome of pomegranate 'Taishanhong' was published. It was sequenced by another Hiseq 2500 platform of Illumina, and the clean reads were assembled by ALLPATHS-LG. The repetitive DNA accounted for 51.2% (140.2 Mb) of the genome assembly. The final genome assembled sequence was 274Mb, and 30,903 protein-coding was predicted (Yuan et al., 2018). As is well known, pomegranate fruit is rich in ellagitannin-based compounds that have antioxidant activities (Johanningsmeir and Harris, 2011). The genomic and RNA-Seq analyses made a direction for research of ellagitannin biosynthesis. And more important, pomegranate represents a unique system for studying ovule developmental biology [1]. AGAMOUS(AG)-clade genes are required for specifying the ovule identity (Yuan et al., 2018; Brambilla et al., 2007; Colombo et al., 2008). In the genome pomegranate, there are 237 genes belonging to 12 families that might participate in ovule development, and the pomegranate genome has the largest number of AG-clade genes compared with others (Yuan et al., 2018). It indicated that the expansion of AG-clade genes might play an important role in the development of the pomegranate-specific type of ovules (Yuan et al., 2018). Another important gene family associated with regulating ovule production is CUP-SHAPED COTYLEDON (CUC) (Duszynska et al., 2013). And in the genome of pomegranate, it has a greater number too. The expansion of CUC might play a key role in the progress of the production of a large number of ovules (Duszynska et al., 2013).

Combining NGS data with SOAPdenovo or ALLPATHS-LG assemble strategy, the high-quality genomes of pomegranate were assembled, will contribute to investigation into its adaptation and biosynthesis of secondary metabolites and it will pave the way for elucidating many other desirable traits in this nutritious fruit crop with medicinal properties

3.1 The Genome Assembled by TGS

In 2020, another genome of a soft-seeded pomegranate cultivar 'Tunisia' was published. Canu was used for *de novo* assembly of the genome with Pacbio clean reads, and the final genome size was 320.31 Mb with eight chromosomes which was built by Hi-C technology and verified by the genetic map. It might need further analysis to explain the difference in the number of chromosomes with others in future studies (Alkan et al., 2011). The size of contig N50 is the largest among those three genome editions, it is 4.49 Mb. The advantage of TGS was obvious. In this study, 33,594 protein-coding gene models were predicted according to a combination of *ab initio*, homology-based, and transcript evidence gathered from mixed-tissue RNA sequences (Luo et al., 2020). The analysis of seed hardness revealed that the hard-seeded varieties formed a subclade and the soft- and semi-soft-seeded varieties formed another subclade [8], the linkage disequilibrium decay (LD) of the hard-seeded group decayed significantly faster than the LD of the soft-seeded group ($P < 0.01$), a principal component analysis (PCA) revealed a similar population genetic structure, with hard-seeded cultivars forming a tight cluster separated from the others (Luo et al., 2020). Based on the analysis, geographical barriers limit hybridization between soft- and hard-seeded, as the hard-seeded populations were from China, whereas the soft-seeded populations originated from the United States of America, Italy, Tunisia, and other countries (Luo et al., 2020).

4. Prospects of Pomegranate Genomic Development

Nowadays, several sequencing technologies, such as Nanopore and PacBio, along with their high-accuracy variant called PacBio HiFi, have proven to be valuable for genome sequencing and *de novo* assembly. In *de novo* chromosome-scale assembly, additional information is often utilized to improve the contiguity and order of assembled sequences. Hi-C, a method that captures chromatin interactions, can be employed to generate proximity-based data for scaffolding contigs and assembling them into chromosome-level structures (Awad and Gan, 2020). Despite advancements in sequencing technologies and assembly methods, genome assemblies can still contain gaps and mis-assemblies. Genome assembly is a complex process, and challenges such as repetitive regions, sequencing errors, and limitations of the assembly algorithms can contribute to these issues (Awad and Gan, 2020; Jibran et al., 2018; Staňková et al., 2016). A high-quality genome is needed to promote genetic research. GALA (Gap-free long-read assembler) is a strategy developed in 2020 (Awad and Gan, 2020), which might avoid the gap within chromosomes in the progress of *de novo* assembly. So, the application of GALA is needed in pomegranate genome assembly in the future, to promote its genetic research.

Besides the nuclear genome (Yuan et al., 2018; Qin et al., 2017; Luo et al., 2020), the mitochondrial and chloroplast genomes are important components to consider in genomic studies. These organelle genomes provide valuable insights into various biological processes and evolutionary relationships. The chloroplast is responsible for photosynthesis and contains genes involved in this process. So, the chloroplast genome is particularly useful for studying plant evolution, phylogenetics, and understanding the genetic basis of specific traits related to photosynthesis. The chloroplast genome of pomegranate, as mentioned, is a significant achievement and provides a valuable resource for further research in this area [40]. Each organelle genome contributes unique information and insights into different aspects of biology. Therefore, the mitochondrial genome is appealed and it can provide more information to better understand this species from genomic perspective. Understanding the complete set of genomes, including the nuclear, mitochondrial, and chloroplast genomes, allows for a more comprehensive understanding of the genetic makeup and evolutionary history of the species.

References

Alkan, C., Sajjadian, S. and Eichler, E.E. 2011. Limitations of next-generation genome sequence assembly. *JNM.*, *8*(1): 61.

Andrews, S. 2010. FastQC: A quality control tool for high throughput sequence data. *In*: *Babraham Bioinformatics*. Babraham Institute, Cambridge, United Kingdom.

Awad, M., Gan, X. 2020. GALA: Gap-free chromosome-scale assembly with long reads. *JB*.bioRxiv (Preprint)

. Bolger, A.M., Lohse, M. and Usadel, B. 2014. Trimmomatic: A flexible trimmer for Illumina sequence data. *J. Bioinformatics*, *30*(15): 2114–20.

Brambilla, V., Battaglia, R., Colombo, M., Masiero, S., Bencivenga, S., Kater, M.M. and Colombo, L. 2007. Genetic and molecular interactions between BELL1 and MADS box factors support ovule development in Arabidopsis. *JTPC.*, *19*(8): 2544–56.

Chen, Y., Nie, F., Xie, S-Q., Zheng, Y-F., Dai, Q., Bray, T., Wang, Y-X., Xing, J-F., Huang, Z-J. and Wang, D-P. 2021. Efficient assembly of nanopore reads via highly accurate and intact error correction. *JNC*, *12*(1): 1–10.

Cheng, H., Concepcion, G.T., Feng, X., Zhang, H. and Li, H. 2021. Haplotype-resolved *de novo* assembly using phased assembly graphs with hifiasm. *JNM*, *18*(2): 170–75.

Chin, C-S., Peluso, P., Sedlazeck, F.J., Nattestad, M., Concepcion, G.T., Clum, A., Dunn, C., O'Malley, R., Figueroa-Balderas, R. and Morales-Cruz. 2016. Phased diploid genome assembly with single-molecule real-time sequencing. *AJNM.*, *13*(12): 1050–54.

Colombo, L., Battaglia, R. and Kater, M.M. 2008. Arabidopsis ovule development and its evolutionary conservation. *TIPS.*, *13*(8): 444–50.

Del Angel, V.D., Hjerde, E., Sterck, L., Capella-Gutierrez, S., Notredame, C., Pettersson, O.V., Amselem, J., Bouri, L., Bocs, S. and Klopp, C. 2018. Ten steps to get started in Genome Assembly and Annotation. *JF*, *7*.

Duszynska, D., McKeown, P.C., Juenger, T.E., Pietraszewska-Bogiel, A., Geelen, D. and Spillane, C. 2013. Gamete fertility and ovule number variation in selfed reciprocal F 1 hybrid triploid plants are heritable and display epigenetic parent of origin effects. *JNP*, *198*(1): 71–81.

Eid, J., Fehr, A., Gray, J., Luong, K., Lyle, J., Otto, G., Peluso, P., Rank, D., Baybayan, P. and Bettman. 2009. Real-time DNA sequencing from single polymerase molecules. *BJS*, *323*(5910): 133–38.

Halvorsen, B.L., Holte, K., Myhrstad, M.C., Barikmo, I., Hvattum, E., Remberg, S.F., Wold, A-B., Haffner, K., Baugerød, H. and Andersen, L.F. 2002. A systematic screening of total antioxidants in dietary plants. *J. of Nutrition*, *132*(3): 461–71.

Hon, T., Mars, K., Young, G., Tsai, Y-C., Karalius, J.W., Landolin, J.M., Maurer, N., Kudrna, D., Hardigan, M.A. and Steiner, C.C. 2020. Highly accurate long-read HiFi sequencing data for five complex genomes. *JSD*, *7*(1): 1–11.

. Hong, H., Zhang, W., Shen, J., Su, Z., Ning, B., Han, T., Perkins, R., Shi, L. and Tong, W. 2013. Critical role of bioinformatics in translating huge amounts of next-generation sequencing data into personalized medicine. *JSCLS*, *56*(2): 110–18.

Li, S. 1998. *Compendium of materia Medica mssence. In*: Beijing: Science Press.

Jibran, R., Dzierzon, H., Bassil, N., Bushakra, J.M., Edger, P.P., Sullivan, S., Finn, C.E., Dossett, M., Vining, K.J. and VanBuren, R. 2018. Chromosome-scale scaffolding of the black raspberry (*Rubus occidentalis* L.) genome based on chromatin interaction data. *JHR*, *5*(1): 1–11.

. Johanningsmeier, S.D. and Harris GK. 2011. Pomegranate as a functional food and nutraceutical source. *Annual Review of Food Science and Technology*, *2*: 181–201.

Kim, N.D., Mehta, R., Yu, W., Neeman, I., Livney, T., Amichay, A., Poirier, D., Nicholls, P., Kirby, A., Jiang, W., et al., 2002. Chemopreventive and adjuvant therapeutic potential of pomegranate (*Punica granatum*) for human breast cancer. *JBCR.*, *71*(3): 203–17.

Kolmogorov, M., Yuan, J., Lin, Y. and Pevzner, P.A. 2019. Assembly of long, error-prone reads using repeat graphs. *JNB*, *37*(5): 540–46.

Koren, S., Walenz, B.P., Berlin, K., Miller, J.R., Bergman, N.H. and Phillippy, A.M. 2017. Canu: Scalable and accurate long-read assembly via adaptive k-mer weighting and repeat separation. *Gen. Rev.*, *27*(5): 722–36.

Kulski, J.K. 2016. Next-generation sequencing: An overview of the history, tools, and 'omic' applications. *NGS: Applications, Challenges*, *2016*: 3–60.

Li, H. 2016. Minimap and miniasm: Fast mapping and *de novo* assembly for noisy long sequences. *JB*, *32*(14): 2103–10.

Liu, H., Wu, S., Li, A. and Ruan, J. 2020. *SMARTdenovo: A de novo Assembler Using Long Noisy Reads*. National Library of Medicine.

Luo, X., Li, H., Wu, Z., Yao, W., Zhao, P., Cao, D., Yu, H., Li, K., Poudel, K. and Zhao D. 2020. The pomegranate (*Punica granatum* L.) draft genome dissects genetic divergence between soft and hard seeded cultivars. *PBJ*, *18*(4): 955–68.

. Metzker, M.L. 2010. Sequencing technologies: The next generation. *Nat. Review Genetics*, *11*(1): 31–46.

Nurk, S., Walenz, B.P., Rhie, A., Vollger, M.R., Logsdon, G.A., Grothe, R., Miga, K.H., Eichler, E.E., Phillippy, A.M. and Koren, S. 2020. HiCanu: Accurate assembly of segmental duplications, satellites, and allelic variants from high-fidelity long reads. *JGR*, *30*(9): 1291–1305.

Qin, G., Xu, C., Ming, R., Tang, H., Guyot, R., Kramer, E.M., Hu, Y., Yi, X., Qi, Y. and Xu X. 2017. The pomegranate (*Punica granatum* L.) genome and the genomics of punicalagin biosynthesis. *JTPJ*, *91*(6): 1108–28.

Ruan, J. and Li, H. 2020. Fast and accurate long-read assembly with wtdbg2., *JNM*, *17*(2): 155–58.

Sánchez-Lamar, A., Fonseca, G., Fuentes, J.L., Cozzi, R., Cundari, E., Fiore, M., Ricordy, R., Perticone, P., Degrassi, F. and De Salvia, R.J. 2008. Assessment of the genotoxic risk of *Punica granatum* L.(Punicaceae) whole fruit extracts. *JoE*, *115*(3): 416–22.

Sanger, F. and Coulson, A.R. 1975. A rapid method for determining sequences in DNA by primed synthesis with DNA polymerase. *J. Mol. Biol.*, *94*(3): 441–48.

Sanger, F., Coulson, A.R., Hong, G., Hill, D. and Petersen, G.1982. Nucleotide sequence of bacteriophage λ DNA. *1. J. Mol. Bio.*, *62*(4): 729–73.

. Schadt, E.E., Turner, S. and Kasarskis, A . 2010. A window into third-generation sequencing. *Hum. Mol. Gen.*, *19*(R2): R227–R240.

Staden, R. 1979. A strategy of DNA sequencing employing computer programs. *JNAR*, *6*(7): 2601–10.

Staňková, H., Hastie, A.R., Chan, S., Vrána, J., Tulpová, Z., Kubaláková, M., Visendi, P., Hayashi, S., Luo, M. and Batley, J. 2016. BioNano genome mapping of individual chromosomes supports physical mapping and sequence assembly in complex plant genomes. *JPB*, *14*(7): 1523–31.

Vaser, R. and Sikic, M. 2020. Jb: Raven: A *de novo* genome assembler for long reads.

Wenger A.M., Peluso P., Rowell W.J., Chang P-C, Hall R.J., Concepcion G.T., Ebler J., Fungtammasan A., Kolesnikov A., Olson N.D. 2019. Accurate circular consensus long-read sequencing improves variant detection and assembly of a human genome. *JNB*, *37*(10): 1155–62.

Xiao, C-L., Chen, Y., Xie, S-Q., Chen, K-N., Wang, Y., Han, Y., Luo, F. and Xie, Z. 2017. MECAT: Fast mapping, error correction, and *de novo* assembly for single-molecule sequencing reads. *JNM, 14*(11): 1072.

Yang, W., Zou, J., Yu, Y., Long, W. and Li, S. 2021. Repeats in mitochondrial and chloroplast genomes characterize the ecotypes of the *Oryza*. *JMB, 41*(1): 1–11.

Yuan, Z., Fang, Y., Zhang, T., Fei, Z., Han, F., Liu, C., Liu, M., Xiao, W., Zhang, W. and Wu, S. 2018. The pomegranate (*Punica granatum* L.) genome provides insights into fruit quality and ovule developmental biology. *JPBJ, 16*(7): 1363–74.

. . . . 40.

Multiomics in Pomegranate

Ran Wan[1] and *Jiangli Shi*[1*]

Multiomics data with biological traits allows a systemic exploration of the regulation network of the biological organization and the complex traits with external environments at various levels. Pomegranate (*Punica granatum* L.) is an ancient, important, and popular fruit crop. As pomegranate genome data published, multiomics have been extensively used for elucidating puzzling issues, e.g., growth and development, fruit quality, stress responses. Here, we have summarized recent significant progress with various omics in pomegranate seed hardness, juice quality, the color of aril and husk, abiotic and biotic stresses, postharvest technique, and so on. This chapter summarizes abundant valuable biological information under certain environments or treatments, clarifies the correlation of certain traits with genes, transcripts, metabolites, or proteins, and reveals the related regulation mechanism, aiming to speed up the development of the pomegranate industry.

1. Introduction

Multiomics including genome, proteome, transcriptome, epigenome, metabolome, and microbiome, focus on the complex biological big data so as to find novel associations, which have greatly contributed to elucidating the complexity of genetic manipulation and defense mechanisms in plants, and speeded up the breeding programs to improve various resistances and fruit quality of horticulture crops. With the innovation of high-throughput sequencing technology, omics sequencing has been widely used in plants, because of its low cost, rich and comprehensive information, and high practicability. Transcriptomics is defined as the study on

[1] Jiangli Shi, College of Horticulture, Henan Agricultural University, 450002, Zhengzhou, China.

* Corresponding author: sjli30@henau.edu.cn

gene transcription and transcriptional regulation, and is important for exploring the expression profiles of the ribonucleic acid (RNA) molecules (called transcripts) expressed in some given entity, and analyzing the locations, structures, and functions of new transcripts in cells. Proteomics provides the protein-interaction mechanism and function analysis in the genome, cells or tissues of organisms under various environmental stresses. Proteomics can identify a large number of proteins *in vivo*, and then clarify the post-translational modifications and protein expression differences. Metabolomics is an important part of system biology after genomics, transcriptomics, and proteomics, and has become the research hot spot among omics. After the reaction of endogenous substances or the effect of environmental factors, the detected metabolites can display the characteristics of the closest phenotyping, and reveal the mechanism of the related-genes expressions. Additionally, metabolomic is the ultimate expression of the overall function or state of a biological system (AbuQamar et al., 2016).

Different omics reflect the transcription, translation, and metabolism in horticultural plant genes at various levels, so as to achieve more valuable bioinformation and better elucidate the regulation process for various physiological phenomena of horticultural plants. Pomegranate (*Punica granatum* L.), as an ancient cultivation history, has become an emerging profitable fruit crop due to its attractive features, higher medicinal and nutrition values. However, compared to other fruit crops, such as orange (*Citrus sinensis*), apple (*Malus domestica*), grape (*Vitis vinifera*), further elucidation of genetics and evolution of some interesting traits had been restricted for many years before its genomic information was released (Wang et al., 2021). Fortunately, the pomegranate genome sequencing has been done in recent several years, such as hard-seed cultivar 'Dabenzi' (Qin et al., 2017) hard-seed cultivar 'Taishanhong' (Yuan et al., 2018), and soft-seed cultivar 'Tunisia' (Luo, et al., 2018b, 2020). Here, we highlight the current progresses on growth and development, fruit qualities, and stress responses in pomegranate using multiple omic technology.

2. Pomegranate Growth and Development

2.1 Seed

Seed hardness is one of the important quality characteristics of fresh pomegranate fruit, which determines consumer acceptance. Generally, lower the seed hardness, higher the edibility and better taste. Pomegranate can be roughly divided into soft-seed cultivar e.g., 'Tunisia' and 'Mollar', semi-soft-seed cultivar and hard-seed cultivar. Seed development is a complex physiological and biochemical process. Several omics studies have demonstrated the related molecular mechanism of the soft-seed formation of pomegranate, which is obviously different from the hard-seed ones.

Several researches have been conducted non-targeting metabolic, miRNA, and transcriptomic profiling of the seeds of three typical hard-seed 'Dabenzi', 'Sanbai', and 'Yudazi' (>4.2 kg/cm^2) along with the popular soft-seed 'Tunisia' (<3.67 kg/cm^2) in China at different developmental stages (Qin et al., 2020; Shi et al., 2022a; Xue et al., 2017). These researches all pointed out that lignin biosynthesis is the key determinant process for pomegranate seed hardness. Lignin metabolic pathway displayed obvious difference between soft- and hard-seed pomegranates. Compared with soft seeds, hard seeds always have a higher level of monolignols, including coniferyl alcohol and sinapyl alcohol, meanwhile they have a higher gene expression level, such as functional genes, *PgCCR*, *PgCCoA-OMT* and *PgPAL*, and some transcriptional regulators *PgMYB*, *PgWRKY* and *PgNAC* (Qin et al., 2020; Xue et al., 2017). Utilizing GWAS of the cultivars of different hardness, several seed-hardness associated loci were identified, including *SUC6/8*, *FOXO*, and *MAPK* genes (Luo et al., 2020). Regarding transcription factors, *PgNAC1*, *PgWRKY*, and *PgMYC* were targeted by differentially expressed mdm-miR164e and mdm-miR172b, affecting the formation of seed hardness (Luo et al., 2018b). Moreover, hemicellulose constituents, flavonoids and some genes, for example, *PgCSE4/7/8* and *PgNAC66,* also might contribute to the pomegranate seed hardness (Xue et al., 2017; Qin et al., 2020). Thus, seed hardness is the result of a complex biological process regulated by a multilevel network in pomegranate. The above results provided some candidate functional genes and key regulators, which will be important issues in future for dissecting the mechanism of soft-seed pomegranates.

Seed germination begins with water uptake by the quiescent dry seed and is composed of radicle protrusion through the embryo-surrounding tissues. This process participates in a series of orderly physiological and morphogenetic processes involving seed energy conversion, nutrient consumption, and changes in metabolites. Using an integrative system biology of metabolomic and transcriptomic profile during pomegranate seed germination, the functionality and complexity of the physiological and morphogenetic processes as well as gene expression and metabolic differences were analyzed during seed germination stages (Fu et al., 2021). The total 40 differentially accumulated metabolites (DAMs) were identified from 489 detected metabolites and 6,984 genes significantly differentially expressed throughout the whole germination process. According to the metabolomics profiling, most of the DAMs were primary metabolites, including various amino acids (including glycine, tyrosine, proline, serine, and isoleucine), saccharides (raffinose, mannose, and fructose lactulose), the secondary metabolites in small amounts, including alcohols such as sorbitol and inositol, galactinol, 24, 25-dihydrolanosterol, lipids, and trehalose 6-phosphate. The positive correlation between two modules of Stage 1 and Stage 4 showed that *CDL15_Pgr011515*, *CDL15_Pgr004967*, and *CDL15_Pgr025596*, were predicted to be hub genes at the early stage (Stage 1 in Figure 1A), while *TCP11* and homeobox family members were involved in multiple metabolic pathways in the final stage (Stage 4 in Figure 1D). Moreover, during pomegranate seed germination, key expressed genes and DAMs highlight the roles of flavonoid metabolism during pomegranate seed germination.

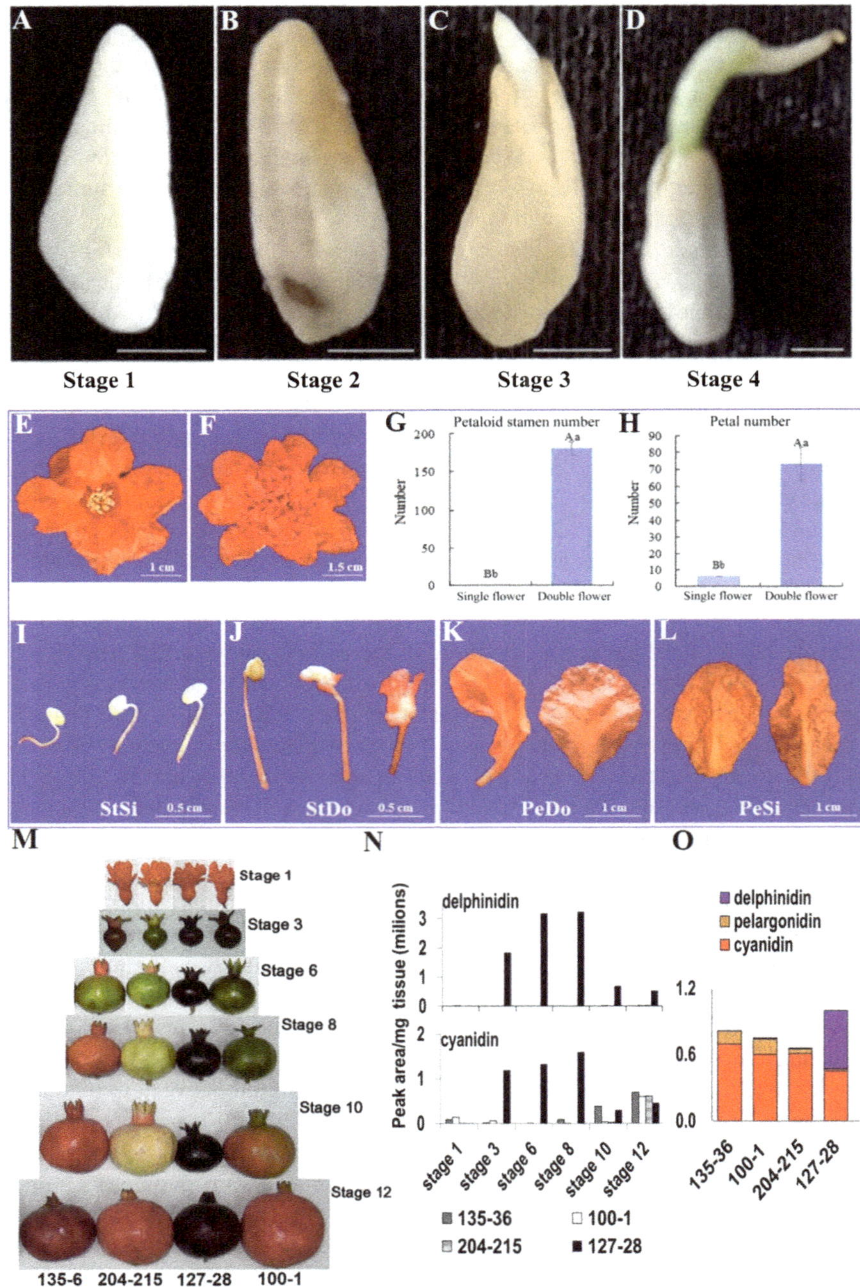

Fig. 1 Pomegranate tissues. **A**. Stage 1, imbibition stage (when the seed absorbs water and reaches saturation), which is often reached after 48 h of water absorption. **B**. Stage 2, preliminary

Contd.

An attractive feature in the pomegranate fruit is that more than one hundred ovules grow in one pomegranate ovary. They develop into seeds with arils, which consist of epidermal cells derived from the integument (Yuan et al., 2018). Integrated analysis of genomic and transcriptomic provided insights into the molecular mechanisms underlying the unique ovule development processes. The pomegranate genome has 237 candidate genes belonging to 12 families associated with ovule development. The pomegranate-specific ovule development and the polycaryoptic phenotype can likely be attributed to the expansions of the AG and CUC families and the contraction and inactivation of the BEL1 family (Yuan et al., 2018).

2.2 Flower

As we know, pomegranate has two types of flowers on the same plant: functional male flowers (FMF) and bisexual flowers (BF) (Chen et al., 2017). BF are female-fertile flowers that can set fruits. FMF are female-sterile flowers that fail to set fruit till eventually dropping, due to the abnormal ovule development. Chen et al. (2017) found that the key stage for the termination of FMF ovule development was when the bud vertical diameter was 5.0–13.0 mm through comparing ovule development in FMF and BF, using scanning electron microscopy. The transcriptomic analysis of 10-year-old "Tunisian soft seed" pomegranate found that the *INNER OUTER* (*INO/YABBY4*) and *AINTEGUMENTA* (*ANT*) homolog genes, as well as their regulator genes, such as *AGAMOUS*-like (*AG-like*) and *SPOROCYTELESS* (*SPL*) homolog genes might be involved in the development of the integument for their

Contd.

stage (when the protoplasm and seed cell wall are hydrated and the protoplasm changes from the gel state to the solid state (during this stage, various enzymes begin to become activated, and respiration and metabolism sharply increase). **C**. Stage 3, emergence stage (when the embryo breaks through the seed coat. **D**. Stage 4, germination stage (before the first cotyledon is unfolded and after the stem formation reaches a length of 3 mm. Bar = 0.2 cm. **E**. Morphology of single-petal flower. **F**. Morphology of double-petal flower. **G**. Petaloid stamen number of single and double-petal flower. **H**. Petal number of single and double-petal flowers. **I**. Stamen morphology of single-petal flower. **J**. Stamen morphology of double-petal flower, including petaloid stamens. **K**. Petal morphology of double-petal flower, in which the left side is transitional petal, and the right side is normal petal. **L**. Morphology of single petals. Lowercase letters represent the difference is significant ($p < 0.05$), uppercase letters represent the difference is highly significant ($p < 0.01$). **M**. Fruits of four pomegranate varieties at six different developmental stages, from flower (Stage 1) to fully mature fruit (Stage 12). **N**. Total level of delphinidin and its derivatives (purple color) and of cyanidin and its derivatives (red color) in the peel during the various developmental stages of the fruit in four pomegranate varieties. **O**. Anthocyanin composition in the peel of ripe fruit (Stage 12). The varieties shown are P.G.100–101, P.G.135–136, P.G.204-215, and P.G.127–128 ('Black') [*Journal of Integrative Agriculture*, 2021, *20*(1): 132–146. *Frontiers in Plant Science*, 2021, *12*: 642019. *Plants*, 2023, *12*: 2402].

down-regulation in FMF at the key stage of ovule development cessation. Moreover, the ethylene response signal genes, *ETR* (ethylene-resistant) and *ERF1/2* (ethylene-responsive factor) might affect the process.

Furthermore, miRNAs played an important role in FMF through miRNA sequencing (Chen et al., 2020). *Pg-miR858b* and *Pg-miRN01* strongly associated with pistil development at the early stage. *Pg-miR444b.1*, *Pg-miRN11*, *Pg-miR166a/165a-3p*, and *Pg-miR952b* may influence pomegranate integument development at later stages. *Pg-miR166a-3p* may affect pomegranate female development by interacting with its targets, *PgPHB (PHABULOSA)* and *PgPHV (PHAVOLUTA)*. The development termination and failure of further elongation of ovule integument in functional male flower might be involved in the lower expression of *INO (INNER NO OUTTER)*.

Petaloidy leads to a plump floral pattern and increases the landscape value of ornamental pomegranates (Figure 1E–L) (Huo et al., 2023). Transcriptomic and proteomic sequencing of the stamens and petals were conducted in the single-petal and double petal flowers of pomegranate (Huo et al., 2023). Totally, 24,567 genes (including 2,896 new genes) and 5,865 proteins were identified, of which, 5,721 were quantified at both transcriptional and translational levels. Additionally, the hormone-related DEGs/DAPs, PgJAR1, PgILR1, and PgLAX2, and PgARF, and transcription factors DEGs/DAPs, EREBP, LOB, MEF2, MYB, C3H, and trihelix, might promote petal doubling. Furthermore, combined with the correlation analysis, the enrichment results of transcriptome and proteome showed that cell wall metabolism, jasmonic acid signal transduction, redox balance, and transmembrane transport participated in petaloidy.

2.3 Fruit Peel

2.3.1 Peel color

The peel color of pomegranate fruits is a major factor affecting consumer acceptance and marketability. Thus, one of the main objectives of pomegranate breeders is to enhance peel color. Fruit peel color is a complex trait that is involved in physiological, biochemical, and molecular processes (Harel-Beja et al., 2019; Zhang et al., 2020). To date, the biological and genetic factors regulating fruit peel color have been studied extensively in some fruit plants, such as grape, jujubes, apples, pears, and so on (Lijavetzky et al., 2006; Shi et al., 2020; Liu et al., 2021b). The genes on the anthocyanin biosynthesis pathway were identified in many plant species and most showed high homology in their encoded amino acid sequences. However, little is known about the development of peel color in pomegranate. Anthocyanins mainly consist of cyanidin and delphinidin glucosides, and affect the coloration of the pomegranate peel and arils. Additionally, there is a high variation in the anthocyanin contents among different pomegranate varieties and cultivation conditions (Zhang et al., 2020).

To reveal the complex molecular network controlling fruit peel color in pomegranates, iTRAQ-based proteome and RNA sequencing-based transcriptome were carried out using 'Tunisia' (red fruit, T) and 'White' (white fruit, S) pomegranate cultivars at two stages of fruit development: 60 days (P1) and 120 days (P2) after flowering (Luo et al., 2018a). In total, 237,421 spectra were generated, and 18,407 unique peptides and 5,357 proteins were identified through iTRAQ and LC-MS/MS analysis. Among the proteins, 888 and 1,969 DAPs were found in SP1_TP1 and SP2_TP2, respectively. A total of 552 shared DAPs and 1,684 shared DEGs were identified during the comparison between SP1_TP1 and SP2_TP2. More DAPs and DEGs were detected between the 'Tunisia' and 'White' cultivars at the fruit ripening stage than at the fruit coloring stage, greater changes in peel color occurred at the ripening stage. In summary, a total of 27 differentially abundant proteins (increased abundance) and 54 DEGs (16 upregulated and 38 downregulated) were identified, which involved in the biosynthesis of anthocyanins, stilbenoids, diarylheptanoids, gingerols, flavonoids, and phenylpropanoids, and contributed to pomegranate fruit peel color. Otherwise, several candidate proteins and genes corresponded to enzymes related to general reactions (*PAL*, *4CL*, *DFR*, *LDOX/ANS*, *CHS*, and *F3'5'H*) and glycosylation (*GT1* and *UGAT*) of compounds and pigments were linked to the color of pomegranate fruit peel.

Interestingly, in the 'Black' pomegranate variety (P.G.127-28), it was found that it contained exceptionally high levels of anthocyanins in fruit peel from Israel, reaching up to two orders of magnitude higher content as compared to that of other pomegranate varieties (Trainin et al., 2021). The main anthocyanins in pomegranate fruits included 3-glucosides and 3,5-diglucosides of delphinidin, cyanidin, and pelargonidin (Du et al., 1975). Trainin et al. (2021) found that delphinidin was highly abundant in the peel of 'Black' variety (Figure 1M–O). Importantly, the pattern of anthocyanin accumulation in 'Black' variety differed from that of other pomegranates in the fruit peel during fruit development, with higher anthocyanin levels and not dependent on light. Combined transcriptomic and genetic analysis of F_2 and F_3 populations, respectively, a putative anthocyanidin reductase (*ANR*) gene was regarded to be responsible for the different anthocyanin composition and high anthocyanin levels of the 'black' trait in pomegranate, using RNA-Seq and single nucleotide polymorphism (SNP) markers. Only pomegranate varieties exhibiting the 'black' trait carried a base pair deletion toward the end of the gene, causing a frame shift resulting in a shorter protein (Trainin et al., 2021).

Ginzberg and Faigenboim (2022) focused on the fruit skin (the outer red layer of the peel) representing the primary concerns of sufficient red pigment accumulation and minimal cracking. The skin transcriptome of pomegranate cv. 'Wonderful' was performed at three distinct developmental stages (early fruit with dark green skin, mid-growth fruit at the onset of color break, and ripening fruit at harvest), and 29,281 unique transcripts were then obtained (Ginzberg and Faigenboim, 2022). Interestingly, although pomegranate is a non-climacteric fruit, about 18% of the transcription factors were ethylene-response factors (ERFs), which mediated regulating cell wall biogenesis and signaling in the early fruit skin, and cutin biosynthesis, stress responses, and stress-related developmental processes in early and ripening fruit skin. Also, the ripening mechanism of the climacteric nature of pomegranate skin differed from that in strawberry, a model non-climacteric fruit. It was noted that the biosynthetic pathways of important metabolites in pomegranate, hydrolyzable tannins and anthocyanins, were co-upregulated at the ripening stage, in line with the visual enhancement of red coloration (Ginzberg and Faigenboim, 2022). Meanwhile, cuticle- and cell-wall-related genes were differentially expressed between the developmental stages, and were mainly up regulated in the skin of early fruit, with lower expression at mid-growth and ripening stages. The obtained results also inferred that lignification may be involved in skin hardening in the mature fruit (Ginzberg and Faigenboim, 2022).

2.3.2 Peel extracts

Pomegranate peel is the most abundant by-product of the agri-food pomegranate processing chain (Tozzi et al., 2022). Pomegranate peel extracts possessed remarkable antioxidant, antibacterial, anti-inflammatory, and hypolipidemic bioactivities (Li et al., 2015). Through the HPLC fingerprint combined with quantitative analysis, eight monophenols (including gallic acid, punicalagin, catechin, chlorogenic acid, caffeic acid, epicatechin, rutin, and ellagic acid) were simultaneously quantified in pomegranate peels (Li et al., 2015). Spectrophotometric-based assays and the liquid chromatography mass spectrometry (LC-MS)-based approach were used to explore the effects of hot water blanching pre-treatment of 80° for 3 min on yield, bioactive compounds, antioxidants, enzyme inactivation, and antibacterial activity in peel extracts from 'Wonderful', 'Acco', and 'Herskawitz' (Magangana et al., 2022). The results revealed that the POD and PPO enzyme activity decreased, antioxidant compounds increased, importantly, and antibacterial activity was enhanced in pomegranate peel extracts of 'Wonderful', 'Acco', and 'Herskawitz'. A total of 30 chemicals, including five phenolic acids, one organic acids, four flavonoids, 14 ellagitannins, and six other unknown polyphenols, were present in peel extracts from 'Acco', and 'Herskawitz' pomegranate peel extracts, whereas, only 25 metabolites were present in peel extracts from 'Wonderful'. Five chemicals were absent, namely ellagic acid rhamnoside, dehydro-galloyl-hexahydroxydiphenol-hexoside, and unknowns c, d, and f (Magangana et al., 2022).

Furthermore, gene expression and metabolite (anthocyanins and punicalagin) accumulation were evaluated at three stages of fruit development by Harel-Beja et al. (2019). RP-HPLC and transcriptomic sequences showed that the red peel (P.G.116-17 from Israel) and the pink peel (P.G.200-211 from Spanish) contained high and low levels of anthocyanins, respectively. It was reported that the gene expression of *PAL*, *CHI*, *CHS*, and *F3'H*, had a high correlation with cyanidin derivatives. Furthermore, the expression of *DFR*, *ANS*, *UGT*, *CHS*, cinnamate 4-hydroxylase, and *F3H* were highly correlated with both anthocyanins and punicalagin. The gene expression of more than 60 TFs was highly correlated ($r > 0.75$) to anthocyanins and punicalagin, especially two *bHLHs* and one *MYB*.

2.4 Arils

2.4.1 Aril color

"Tunisian soft seed" pomegranate seeds are very soft and edible, rich in a variety of bioactive components, and are popularized worldwide. However, the flavor and color of the "Tunisian soft-seed" pomegranate vary significantly among different planting areas (Yuan et al., 2022a). To explore the DEGs related to the formation of flavor and color quality of pomegranate arils, transcriptome and metabolites of "Tunisian soft seed" pomegranate arils were examined from seven planting areas, including Jianshui, Dali, Lijiang, and Hehuize in Yunnan Province, Huili and Datian in Sichuan Province, and Xingyang in Henan Province (Table 1) (Yuan et al., 2022a). Using integrated genomic and transcriptomic analyses provided insights into the evolution of the anthocyanin biosynthetic pathway. The fruit's unique red color was conferred by anthocyanins (Yuan et al., 2022a). Borochov-Neori et al. (2011) showed that anthocyanin accumulation changed inversely to the season's temperatures by RP-HPLC analysis. Attanayake et al. (2018) focused on the expression of major anthocyanin biosynthetic genes in the peels and arils of a yellow-peeled and a pink-aril pomegranate cultivar in three agro-climatologically different locations in Sri Lanka. It was observed that drier and warmer climates promoted the accumulation of total phenolic content (TPC), antioxidant capacity (AOX), and α, β, and total punicalagin, in both peels and arils, in comparison with wetter and cooler climates. Meanwhile, the expression of pomegranate *DFR*, *F3H*, and *ANS* transcripts in both peels and arils was relatively higher in hotter and drier climates than in cooler and wetter climates. Therefore, it was inferred that drier and warmer environments were more beneficial to the production of healthy biochemical compounds for pomegranate production.

The studies proved that anthocyanin accumulation determines aril color in pomegranate. Nine anthocyanins were major compounds, including cyanidin derivatives, delphinidin derivatives, and pelargonidin derivatives. Meanwhile, cyanidins were generally more abundant, while delphinidin accumulation was enhanced in the cooler season. Monoglucosylated anthocyanins prevailed at cooler temperatures and subsided during seasonal warming with a concomitant increase in diglucoside proportion (Borochov-Neori et al., 2011; Yuan et al., 2022a). In the Huili production area of Sichuan, the genes with highly upregulated expression in redder arils encoded enzymes such as chalcone synthase (CHS), chalcone isomerase (CHI), flavonoid 3-hydroxylase (F3H), flavonoid 3'-hydroxylase (F3'H), dihydroflavonol 4-reductase (DFR), anthocyanidin synthase/leucoanthocyanidin dioxygenase (ANS/LDOX), UDP-glucose: flavonoid glucosyltransferases (UFGT) and anthocyanin O-methyltransferase (AOMT), which were responsible for the aril color transition from white to red. Moreover, the tissue-specific expression patterns of AOMTs could be responsible for anthocyanin accumulation in peel and aril (Yuan et al., 2018; Attanayake et al., 2018; Yuan et al., 2022a). Furthermore, 7 *R2R3-MYB*, 9 *bHLH*, and 13 *WD40* were highly expressed in both peel and aril, suggesting they might regulate anthocyanin production in pomegranate fruit (Yuan et al., 2018).

Table 1 Morphological and color phenotype observation of Tunisian soft-seed pomegranate arils producing in 7 regions in China.

Sample		*Y_DTN*	*Y_LTN*	*Y_JTN*	*Y_QTN*	*S_DHT*	*S_PTN*	*H_HYT*
Color phenotype								
Fitting graph								
Color attributes	L*	26.87 ± 0.10^{d}	34.95 ± 0.09^{b}	27.69 ± 0.42^{d}	36.43 ± 0.99^{a}	16.91 ± 0.33^{e}	27.30 ± 0.27^{d}	31.02 ± 0.05^{c}
	a*	27.40 ± 0.07^{a}	22.63 ± 0.10^{c}	27.12 ± 0.44^{a}	27.47 ± 0.72^{a}	24.71 ± 0.22^{b}	27.70 ± 0.15^{a}	27.22 ± 0.06^{a}
	b*	1.16 ± 0.02^{c}	1.67 ± 0.06^{a}	$1.27 \pm 0.24b^{c}$	-0.77 ± 0.11^{e}	1.48 ± 0.14^{ab}	0.50 ± 0.05^{d}	1.31 ± 0.03^{bc}
	C*	27.43 ± 0.07^{a}	22.69 ± 0.10^{c}	27.15 ± 0.45^{a}	27.48 ± 0.72^{a}	24.75 ± 0.23^{b}	27.70 ± 0.15^{a}	27.25 ± 0.06^{a}
	H*	2.42 ± 0.04^{c}	4.23 ± 0.15^{a}	2.67 ± 0.48^{c}	-1.61 ± 0.27^{e}	3.42 ± 0.30^{b}	1.03 ± 0.10^{d}	2.76 ± 0.05^{c}
	ΔE	38.39 ± 0.05^{e}	41.67 ± 0.05^{b}	38.79 ± 0.03^{d}	45.65 ± 0.37^{a}	29.98 ± 0.02^{f}	38.89 ± 0.10^{d}	41.29 ± 0.03^{c}

Data were shown as the mean ± standard error (n = 6). Y_DTN: BinChuan, Da Li, Yunnan Province; Y_LTN: Jianshui, Honghe, Yunnan Province; Y_JTN: Yongsheng, Lijiang, Yunnan Province; Y_QTN: Qujing, Huize, Yunnan Province; S_DHT: Huili, Daliangshan, Sichuan Province; S_PTN: Datian, Panzhihua, Sichuan Province; H_HYT: Xingyang, Henan Province. Different lowercase letters in same row were statistically significant at $p < 0.05$ (Tukey's test). [*Food Chemistry*, *2022*, *370*: 131270]

2.4.2 *Sugar, acid, and volatiles in aril*

Pomegranate arils from different planting areas or different cultivars in China, India, or Italy also displayed the diversity in sugar and acids (Singh et al., 2019; Hasanpour et al., 2020; Yuan et al., 2022a, 2022b;), which provided good evidences to clear the sugar and acid mechanism in pomegranate. The results demonstrated that the main soluble sugars were fructose and glucose; the main organic acids included oxalic acid, malic acid, citric acid (the highest content), quinic acid, lactic acid, and acetic acid (Singh et al., 2019; Hasanpour et al., 2020; Cirillo et al., 2022; Feng et al., 2022; Yuan et al., 2022). In Italy, 'Mazandaran' pomegranate was found to be different from the other ecotypes, having a high content of citric and succinic acids. Pomegranate cvs. 'Bajestan', 'Ferdows', and 'Yazd' contained comparatively higher amounts of anthocyanins and ellagic acid derivatives than other pomegranate ecotypes (Hasanpour et al., 2020). Moreover, the gene expression of *invertase 2* (*INV2*), *INV1*, *FRK2 (phosphofructokinase 2)*, *FRK7*, *PFK2*, *PFK7*, and *HK1 (hexokinase 1)* was closely correlated with the contents of fructose and glucose during fruit development stage, whereas the gene expression of sucrose synthase 3 (*SUS3*) and *INV1* was negatively correlated with the sucrose content. The gene expression of *MDH* gene and *WRKY42* was closely related to the content of sucrose, malic acid, citric acid, and succinic acid during fruit development stage (Feng et al., 2022). Furthermore, 'Tunisian' pomegranate arils from seven Chinese regions were investigated (Yuan et al., 2022b). The results showed that 40 volatile compounds, including 12 alcohols, eight acids, nine aldehydes, four olefins and others, were identified in arils. Among them, hexanol (35.62–50.4%), (Z)-3-hexen-1-ol (17.7–25.32%), and α-terpineol (3.05–6.42%) accounted for a relatively high content. Five differential accumulated metabolites were significantly enriched in the fatty acid biosynthesis, including palmitic acid, octanoic acid, lauric acid, decanoic acid, and myristic acid. Particularly, hexanoic acid was significantly correlated with 38 DEGs, suggesting that it was an important metabolite for aril flavor.

2.5 *Juice*

Pomegranate presents a remarkable ecotypic variation in fruit characteristics, such as husk color, fruit size, antioxidants' contents, and tastes. Many studies pointed out that pomegranate fruit was rich in bioactive polyphenolics and naturally available phytonutrients, benefiting significantly for human health. Especially, antioxidants including phenolic acids, tannins, flavonols, and anthocyanins, were explored the medical values for cardiovascular diseases, hypertension, many types of cancers aging and diabetes, owing to antibacterial, antifungal, and antiviral properties (Gómez-Bellot et al., 2023).

The content of total anthocyanin, flavonoid, phenolic acid, and tannins also varied from pomegranate cultivars and planting areas. More than 65 punicalagins, ellagic acid derivatives, flavonoids, anthocyanins, and phenylpropanoids were simultaneously detected in four centuries old *Punica granatum* L. ecotypes from northern Italy, and compared with those of *P. granatum* cv. "Dente di Cavallo" (widely cultivated in Italy), using a simple ultra-HPLC (uHPLC) separation and MS^n linear ion trap mass spectrometric characterization (Calani et al., 2013). The results showed that the main phenolic compounds in juices from 12 pomegranate varieties and five pomegranate clones in Spain were determined by HPLC-DAD-ESI-MS, which included 13 anthocyanins and 14 other phenolic compounds. Total phenolic content ranged from 580.8–2551.3 mg/L in pomegranate juice. Anthocyanins accounted for 20–82% of total phenolic content, flavonoids for 1.6–23.6% of total phenolic compounds, while phenolic acids and ellagitannins were in the range 16.4–65.8% (Kalaycioglu and Erim, 2017). Juice of Iranian pomegranate ecotypes from the three pomegranate origins of east Iran, including Kashmar, Bajestan (from Khorasan Razavi province), and Ferdows (from South Khorasan province), were investigated using untargeted 1H-NMR-Based metabolomics technology (Hasanpour et al., 2020). The results revealed that the pomegranates from various geographical regions in Iran produced distinct metabolite profiles in juice. Furthermore, phenolic compounds exhibited the higher content in Bajestan and Ferdows (two connected geographical origins with relatively similar climates), than in Kashmar. Especially, different from the other ecotypes, Mazandaran pomegranate contained a higher content of citric and succinic acids; Bajestan, Ferdows, and Yazd pomegranates contained comparatively higher amounts of anthocyanins and ellagic acid derivatives. Moreover, four phenolic compounds (gallic acid, ellagic acid, protocatechuic acid, and catechin) and four sugars (β-glucose, α-glucose, fructose and sucrose) were identified by proton resonances, while organic acids including succinic, fumaric, citric, malic, and tartaric acids were confirmed by 2D NMR experiments. Citric acid was the predominant organic acid in pomegranate juice.

To comprehensively investigate the metabolic profiles and potential metabolism of essential metabolites, a widely targeted metabolome, integrated with the transcriptome was conducted of juices (edible parts) of pomegranate fruits at 50, 95, and 140 days after flowering (DAF) (Zhao et al., 2023). Totally, 590 metabolites, including 11 sugar and sugar alcohols, 17 common organic acids, 20 essential amino acids, and a variety of flavonoids, were detected in pomegranate juices, in which the flavonoid biosynthesis was the most enriched pathway during fruit development. Notably, the redirection of metabolite flux from catechin and its derivative synthesis to anthocyanin synthesis occurred at the later developmental stages, which might be associated with the increased expression of *DFR*, *ANS*, *UTG75C1* and one anthocyanidin 3-O-glucosyltransferase (BZ1), as well as the decreased expression of one leucoanthocyanidin reductase (*LAR*) (Zhao et al., 2023). Interestingly, drought could affect the metabolites in pomegranate juices. The controlled irrigation stress (irrigation was applied at 25% of the water requirements of the crop during the ripening phase) would increase the production of bioactive compounds by increasing the phenylpropanoids metabolism in pomegranate juices, where 13 polyphenols were identified as upregulated metabolites, such as dihydrokaempferol, quercetrin, quercetin 3-O-(6'-acetyl-glucoside), 6-hydroxydelphinidin 3-glucoside, gallocatechin, and so on (Gómez-Bellot et al., 2023).

3. Pomegranate Postharvest Biotechnology

3.1 Low-temperature Storage

The fruit quality loses rapidly after harvest due to continued metabolic responses and physiological disorders in horticultural crops. The red husk of pomegranate fruits without visual defects is an important quality attribute for marketability. To meet the market demand for higher fruit quality and commercial value, cold storage is performed widely in combination with various treatments. Also, low temperature storage had become one of the most common methods for pomegranate postharvest storage.

Considered that pomegranate fruits are chilling-sensitive, they may develop chilling injury (CI) symptoms, displaying surface pitting on the peel, and internal aril browning when exposed to unfavorable low temperatures. Previous studies showed that harvest time, varieties, husk scald incidence, and low-temperature-conditioning (LTC) all influenced the pomegranate quality during postharvest storage. Several RNA-Seq analysis of pomegranate fruits were conducted to reveal the molecular mechanism of chilling tolerance, husk scald and harvest time during low temperature storage (Kashash et al., 2018, 2019a, 2019b; Belay et al., 2020).

To explore the effect of harvest time on chilling tolerance of pomegranate fruit, and elucidate the molecular mechanisms that govern chilling tolerance of pomegranate fruit, RNA-Seq analysis of inner membrane tissues from 'Wonderful' (relatively stronger chilling tolerance) was performed at early-harvested fruit (harvest day) and late-harvested fruit (a month later) stored at 1°C for 2 weeks (Kashash et al., 2019a). Pair-wise comparisons revealed that 6,853 and 8,000 transcripts were significantly ($p \leq 0.01$) induced or repressed in the early-harvested fruit and the late-harvested fruit, respectively (Kashash et al., 2019a). Transcripts related to several regulatory, metabolic, and stress-adaptation pathways were specifically induced in the late-harvested 'Wonderful' pomegranate fruits, while suppressed in the early-harvested ones. Furthermore, various stress-related transcription factors, such as AP2/ERFs, WRKYs, MYBs, bHLH, homeobox, and HSFs, were involved in the activation of jasmonic acid and ethylene biosynthesis, and signal transduction pathways. The differentially expressed transcripts were related to primary and secondary carbohydrate metabolism, and activated starch degradation and galactinol and raffinose biosynthesis genes (Kashash et al., 2019a). Compared to 'Ganesh' (chilling-sensitive), many transcripts related to various pathways, such as the biosynthesis of jasmonic acid, galactinol, raffinose, phenol and phenylpropanoid, calcium and MAPK signaling, lipid metabolism, and various transcription factors and heat-shock proteins, have been significantly up-regulated (Kashash et al., 2019b).

During the pomegranate fruit storage, scald peels as a physiological disorder, directly affect fruit quality. 'Wonderful' fruits were stored at 7°C, 91 ± 4% of relative humidity for 3 months, followed by 2 weeks shelf-life at 22°C, 66 ± 4% of relative humidity. The healthy- and scald- peels were used for whole transcriptome sequencing, respectively (Belay et al., 2020). The results showed that 652 DEGs were identified between healthy and scald fruit peels. GO analysis showed that 432 genes were classified into molecular functions, 272 in cellular components, and 205 in biological processes. This observation could suggest that no stress was induced in the healthy fruit peel. Comparatively, *Pgr023188* and *Pgr025081* encoded uncharacterized protein and *Pgr007593* encoded glycosyltransferase, with significantly highest fold changes in scald-peels; while *Pgr003448*, *Pgr006024*, and *Pgr023696* demonstrated that they were involved in various iron-binding and oxidoreductase activities, with significantly down-regulation. The results provided an initial foundation and orientation for future studies on pomegranate postharvest handling and storage induced stresses (Belay et al., 2020).

It is very interesting that the postharvest LTC treatment was recently developed for 'Wonderful' pomegranate fruits. To elucidate the molecular mechanisms involved in the fruit responses to LTC, RNA-Seq analysis of inner membrane tissues were divided into four time points: (1) immediately after harvest; (2) after the LTC treatment; (3) after 2 weeks of cold storage at 1°C; and (4) after LTC + 2 weeks at 1°C (Kashash et al., 2018). Pairwise comparisons revealed that a total of 7,798 transcripts were significantly differentially expressed ($P \leq 0.05$) in the chilling treatment (including 2,422 upregulated and 5,376 downregulated), whereas, just 1,997 transcripts significantly differentially expressed in the LTC + chilling treatment (974 upregulated and 1,023 downregulated) (Kashash et al., 2018). Additionally, some downregulated chilling-specific transcripts were classified into carbohydrate, amino acid and lipid metabolism, stress, cell wall, protein and signaling bioprocesses. Although the effects of chilling and LTC on the expression of MYBs and WRKYs were not detected any consistent trend, LTC increased transcript levels of HSPs and chaperones and of BTB/POZ domain-containing proteins for stress-adaptation processes. In summary, the LTC + chilling regulated gene expression patterns related to stress mechanisms, including modification of specific transcription factor, signaling and stress-related genes (Kashash et al., 2018).

3.2 Postharvest Browning

Aril browning is considered as a typical symptom of chilling injury, affecting greatly fruit quality and marketability (Shi et al., 2022b). A comprehensive comparison of aril traits and secondary metabolites was performed between healthy and browning arils using widely targeted secondary metabolomics (Shi et al., 2022b). A total of 399 metabolites were identified in pomegranate arils, and included 75 up-accumulated and 14 down-accumulated in browning arils. The results observed that the colored pigments exhibited respective change profiles in browning arils during the cold storage. Especially, cyanidin-3-O-arabinoside was the first time to be detected in arils, and was largely correlated with the browning aril color among 7 anthocyanins derivatives in pomegranate. During the cold storage, the key oxidase to aril browning was PPO rather than POD. Otherwise, p-coumaric acid that controls melanin formation via inhibiting tyrosinase activity, may affect greatly pomegranate aril browning. In summary, aril browning was mainly attributed to water loss, the oxidization of (PPO), and hydrolysis reaction (Shi et al., 2022b).

Symptoms of peel browning include pitting, husk scald, some softening, a higher sensitivity to decay, internal seed discoloration, and browning of chilling injury in postharvest hard-, semi soft- and soft-pomegranate fruit (Qi et al., 2022; Valdenegro et al., 2022; Wan et al., 2023; Kashash et al., 2018). Combined with the physiological, the main phenolic metabolites and genetic changes, peel browning was considered closely associated with low antioxidant ability, due to increasing PPO and POD activities and decreasing APX and CAT activities and also associated with declined antioxidants, such as anthocyanins, phenolic acids, flavonoids as well as ascorbic acid and glutathione contents (Liu et al., 2021a; Qi et al., 2022). DEGs in pomegranate peel under different browning degrees were enriched in the metabolic processes of phenolic compounds, lipids, ascorbic acid, glutathione, sugar, starch, energy-related, ethylene biosynthesis, and signal transduction pathways (Qi et al., 2022).

4. Abiotic and Biotic Stresses

Plants often live in adverse environmental conditions, exposed to various stresses, such as heat, cold, heavy metals, salt, radiation, poor lighting, nutrient deficiency, drought or flooding, and pathogens (Chen et al., 2021; Manghwar et al., 2022). To adapt to unfavorable environments, plants have evolved specialized molecular mechanisms that serve to balance the trade-off between abiotic stress responses and growth (Chen et al., 2021). Plants utilize various signaling molecules, including hormones for mediating the plant response to stresses (Manghwar et al., 2022). Multiomics, as powerful tools, has been largely used to explore plant responses against these stresses (AbuQamar et al., 2016).

4.1 Salt

Soil salinization is defined as the excess or deposition of salt ions in land, directly influenced growth and development of horticultural plants, meanwhile, plants can adapt to saline environment by the regulation of multiple genes. Approximately 20% of the world's irrigated agricultural lands are adversely affected by soil salinization (Zhao et al., 2021). At present, pomegranate is widely grown in arid and semi-arid soils, with constant soil salinization (Adiba et al., 2023; Liu et al., 2019). Studies illuminating the mechanism of the response to salt stress in pomegranate will provide valuable molecular information for selecting salt-tolerant genes in pomegranate breeding.

To elucidate the molecular response to salt stress on mRNA levels, 18 cDNA libraries of pomegranate roots and leaves from 0 (controls), 3, and 6 days were constructed after 200 mM NaCl treatment (Liu et al., 2019). The results indicated that 34,047 genes by mapping to genome, and 2,255 DEGs were identified, including 1,080 upregulated and 1,175 downregulated genes. Ions/metal ions binding-related genes were first suppressed and then recovered in leaves, while ion transport and oxidation-reduction process were restricted under salt stress. The ABA associated pathway response to salt threat in pomegranate defense was attached for many related genes that expressed distinctly both in roots and leaves. Three *PYLs*, as ABA receptors, were significantly downregulated in roots and leaves under salt stress, while seven *PP2Cs* were upregulated, and five *PP2Cs* downregulated in roots or leaves. Compared with controls, the most abundant of differential expression transcriptional factors under salt stress compared to controls included *NAC*, *ERF*, *MYB*-related, *C2H2*, MYB, *bHLH*, *GRAS*, *WRKY*, and *bZIP* family genes (Liu et al., 2019). Calzone et al. (2023) also evaluated the effect of NaCl treatment on two pomegranate cultivars ('Wonderful' and 'Parfianka') using the profiles of secondary metabolites, and the treatments were set as the moderate levels of salt stress (100 mM NaCl for 35 consecutive days) and sequentially with a realistic O_3 treatment (100 ppb for 5 h). It was observed that NaCl or O_3 induced the accumulation of cinnamic acid derivatives (more than 3-fold higher than controls), exhibiting a chemical composition plasticity against oxidative stress in 'Wonderful' leaves. Accumulation of ellagitannins and punicalagins under salt stress can help 'Wonderful' leaves coping with subsequent O^3 treatment. On the other hand, 'Parfianka' leaves had a constitutive amount of phenolics/polyphenolics during the whole treatment period, suggesting that 'Parffanka' was more salt- and O_3-tolerant than 'Wonderful'.

4.2 Drought

Water is necessary for plant growth and development. However, drought in soil affects plant growth and development, fruit yield, and quality, even survival (Qian et al., 2020). Precipitation has gradually decreased worldwide in recent years, while evaporation has increased as a result of rising temperatures. Thus, drought has become recognized as one of the major abiotic stress factors resulting in yield reduction or even crop extinction in agricultural production (Qian et al., 2020). Pomegranate is mainly cultivated in semi-arid areas. Thus, understanding the response mechanisms to drought stress is of great importance (Catola et al., 2016). Previous studies reported that ABA can effectively trigger pomegranate tolerance to drought stress, through elevating antioxidant enzyme activities and GSH/GSSG ratio (Qian et al., 2020). According to the comparative transcriptomic analysis of ABA treated and untreated pomegranate leaves, ABA receptor *PYL4* gene was upregulated, while its downstream regulating target *PP2C* gene was downregulated after ABA treatment. The results also implicated the possible roles of genes associated with brassinosteroid (BR) metabolic process (6 genes), MAPK signaling pathway (10 genes), peroxisome (4 genes), riboflavin metabolism, and carotenoid biosynthesis responding to drought (Qian et al., 2020).

Nevertheless, exploiting water scarcity conditions could be a viable approach to improve the quantitative and qualitative attributes of pomegranate seed oil

(Adiba et al., 2023). NMR and FTIR (Fourier transform infrared spectroscopy) technologies were used to investigate how sustained deficit irrigation (SDI-50), equivalent to 50% of crop evapotranspiration, influences pomegranate seed oil (PSO) attributes such as phenols, flavonoids, and tannins content, and the seeds' lipochemical fingerprints compared to fully irrigated trees of the five local clones "Grenade Jaune", 'Djebali', "Ounk Hmam" 'Sefri', and 'Gjeibi' as well as the two exotic varieties "Zheri Precoce" (Tunisia) and "Mollar Osin Hueso" (China) at fully ripening stages (Adiba et al., 2023). PSO yield composed of total phenol, total tannins, was significantly increased by SDI-50 treatment in five of seven detected cultivars compared to control. The highest record was in "Zheri Precoce" cultivar, which was twice that of the control. SDI-50 also induced a substantial increase in total phenolic content, with an average increase of 7.5%, which was correlated with an increase in antioxidant activity. Noticeably, water stress appears to increase the proportion of certain fatty acids, such as oleic acid and linoleic acid, which are associated with improved oil quality and health benefits (Adiba et al., 2023).

4.3 Cold

Since "Tunisian soft seed" pomegranate is susceptible to cold, freezing injury can cause cold damage at below –10°C for more than half a day in the winter or low temperatures in early spring (Guan et al., 2022). Subsequently, the growth of pomegranate buds and new shoots were retarded when the temperature was below 3°C in spring in China. Thus, cold weather causes at least 20–30% yield losses every year (Guan et al., 2022). At present, some researches aimed to develop the mechanism for improving cold resistance in pomegranate.

Based on the transcriptome analysis of "Tunisian soft-seed" pomegranate under normal (control) and low-temperature conditions (6 and 0°C), Guan et al. (2022) observed the key genes and crucial biological pathways involved in response to cold stress (Guan et al., 2022). The total of 7,772 DEGs (3,465 upregulated and 4,307 downregulated) were found in 6°C vs CK, while 3,246 DEGs (1,682 upregulated and 1,564 downregulated) in 0°C vs CK, with 1,561 common DEGs. A large number transcriptional factors such as bHLH, NAC, ERF, WRKY, MYB, and bZIP participated in the response to cold stress in pomegranate (Guan et al., 2022). The response to cold stress in pomegranate is vitally interrelated in photosynthesis, photosynthesis-antenna proteins, and carbon fixation in photosynthetic organisms, because 14 *LHCA* and *LHCB* (light-harvesting complex I and II chlorophyll a/b binding protein) genes and dozens of genes related to photosystem and carbon fixation were downregulated several folds under cold stress (Guan et al., 2022). Also, many DEGs involved in the accumulation of soluble sugars were particularly upregulated, e.g., genes encoding α/β-amylase (α/βAMY), β-glucosidase (bglU, bglX, and bglB), fructofuranosidase (sacA), and hexokinase (HK) as well as trehalose phosphate synthase (TPS), trehalose phosphatase (TP), UDP-glucose 4-epimerase (galE), UTP-glucose-1-phosphate uridylyltransferase (UGP2). Collectively, the results provide useful information to understand the molecular mechanism of pomegranate response to cold stress and also lay a foundation for selecting candidate genes related to cold tolerance molecular breeding in pomegranate (Guan et al., 2022).

4.4 Bacterial Blight

The bacterial blight-causing pathogen in pomegranate, *Xanthomonas. axonopodis* pv. punicae, was identified in 1952, but was not considered a serious threat at that time. Gradually, the disease has become a serious threat and outbreaks have reached an epiphytotic level in several parts of central India, leading to the loss of yield and quality. The blight affects all above-ground parts of the plant of which the fruits are the most vulnerable, producing water-soaked lesions, and then necrotic dark brown/blackish brown spots/lesions, and at last large blighted areas with dried silvery bacterial ooze (Singh et al., 2020).

The transcriptome of pomegranate leaves and fruit was investigated to the defense response to *X. axonopodis* pv. *punicae* using the susceptible 'Bhagawa' and the moderately resistant genotype (IC 524207) at three progressive infection stages from hills of Himalayas in India (Singh et al., 2020). A total of 34,626 unigenes with size >2 kb was obtained, and 85.3% unigenes were annotated in at least one of the seven databases examined. GO terms revealed that the DEGs were significantly enriched oxidation reduction biological process, protein binding and oxidoreductase activity. Comparative analysis of gene-expression signatures showed that the more upregulated DEGs were involved in IC 524207, and more rapid response to pathogen infection occurred in the resistant genotype, compared to susceptible cultivar. Compared with resistant genotype, the genes related to phenylpropanoid biosynthesis, photosynthesis and starch sugar metabolism, known as energy and defense associated pathways, were downregulated in susceptible cultivar. The obtained results indicated that the profiles of DEGs in leaves and fruit could illuminate the reasons for the response to blight infection in pomegranate.

References

AbuQamar, S.F., Moustafa, K. and Tran, L.S.P. 2016. 'Omics' and plant responses to Botrytis cinerea. *Frontiers in Plant Science*, *7*. doi:10.3389/fpls.2016.01658.

Adiba, A., Razouk, R., Haddioui, A., Ouaabou, R., Hamdani, A., Kouighat, M. and Hssaini, L. 2023. FTIR spectroscopy-based lipochemical fingerprints involved in pomegranate response to water stress. *Heliyon*, *9*(6): e16687. doi:10.1016/j.heliyon.2023.e16687.

Attanayake, R., Eeswaran, R., Rajapaksha, R., Weerakkody, P. and Bandaranayake, P. C.G. 2018. Biochemical composition and expression of anthocyanin biosynthetic genes of a yellow peeled and pinkish ariled pomegranate (*Punica granatum* L.) cultivar are differentially regulated in response to agro-climatic conditions. *Journal of Agricultureal Food Chemistry*, *66*(33): 8761–71. doi:10.1021/acs.jafc.8b02909.

Belay, Z.A., Caleb, O.J., Vorster, A., van Heerden, C. and Opara, U.L. 2020. Transcriptomic changes associated with husk scald incidence on pomegranate fruit peel during cold storage. *Food Research International*, *135*: 109285. doi:10.1016/j.foodres.2020.109285.

Borochov-Neori, H., Judeinstein, S., Harari, M., Bar-Ya'akov, I., Patil, B.S., Lurie, S. and Holland, D. 2011. Climate effects on anthocyanin accumulation and composition in the pomegranate (*Punica granatum* L.) fruit arils. *J. Agric Food Chem*, *59*(10): 5325–34. doi:10.1021/jf2003688.

Calani, L., Beghe, D., Mena, P., Del Rio, D., Bruni, R., Fabbri, A. and Galaverna, G. (2013). Ultra-HPLC-MS(n) (Poly) phenolic profiling and chemometric analysis of juices from ancient *Punica granatum* L. cultivars: A non-targeted approach. *Journal of Agricultural Food Chemistry*, *61*(23): 5600–5609. doi:10.1021/jf400387c.

Calzone, A., Tonelli, M., Cotrozzi, L., Lorenzini, G., Nali, C. and Pellegrini, E. 2023. Significance of phenylpropanoid pathways in the response of two pomegranate cultivars to salinity and ozone stress. *Environmental and Experimental Botany*, *208*. doi:10.1016/j.envexpbot.2023.105249.

Catola, S., Marino, G., Emiliani, G., Huseynova, T., Musayev, M., Akparov, Z. and Maserti, B.E. 2016. Physiological and metabolomic analysis of *Punica granatum* (L.) under drought stress. *Planta*, *243*(2): 441–449. doi:10.1007/s00425-015-2414-1

Chen, H., Bullock, D.A., Jr., Alonso, J.M. and Stepanova, A.N. 2021. To fight or to grow: The balancing role of ethylene in plant abiotic stress responses. *Plants* (*Basel*), *11*(1). doi:10.3390/plants11010033.

Chen, L., Zhang, J., Li, H., Niu, J., Xue, H., Liu, B. and Cao, S. 2017. Transcriptomic analysis reveals candidate genes for female sterility in pomegranate flowers. *Frontiers in Plant Science*, *8*: 1430. doi:10.3389/fpls.2017.01430.

Du, C.T., Wang, P.L. and Francis, F.J. 1975. Anthocyanins of pomegranate, *Punica granatum*. *Journal of Food Science*, *40*(2): 417–18. doi:10.1111/j.1365-2621.1975.tb02217.x.

Feng, L., Wang, C., Yang, X., Jiao, Q. and Yin, Y. 2022. Transcriptomics and metabolomics analyses identified key genes associated with sugar and acid metabolism in sweet and sour pomegranate cultivars during the developmental period. *Plant Physiol Biochem.*, *181*: 12–22. doi:10.1016/j.plaphy.2022.04.007.

Fu, F.F., Peng, Y.S., Wang, G.B., El-Kassaby, Y.A. and Cao, F.L. 2021. Integrative analysis of the metabolome and transcriptome reveals seed germination mechanism in *Punica granatum* L. *Journal of Integrative Agriculture*, *20*(1): 132–46. doi:10.1016/s2095-3119(20)63399-8.

Ginzberg, I. and Faigenboim, A. 2022. Ripening of pomegranate skin as revealed by developmental transcriptomics. *Cells*, *11*(14). doi:10.3390/cells11142215.

Gómez-Bellot, M.J., Garcia, C.J., Parra, A., Vallejo, F. and Ortuño, M.F. 2023. Influence of drought stress on increasing bioactive compounds of pomegranate (*Punica granatum* L.) juice. Exploratory study using LC–MS-based untargeted metabolomics approach. *European Food Research and Technology*, *249*(11): 2947–56. doi:10.1007/s00217-023-04340-8.

Guan, S., Chai, Y., Hao, Q., Ma, Y., Wan, W., Liu, H. and Diao, M. 2022. Transcriptomic and physiological analysis reveals crucial biological pathways associated with low-temperature stress in Tunisian soft-seed pomegranate (*Punica granatum* L.). *Journal of Plant Interactions*, *18*(1). doi:10.1080/17429145.2022.2152887.

Harel-Beja, R., Tian, L., Freilich, S., Habashi, R., Borochov-Neori, H., Lahav, T. and Holland, D. 2019. Gene expression and metabolite profiling analyses of developing pomegranate fruit peel reveal interactions between anthocyanin and punicalagin production. *Tree Genetics and Genomes*, *15*(2). doi:10.1007/s11295-019-1329-6.

Hasanpour, M., Saberi, S. and Iranshahi, M. 2020. Metabolic profiling and untargeted 1H-NMR-Based metabolomics study of different Iranian pomegranate (*Punica granatum*) ecotypes. *Planta Med.*, *86*(3): 212–19. doi:10.1055/a-1038-6592.

Huo, Y., Yang, H., Ding, W., Huang, T., Yuan, Z. and Zhu, Z. 2023. Combined transcriptome and proteome analysis provides insights into petaloidy in pomegranate. *Plants*, *12*: 2402.

Kalaycioglu, Z. and Erim, F.B. 2017. Total phenolic contents, antioxidant activities, and bioactive ingredients of juices from pomegranate cultivars worldwide. *Food Chemistry*, *221*: 496–507. doi:10.1016/j.foodchem.2016.10.084.

Kashash, Y., Doron-Faigenboim, A., Bar-Ya'akov, I., Hatib, K., Beja, R., Trainin, T. and Porat, R. 2019a. Diversity among pomegranate varieties in chilling tolerance and transcriptome responses to cold storage. *Journal of Agricultural and Food Chemistry*, *67*(2): 760–71. doi:10.1021/acs.jafc.8b06321.

Kashash, Y., Doron-Faigenboim, A., Holland, D. and Porat, R. 2019b. Effects of harvest time on chilling tolerance and the transcriptome of 'Wonderful' pomegranate fruit. *Postharvest Biology and Technology*, *147*: 10–19. doi:10.1016/j.postharvbio.2018.09.005.

Kashash, Y., Doron-Faigenboim, A., Holland, D. and Porat, R. 2018. Effects of low-temperature conditioning and cold storage on development of chilling injuries and the transcriptome of 'Wonderful' pomegranate fruit. *International Journal of Food Science and Technology, 53*(9): 2064–76. doi:10.1111/ijfs.13793.

Li, J., He, X., Li, M., Zhao, W., Liu, L. and Kong, X. 2015. Chemical fingerprint and quantitative analysis for quality control of polyphenols extracted from pomegranate peel by HPLC. *Food Chemistry, 176*: 7–11. doi:10.1016/j.foodchem.2014.12.040.

Liu, C., Zhang, Z., Dang, Z., Xu, J. and Ren, X. 2021a. New insights on phenolic compound metabolism in pomegranate fruit during storage. *Scientia Horticulturae, 285*. doi:10.1016/j.scienta.2021.110138.

Liu, C., Zhao, Y., Zhao, X., Wang, J., Gu, M. and Yuan, Z. 2019. Transcriptomic profiling of pomegranate provides insights into salt tolerance. *Agronomy, 10*(1): doi:10.3390/agronomy10010044.

Liu, H., Liu, Z., Wu, Y., Zheng, L. and Zhang, G. 2021b. Regulatory mechanisms of anthocyanin biosynthesis in apple and pear. *International Journal of Molecular Science, 22*: 8441. https://doi.org/10.3390/ijms22168441.

Lijavetzky, D., Ruiz-Garcia, L., Cabezas, J.A., De Andres, M.T., Bravo, G., ... and Martinez-Zapater, J.M. 2006. Molecular genetics of berry colour variation in table grape. *Molecular Genetics Genomics, 276*: 427–35. doi.org/10.1007/s00438-006-0149-1.

Luo, X., Cao, D., Li, H., Zhao, D., Xue, H., Niu, J. and Cao, S. 2018a. Complementary iTRAQ-based proteomic and RNA sequencing-based transcriptomic analyses reveal a complex network regulating pomegranate (*Punica granatum* L.) fruit peel color. *Scientific Reports, 8*(1): 12362. doi:10.1038/s41598-018-30088-3.

Luo, X., Cao, D., Zhang, J., Chen, L., Xia, X., Li, H. and Cao, S. 2018b. Integrated microRNA and mRNA expression profiling reveals a complex network regulating pomegranate (*Punica granatum* L.) seed hardness. *Scientific Reports, 8*(1): 9292. doi:10.1038/s41598-018-27664-y.

Luo, X., Li, H.X., Wu, Z.K., Yao, W., Zhao, P., Cao, D. and Cao, S.Y. 2020. The pomegranate (*Punica granatum* L.) draft genome dissects genetic divergence between soft- and hard-seeded cultivars. *Plant Biotechnology Journal, 18*(4): 955–68. doi:10.1111/pbi.13260.

Magangana, T.P., Makunga, N.P., Fawole, O.A., Stander, M.A. and Opara, U.L. 2022. Antioxidant, antimicrobial, and metabolomic characterization of blanched pomegranate peel extracts: Effect of cultivar. *Molecules, 27*(9). doi:10.3390/molecules27092979.

Manghwar, H., Hussain, A., Ali, Q. and Liu, F. 2022. Brassinosteroids (BRs) role in plant development and coping with different stresses. *International Journal of Molecular Science, 23*(3). doi:10.3390/ijms23031012.

Qi, X., Zhao, J., Jia, Z., Cao, Z., Liu, C., Li, J. and Qin, G. 2022. Potential metabolic pathways and related processes involved in pericarp browning for postharvest pomegranate fruits. *Horticulturae, 8*(10). doi:10.3390/horticulturae8100924.

Qian, J.J., Zhang, X.P., Yan, Y., Wang, N., Ge, W.Q., Zhou, Q. and Yang, Y.C. 2020. Unravelling the molecular mechanisms of abscisic acid-mediated drought-stress alleviation in pomegranate (*Punica granatum* L.). *Plant Physiology and Biochemistry, 157*: 211–18. doi:10.1016/j.plaphy.2020.10.020.

Qin, G., Li J., Liu, Chun., Chne, C., Jia, B. and Xu, Y. (2021). Risk assessment of fungicide pesticide residues in vegetables and fruits in the mid-western region of China. *Journal of Fruit Science, 38*(5): 806-16. doi:10.13925/j.cnki.gsxb.20200389.

Qin, G., Liu, C., Li, J., Qi, Y., Gao, Z., Zhang, X. and Xu, Y. 2020. Diversity of metabolite accumulation patterns in inner and outer seed coats of pomegranate: Exploring their relationship with genetic mechanisms of seed coat development. *Horticulture Research, 7*: 10. doi:10.1038/s41438-019-0233-4.

Qin, G.H., Xu, C.Y., Ming, R., Tang, H.B., Guyot, R., Kramer, E.M. and Xu, Y.L. 2017. The pomegranate (*Punica granatum* L.) genome and the genomics of punicalagin biosynthesis. *Plant Journal, 91*(6): 1108–28. doi:10.1111/tpj.13625.

Shi, J., Wang, S., Tong, R., Wang, S., Chen, Y., Wu, W. and Zheng, X. 2022a. Widely targeted secondary metabolomics explored pomegranate aril browning during cold storage. *Postharvest Biology and Technology, 186*. doi:10.1016/j.postharvbio.2022.111839.

Shi, J., Wang, S., Wang, L., Tong, R., Wang, S., Hu, Q., Wan, R., Jian, Z., Zheng, X. and Chen, Y. 2022b. Comparison analysis of key genes expression and metabolites with differential accumulation in lignin biosynthesis from soft- and hard-seed pomegranates. *Journal of Henan Agricultural University*, *56*(1): 61–69. doi:10.16445/j.cnki.1000-2340.20220122.000.

Shi, Q., Du, J., Zhu, D., Li, X. and Li, X. 2020. Metabolomic and transcriptomic analyses of anthocyanin biosynthesis mechanisms in the color mutant Ziziphus jujuba cv. Tailihong. *J. Agric. Food Chem.*, Dec. 23; *68*(51): 15186–98. doi: 10.1021/acs.jafc.0c05334.

Singh, N.V., Parashuram, S., Sharma, J., Potlannagari, R.S., Karuppannan, D.B., Pal, R.K. and Reddy, U.K. 2020. Comparative transcriptome profiling of pomegranate genotypes having resistance and susceptible reaction to *Xanthomonas axonopodis* pv. punicae. *Saudi J. Biol. Sci.*, *27*(12), 3514–28. doi:10.1016/j.sjbs.2020.07.023.

Singh, S.P., Pal, R.K., Saini, M.K., Singh, J., Gaikwad, N., Parashuram, S. and Kaur, C. 2019. Targeted metabolite profiling to gain chemometric insight into Indian pomegranate cultivars and elite germplasm. *Journal of Science of Food and Agriculture*, *99*(11): 5073–82. doi:10.1002/jsfa.9751.

Tozzi, F., Núñez-Gómez, D., Legua, P., Del Bubba, M., Giordani, E. and Melgarejo, P. 2022. Qualitative and varietal characterization of pomegranate peel: High-value co-product or waste of production? *Scientia Horticulturae*, *291*. doi:10.1016/j.scienta.2021.110601.

Trainin, T., Harel-Beja, R., Bar-Ya'akov, I., Ben-Simhon, Z., Yahalomi, R., Borochov-Neori, H., Ophir, R., Sherman, A., Doron-Faigenboim, A. and Holland, D. 2021. Fine mapping of the 'black' peel color in pomegranate (*Punica granatum* L.) strongly suggests that a mutation in the anthocyanidin reductase (ANR) gene is responsible for the trait. *Front. Plant Sci.*, *12*: 642019. doi: 10.3389/fpls.2021.642019.

Valdenegro, M., Fuentes, L., Bernales, M., Huidobro, C., Monsalve, L., Hernandez, I. and Simpson, R. 2022. Antioxidant and fatty acid changes in pomegranate peel with induced chilling injury and browning by ethylene during long storage times. *Frontiers in Plant Science*, *13*: 771094. doi:10.3389/fpls.2022.771094.

Wan, R., Song, J., Lv, Z., Qi, X., Feng, Z., Yang, Z., Cao, X., Shi, J., Jian, Z., Tong, R., et al., 2023. Effects of 1-MCP treatment on postharvest fruit of five pomegranate varieties during low-temperature storage. *Horticulturae*, *9*: 1031. https://doi.org/10.3390/horticulturae9091031.

Wang Y., Guo L., Zhao X., et al. Advances in mechanisms and omics pertaining to fruit cracking in horticultural plants. Agronomy 2021, 11: 1045.

Xue, H., Cao, S., Li, H., Zhang, J., Niu, J., Chen, L. and Zhao, D. 2017. *De novo* transcriptome assembly and quantification reveal differentially expressed genes between soft-seed and hard-seed pomegranate (*Punica granatum* L.). *PLoS One*, *12*(6): e0178809. doi:10.1371/journal.pone.0178809.

Yuan, L., Niu, H., Yun, Y., Tian, J., Lao, F., Liao, X. and Zhou, L. 2022a. Analysis of coloration characteristics of Tunisian soft-seed pomegranate arils based on transcriptome and metabolome. *Food Chemistry*, *370*: 131270. doi:10.1016/j.foodchem.2021.131270.

Yuan, L., Yun, Y., Tian, J., Gao, Z., Xu, Z., Liao, X. and Zhou, L. 2022b. Transcription profile analysis for biosynthesis of flavor volatiles of Tunisian soft-seed pomegranate arils. *Food Research International*, *156*: 111304. doi:10.1016/j.foodres.2022.111304.

Yuan, Z.H., Fang, Y.M., Zhang, T.K., Fei, Z.J., Han, F.M., Liu, C.Y., . . . and Zheng, H.K. 2018. The pomegranate (*Punica granatum* L.) genome provides insights into fruit quality and ovule developmental biology. *Plant Biotechnology Journal*, *16*(7): 1363–74. doi:10.1111/pbi.12875.

Zhang, X., Zhao, Y., Ren, Y., Wang, Y. and Yuan, Z. 2020. Fruit breeding in regard to color and seed hardness: A genomic view from pomegranate. *Agronomy*, *10*(7). doi:10.3390/agronomy10070991.

Zhao, S., Zhang, Q., Liu, M., Zhou, H., Ma, C. and Wang, P. 2021. Regulation of plant responses to salt stress. *International Journal of Molecular Science*, *22*(9). doi:10.3390/ijms22094609.

Zhao, J., Qi, X., Li, J., Cao, Z., Liu, X., Yu, Q., Xu, Y. and Qin, G. 2023. Metabolic profiles of pomegranate juices during fruit development and the redirection of flavonoid metabolism. *Horticulturae*, *9*: 881. https://doi.org/10.3390/horticulturae9080881.

5

Chloroplast Genome of Pomegranate

Ming Yan[1] and *Zhaohe Yuan*[1*]

The chloroplast genomes of pomegranates exhibit the characteristic quadripartite structure, comprising pairs of inverted repeats interspersed with a small single-copy region and a large single-copy region. The length of these chloroplast genome sequences falls within the range of 158,638 bp–158,655 bp. Consistently, there are 113 genes, maintaining the same gene order, encompassing 79 protein-coding genes, 30 tRNA genes, and 4 rRNA genes. With an overall GC content of 36.92%, the pomegranate chloroplast genomes demonstrate minimal sequence diversity. This low diversity suggests that chloroplast genome sequences may not be optimal for exploring the genetic variations among pomegranate genotypes. The findings underscore the need for alternative approaches in investigating the genetic diversity of pomegranate cultivars.

1. Introduction

Chloroplasts are the photosynthetic organelles of the plant cells, which are derived from free-living cyanobacteria through endosymbiosis (McFadden et al., 2004). Apart from playing key roles in photosynthesis, chloroplasts are also responsible for other aspects of plant physiology and development (Daniell et al., 2016). A new study found that chloroplast retrograde signaling can regulate nuclear alternative splicing of a subset of *Arabidopsis thaliana* transcripts (Petrillo et al., 2014; Herz et al., 2019). Interestingly, researchers have found that chloroplasts play diverse roles in plant defense, including contributing to the production of defense compounds (Savage et al., 2021). Chloroplasts contain their own genome, the chloroplast DNA

1 College of Forestry, Nanjing Forestry University, Nanjing 210037, China.
* Corresponding author: zhyuan88@hotmail.com

(cpDNA), which is highly conserved in genomic structure, gene content, and gene order. Cp genomes have been proved to be an effective biological tool for rapid and accurate species recognition as super-barcode (Chen et al., 2018; Yang, Feng et al., 2019). With the advent of high-throughput sequencing technology, the increasing number of chloroplast genomes of fruit crops has been published (Redwan et al., 2015, Cheng et al., 2017). Before the development of next-generation sequencing technology, cp genome assembly was usually based on conventional primer walking strategies (Jansen et al., 2005), which are laborious and costly. It is now convenient to obtain complete a cp genome by using a genomic DNA extracted from the leaf tissue, because a large number of cp genomes are present in the sample (Twyford and Ness, 2017). Based on homology to cp from related species, these reads from cp can be assembled into circle genomes. Many bioinformatic tools have been developed to recover the cp genome sequence from the total genomic DNA, such as NOVOPlasty (Dierckxsens et al., 2016), chloroExtractor (Ankenbrand et al., 2018), GetOrganelle (Jin et al., 2020).

Pomegranate (*Punica granatum* L.) is an important perennial fruit crop of tropical and subtropical regions of the world. Abdul et al. (2018) sequenced the first complete chloroplast genomes of *P. granatum* (cultivar Helow) cultivated in the mountains of Jabal Al-Akhdar, Oman and described its comparative phylogenetic assortment with chloroplast genomes of related species (Khan et al., 2018). Thirteen complete chloroplast genome sequences of pomegranate accessions have been included in NCBI Genbank databases as of October 2022. All the 13 chloroplast genomes shared the typical quadripartite structure, with a pair of inverted repeats each separated by a small single copy region and a large single copy region (Figure 1). The length of chloroplast genome sequences ranged from 158,638 bp–158,655 bp. There are identical sets of 113 genes with the same gene order, including 79 protein-coding, 30 tRNA, and four rRNA genes. The overall GC content was 36.92%.

2. Sequence Diversity of Pomegranate Chloroplast Genomes

Chloroplast DNA has already been used in accessing the genetic diversity and phylogenetic structure at an intraspecies level. For instance, hypervariable regions of chloroplast DNA such as *atpB-rbcL*, *trnL-trnF*, and *rps16-trnQ* were used to assess the genetic diversity of Tunisian apricot accessions (Batnini et al., 2019). Also, chloroplast microsatellite loci were used to investigate the genetic diversity of Iranian pomegranate genotypes (Norouzi et al., 2012). Yan et al. (2019) investigated the sequence diversity of pomegranate chloroplast genomes (Yan et al., 2019). The results showed that the sequence diversity of pomegranate chloroplast genomes was extremely low and there were no parsimony variable sites in the alignment of chloroplasts of the five pomegranate accessions. Therefore, chloroplast genome sequences might not be appropriate for investigating the genetic diversity of pomegranate genotypes.

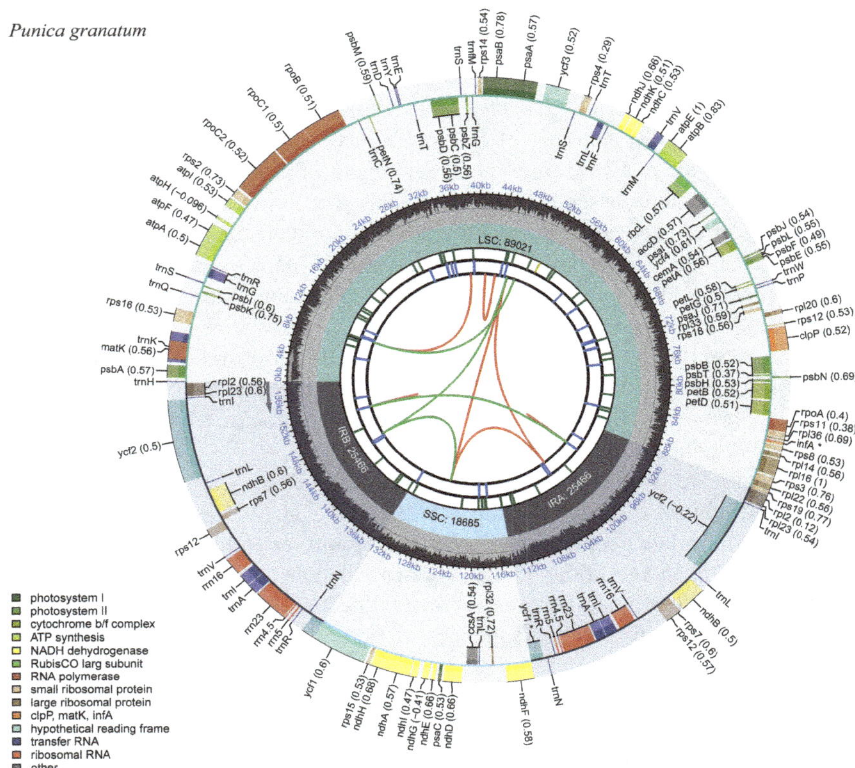

Fig. 1 Chloroplast genome maps of P. granatum. Genes drawn outside the outer circle are transcribed clockwise, and those inside are transcribed counter-clockwise. Genes belonging to different functional groups are color-coded. The length of LSC, SSC, and IRs shown is the mean value.

3. Evolutionary Implications

The significance of complete chloroplast genomes extends beyond their structural understanding, as they serve as crucial tools for unraveling phylogenetic relationships across various taxonomic levels. Recent advancements in phylogenetic analyses, particularly those utilizing protein-coding genes within the chloroplast genome, have played a pivotal role in establishing a comprehensive phylogenetic framework for Viridiplantae. This framework holds substantial value for evolutionary biologists and ecologists alike, offering insights into the broader evolutionary context of plant life.

Within the taxonomic order Myrtales, the genus *Punica* has stirred a debate regarding its familial placement, with suggestions ranging from Punicaceae to Lythraceae and Myrtaceae. Notably, recent phylogenomic analyses, employing 106 single-copy nuclear genes, conclusively support the classification of pomegranate

Table 1 List of the complete pomegranate genomes available in NCBI database and their size.

Cultivar	*Accession Number*	*Length (bp)*	*Publish*
yeba wild pomegranate	MW387436	158638	NA
Bhagwa	MK635347	158641	(Yan, Zhao et al., 2019)
Nana	MK603513	158638	(Yan, Zhao et al., 2019)
Tunisia	MK603512	158639	(Yan, Zhao et al., 2019)
Taishanhong	MK603511	158638	(Yan, Zhao et al., 2019)
Pleniflora	MZ736606	158655	NA
Luqing 1	MN630638	158638	(Feng, Yang et al., 2020)
Daihong 1	MN630639	158651	NA
Sanbaitian	MN630637	158651	NA
Helow	MG878386	158630	(Khan, Asaf et al., 2018)
NA	MT600023	158645	NA
NA	NC_035240/ KY635833	158633	(Rabah, Lee et al., 2017)
NA	MN833212	158639	NA

Note: NA represents there is no cultivar name or publish paper.

within the Lythraceae family rather than the previously proposed monogeneric Punicaeceae family.

To elucidate the precise phylogenetic position of pomegranate within Myrtales and delve deeper into the relationships within this order, Yan et al. (2019) conducted a phylogenetic analysis utilizing complete chloroplast genome sequences. A diverse set of 85 species representing five families from the order Myrtales was meticulously selected for this investigation. The outcomes of the analysis unequivocally revealed that all pomegranate accessions clustered within a single clade, intricately interlinked with other closely related species within the Lythraceae family.

This comprehensive exploration of the pomegranate's chloroplast genome not only contributes to resolving the long-standing debate surrounding its familial affiliation, but also enriches our understanding of the broader evolutionary dynamics within Myrtales. The utilization of complete chloroplast genomes as phylogenetic markers emerges as a powerful approach, facilitating nuanced insights into the intricate relationships shaping the evolutionary tapestry of plant taxa.

Conclusions

In conclusion, this chapter provides a comprehensive overview of the pomegranate chloroplast genomes, with a focus on 13 complete chloroplast genome sequences deposited in NCBI GenBank databases as of October 2022. The consistent quadripartite structure among these genomes, featuring pairs of inverted repeats separated by small and large single-copy regions, is indicative of a shared genomic framework. The length of these chloroplast genomes ranges from 158,638 bp–158,655 bp, and they exhibit identical sets of 113 genes with a consistent gene order, encompassing 79 protein-coding genes, 30 tRNA genes, and 4 rRNA genes. The overall GC content is measured at 36.92%. Notably, the observed extremely low sequence diversity in pomegranate chloroplast genomes underscores the limitations of employing these sequences for assessing the genetic diversity of pomegranate genotypes.

Moreover, the phylogenomic analysis, utilizing complete chloroplast genome sequences, affirms the placement of pomegranate within the Lythraceae family. This robust phylogenetic support enhances our understanding of the evolutionary relationships within this botanical family. As genomic research continues to evolve, these findings contribute valuable insights into the genomics of pomegranates, laying the groundwork for further exploration and applications in fields such as plant breeding and biodiversity conservation.

References

Ankenbrand, M.J., S. Pfaff, N. Terhoeven, M. Qureischi, M. Gündel, C.L. Weiß, T. Hackl and F. Förster. 2018. ChloroExtractor: Extraction and assembly of the chloroplast genome from whole genome shotgun data. *Journal of Open Source Software*, *3*(21): 464.

Batnini, M.A., H. Bourguiba, N. Trifi-Farah and L. Krichen. 2019. Molecular diversity and phylogeny of Tunisian *Prunus armeniaca* L. by evaluating three candidate barcodes of the chloroplast genome. *Scientia Horticulturae*, *245*: 99–106.

Chen, X., J. Zhou, Y. Cui, Y. Wang, B. Duan and H. Yao. 2018. Identification of Ligularia Herbs Using the Complete Chloroplast Genome as a Super-Barcode. *Frontiers in Pharmacology*, *9*.

Cheng, H., J. Li, H. Zhang, B. Cai, Z. Gao, Y. Qiao and L. Mi. 2017. The complete chloroplast genome sequence of strawberry (*Fragaria ananassa* Duch.) and comparison with related species of Rosaceae. *PeerJ.*, *5*: e3919.

Daniell, H., C.-S. Lin, M. Yu and W.-J. Chang. 2016. Chloroplast genomes: Diversity, evolution, and applications in genetic engineering. *Genome Biology*, *171*: 1–29.

Dierckxsens, N., P. Mardulyn and G. Smits. 2016. NOVOPlasty: *De novo* assembly of organelle genomes from whole genome data. *Nucleic Acids Research*, *45*(4): e18–e18.

Feng, L., X. Yang, Q. Jiao, C. Wang, Y. Yin and J. Tao. 2020. Characterization of the complete chloroplast genome of *Punica granatum* 'Luqing1'. *Mitochondrial DNA Part B*, *5*(3): 2578–79.

Herz, M.A. G., M.G. Kubaczka, G. Brzyżek, L. Servi, M. Krzyszton, C. Simpson, J. Brown, S. Swiezewski, E. Petrillo and A.R. Kornblihtt. 2019. Light regulates plant alternative splicing through the control of transcriptional elongation. *Molecular Cell*, *73*(5): 1066–74. e1063.

Jansen, R.K., L.A. Raubeson, J.L. Boore, C.W. Depamphilis, T.W. Chumley, R.C. Haberle, S.K. Wyman, A.J. Alverson, R. Peery and S.J. Herman 2005. Methods for obtaining and analyzing whole chloroplast genome sequences. *In*: *Methods in Enzymology*, *395*: 348–84. Elsevier, New York, USA.

Jin, J.-J., W.-B. Yu, J.-B. Yang, Y. Song, C.W. dePamphilis, T.-S. Yi and D.-Z. Li. 2020. GetOrganelle: A fast and versatile toolkit for accurate *de novo* assembly of organelle genomes. *Genome Biology*, *21*(1): 241.

Khan, A.L., S. Asaf, I.-J. Lee, A. Al-Harrasi and A. Al-Rawahi. 2018. First reported chloroplast genome sequence of *Punica granatum* (cultivar Helow) from Jabal Al-Akhdar, Oman: Phylogenetic comparative assortment with Lagerstroemia. *Genetica*, *146*(6): 461–74.

McFadden, G.I. and G.G. van Dooren 2004. Evolution: Red algal genome affirms a common origin of all plastids. *Current Biology*, *14*(13): R514–R516.

Norouzi, M., M. Talebi and B.-E. Sayed-Tabatabaei. 2012. Chloroplast microsatellite diversity and population genetic structure of Iranian pomegranate (*Punica granatum* L.) genotypes. *Scientia Horticulturae, 137*: 114–20.

Petrillo, E., M.A. Godoy Herz, A. Fuchs, D. Reifer, J. Fuller, M.J. Yanovsky, C. Simpson, J.W. Brown, A. Barta and M. Kalyna. 2014. A chloroplast retrograde signal regulates nuclear alternative splicing. *Science, 344*(6182): 427–30.

Rabah, S.O., C. Lee, N.H. Hajrah, R.M. Makki, H.F. Alharby, A.M. Alhebshi, J.S.M. Sabir, R.K. Jansen and T.A. Ruhlman. 2017. Plastome Sequencing of Ten Nonmodel Crop Species Uncovers a Large Insertion of Mitochondrial DNA in Cashew. *The Plant Genome, 10*(3).

Redwan, R., A. Saidin and S. Kumar. 2015. Complete chloroplast genome sequence of MD-2 pineapple and its comparative analysis among nine other plants from the subclass Commelinidae. *BMC Plant Biology, 15*(1): 1–20.

Savage, Z., C. Duggan, A. Toufexi, P. Pandey, Y. Liang, M.E. Segretin, L.H. Yuen, D.C.A. Gaboriau, A.Y. Leary, Y. Tumtas, V. Khandare, A.D. Ward, S.W. Botchway, B.C. Bateman, I. Pan, M. Schattat, I. Sparkes and T.O. Bozkurt. 2021. Chloroplasts alter their morphology and accumulate at the pathogen interface during infection by *Phytophthora infestans. The Plant Journal, 107*(6): 1771–87.

Twyford, A.D. and R.W. Ness. 2017. Strategies for complete plastid genome sequencing. *Molecular Ecology Resources, 17*(5): 858–68.

Yan, M., X. Zhao, Y. Zhao, Y. Ren and Z. Yuan. 2019. The complete chloroplast genome sequence of pomegranate 'Bhagwa'. *Mitochondrial DNA Part B, 4*(1): 1967–68.

Yan, M., X. Zhao, J. Zhou, Y. Huo, Y. Ding and Z. Yuan 2019. The Complete Chloroplast Genomes of *Punica granatum* and a Comparison with Other Species in Lythraceae. *International Journal of Molecular Sciences, 20*. doi:10.3390/ijms20122886.

Yang, J., L. Feng, M. Yue, Y.-L. He, G.-F. Zhao and Z.-H. Li. 2019. Species delimitation and interspecific relationships of the endangered herb genus Notopterygium inferred from multilocus variations. *Molecular Phylogenetics and Evolution, 133*: 142–51.

Molecular Genomics of Pomegranate Floral Organ Differentiation

Yujie Zhao[1] and *Zhaohe Yuan*[2*]

Pomegranate flowers are of two types on the same tree: bisexual flowers and functional male flowers. The drop of functional male flowers after flowering caused by ovule sterility seriously affects the fruit yield, which is being a bottleneck restricting industrial development. We know that gene regulatory networks involving transcription factors and hormonal communication regulate floral organs initiation and development. Previous research found that MADS-box, YABBY, TALE, etc., transcription factors were involved in floral organs development. This paper will examine some of the key choices as they relate specifically to pomegranate reproductive biology. The impact of these factors on floral development will be discussed. The MADS-box, YABBY, and TALE gene family have been identified in the pomegranate genome, and research will be primarily key genes' functional.

1. Introduction

Flowers are unique to angiosperms. A typical flower possesses four types of organs, i.e., sepals, petals, stamens and carpels, which are arranged on the receptacle from the outside to the center, respectively. We now know that, no matter how diverse the floral organs are, they all experience at least four main developmental processes: initiation, identity determination, morphogenesis, and maturation (Shan et al., 2019).

[1] College of Horticulture, Henan Agricultural University, Zhengzhou 450002, China.

[2] College of Forestry, Nanjing Forestry University, Nanjing 210037, China.

* Corresponding author: zhyuan88@hotmail.com

Initiation of floral organs: Shortly after the formation of a floral meristem, floral organs are initiated in a centripetal sequence on the peripheral regions of the meristem. Each floral organ arises from a small number of cells called founder cells, which subsequently develop into a pre-primordium, or an anlage, and then a recognizable primordium through cell proliferation (Chandler, 2011; Chandler et al., 2011). Once all primordia are initiated, the relative positions and total number of the floral organs, as well as the basic structure, phyllotaxy, and symmetry of the flower, are more or less determined.

Determination of the identities of floral organs: Floral organs can be classified into sepals, petals, stamens, and carpels because they hold their positions and have the attributes of structure and function, or 'identities', that characterize themselves. In fact, even before a floral organ is initiated, the developmental program that determines its identity has been switched on. According to the ABC model of flower development, the four types of floral organs in a typical flower, such as those of *A. thaliana*, are specified by three classes of floral organ identity genes: A, sepals; A + B, petals; B + C, stamens; and C, carpels (Coen and Meyerowitz, 1991).

Morphogenesis of floral organs: Once a floral organ is initiated, it changes its size and shape over time through a process called morphogenesis. Different types of floral organs (for example, sepals, petals, stamens, and carpels) usually take different developmental trajectories, although the basic mechanisms must be the same: they all involve proliferation, expansion and differentiation of cells, as well as establishment of the adaxial-abaxial, proximal-distal and lateral-medial polarities (Irish, 2008; Sauret-Güeto et al., 2013; Walcher-Chevillet and Kramer, 2016; Moyroud and Glover, 2017).

Maturation of floral organs: While morphogenesis is the process through which a tiny primordium develops into a full-sized floral organ, maturation is the process through which the organ gains its visual, olfactory and gustatory traits (for example, color, scent, and taste). Maturation of floral organs is important for the flower because these traits are critical for the success of pollination and/or defenses against florivores and pathogens.

Pomegranate is characterized by having two types of flowers on the same tree: bisexual flowers and functional male flowers. The bisexual flowers have well-formed female (stigma, style and ovary); and male (filaments and anthers) parts and have been referred to as 'fertile', 'vase-shaped', and 'bisexual' flowers (Wetzstein et al., 2011, 2013). Because the bisexual flowers are the type that set fruit, they are commonly referred to as 'female' flowers. The functional male flowers produce well-developed male parts, but the pistil contains reduced female parts (Wetzstein et al., 2011, 2013). Thus, their role is more accurately depicted as functional male flowers, but rather have degenerated female parts. Functional male flowers typically drop and fail to set fruit.

Bisexual flowers had a discoid stigma covered with copious exudate, elongated stigmatic papillae, a single elongate style, and numerous stamens inserted on the inner wall of the calyx tube (Wetzstein et al., 2011; Engin and Gokbayrak, 2017). In contrast, functional male flowers had reduced female parts and exhibited shortened pistils of variable heights. Stigmatic papillae of male flowers had little exudate yet supported pollen germination. However, pollen tubes were rarely observed in styles. Ovules in functional male flowers were rudimentary and exhibited various stages of degeneration (Wetzstein et al., 2011; Engin and Gokbayrak, 2017; Chen et al., 2017). Functional male flowers fall off after flowering and consume a large amount of nutrition.

2. Transcription Factor Regulating the Development of Floral Organ

At present, it is known that MADS-box, YABBY, and other transcriptional regulatory factors play important roles in floral organ development.

2.1 MADS-box Transcription Factor

MADS-box is a critical transcription factor regulating the development of floral organs and plays essential roles in the growth and development of floral transformation, flower meristem determination, the development of male and female gametophytes, and fruit development. On the basis of genetic structures and phylogenetic analysis, the MADS-box family can be divided into two phylogenetically distinct groups: type I and type II (Alvarezbuylla et al., 2000). Most of the well-studied plant genes are type II genes which are considered as the MIKC-type (Smaczniak et al., 2012). The plant-specific MIKC-type MADS-box genes were first identified as floral organ determinant genes in *Arabidopsis thaliana* and *Antirrhinum majus* (Sommer et al., 1990; Yanofsky et al., 1990). Type II genes can be divided into two subfamilies: MIKC* and MIKC^{C} (Arora et al., 2007; Xu et al., 2014). MIKC^{C} subfamily can be divided into 13 branches: AG, AGL6 (AGAMOUS-LIKE, AGL), AGL12, AGL15, AGL17, AP1(APETALA, AP)/FUL (FRUITFULL), AP3/PI (PISTILLATA), FLC (FLOWERING LOCUS C), SOC1 (SUPPRESSOR OF OVEREXPRESSION OF CONSTANS1), SEP (SEPALLATA), SVP (SHORT VEGETATIVE PHASE), BS (B-SISTER), and TM8 (Henschel et al., 2002; Díaz-Riquelme et al., 2009). In *Arabidopsis*, MIKC^{C}-type embraces 12 subfamilies, without Tomato MADS 8 (TM8). So, there is little research on TM8's function (Heijmans et al., 2012).

According to the ABCDE model of flower development, the five class genes of A, B, C, D, and E coordinately regulate the development of sepals, petals, stamens, carpels, and ovules. Arabidopsis has two A-class genes (*AP1* and *AP2*), two B-class genes (*PI* and *AP3*), C/D-class genes (*AG*, *AGL11* and *STK*), and a single E-class gene (*SEP*), of which only *AP2* is not a MADS-box gene (Theisen et al., 2016). The model suggests that A + E control meristem and sepals' formation; A + B + E regulate petals formation; B + C + E control stamens formation; C + E control carpels formation; and C + D + E regulate ovules development (Theisen et al., 2016).

Thirty-six MIKC-type MADS-box genes were identified in the 'Taishanhong' pomegranate genome (Zhao et al., 2020). By utilizing phylogenetic analysis, 36 genes were divided into 14 subfamilies. The gene structure and conserved domain of 36 *PgMADSs* were analyzed, showing that the same subfamilies' gene structures were relatively similar. *Cis*-acting element analysis showed that promoter sequences of PgMADS genes contained multiple hormone response and abiotic stress related elements, suggesting that *PgMADSs* might be closely related to plant hormone signal transduction and adverse situations. It was speculated that *PgMADSs* might be related to the growth and development of pomegranate. Protein interaction network analysis found that *PgMADS15* was located at the core of the interaction network. Tissue-specific expression analysis revealed that the E-class genes (*PgMADS03*, *PgMADS21* and *PgMADS27*) were highly expressed in floral tissues, while *PgMADS29* was not expressed in all tissues, indicating that the functions of the E-class genes were differentiated. *PgMADS15* of the C/D-class was the key gene in the development network of pomegranate flower organs, suggesting that *PgMADS15* might play an essential role in the peel and inner seed coat development of pomegranate. The results in this study will provide a reference for the classification, cloning, and functional research of pomegranate MADS-box genes.

2.2 TALE Transcription Factor

The TALE family plays a vital role in regulating plant growth and development (Mahajan et al., 2012; Sakakibara et al., 2013; Lin et al., 2013; Furumizu et al., 2015), regulating the formation of plant meristems (Arnaud et al., 2014), and the maintenance of organ morphology (Belles-Boix et al., 2006), organ position (Aida et al., 1999), hormone regulation (Shani et al., 2006), signal transduction (Cnops et al., 2006) and tuber formation (Kondhare et al., 2019). Based on protein sequence and evolution, BELL and KNOX belong to the TALE gene family (Arnaud et al., 2014; Ma et al., 2019). Studies have shown that BELL and KNOX proteins specifically recognize and bind to form the BELL-KNOX heterodimer protein (Bhatt et al., 2004), which is essential for the nuclear localization of two transcription factor proteins and the activity of binding target gene (Smith et al., 2002; Kim et al., 2013). TALE can form complexes to regulate ovule development (Brambilla et al., 2007). After binding to the OVATE family protein (OFP), the BELL-KNOX dimer protein is reversely transferred from the nucleus to the cytoplasm to negatively regulate ovule development (Hackbusch et al., 2005). BELL proteins comprise two highly conserved domains: a POX domain (POX is composed of SKY and BEL) and homeodomain. The BELL plays essential roles in ovule development, frond development and fruit development (Byrne et al., 2003; Meng et al., 2018). BEL1 is expressed in the ovule and controlled the ovule integument identity.

The KNOX gene family contains KNOX1, KNOX2, ELK, and homeodomain, except for a novel gene KNATM without the homeodomain (Magnani et al., 2008; Hamant and Pautot, 2010). In addition, KNOX1 and KNOX2 domains merge to form a MEINOX domain. KNOX1 is expressed in the meristem, which is necessary for meristem development and maintenance. Studies have shown that the *KNOX2* gene is involved in regulating the secondary growth of plant cell walls and plays a crucial regulatory role in the development of roots, stems, seed coats, and heartwood (Li et al., 2012; Zhong et al., 2008; Bhargava et al., 2010; Li et al., 2011).

A total of 17 genes of pomegranate TALE family were identified in pomegranate (Wang et al., 2020). *PgTALE* family genes were divided into eight subfamilies (KNOX-I, KNOX-II, KNOX-III, BELL-I, BELL-II, BELL-III, BELL-IV, and BELL-V). All *PgTALEs* had a KNOX domain or a BELL domain, and their structures were conservative. The 1500 bp promoter sequence had multiple *cis*-elements in response to hormones (auxin, gibberellin) and abiotic stress, indicating that most of *PgTALE* were involved in the growth and development of pomegranates and stress. Function prediction and protein-protein network analysis showed that *PgTALE* may participate in regulating the development of apical meristems, flowers, carpels, and ovules (Wang et al., 2020). Analysis of gene expression patterns showed that *PgTALE*s had a particular tissue expression specificity. In conclusion, the knowledge of the TALE gene gained in pomegranate may be applied to other fruit as well.

BELL1 (BEL1), a homeodomain transcription factor, has been reported to be one of the major factors controlling ovule pattering, in particular determining identity and development of the integuments (Bencivenga et al., 2012). Real-time quantitative PCR analysis suggested that the expression level of *PgBEL1* in bisexual flowers was higher than that in functional male flowers at P1-P4 periods (bud vertical diameter < 12.0 mm). *PgBEL1* was highly expressed in leaf, it was 3.2 times of that in calyx while 1.2 times of that in stem, respectively. *PgBEL1* was low expressed in shoot apex. The expression level of *PgBEL1* in pistil was 1.16 times of that in stamen. At the P2 and P3 periods of pomegranate flower development that bud vertical diameter was 5.1 ~ 10.0 mm, the expression level of *PgWUS* in bisexual flowers was higher than that in functional male flowers. *PgWUS* was the lowest expression in calyx and the highest expression in stem. The expression level of *PgWUS* in pistil was 1.5 times of that in stamen and 1.8 times of that in shoot apex (Zhao et al., 2021). There was a significant difference in the expression levels of *PgWUS* in pistil and stamen.

2.3 YABBY Transcription Factor

The YABBY is unique transcription factor in plants, belonging to the zinc finger protein superfamily. YABBY proteins have two highly conserved domains: N-terminal C2C2 type zinc finger domain and Cterminal YABBY domain (Golz et al., 2004; Sieber et al., 2004). The YABBY transcription factors play significant roles in the regulation of diverse developmental processes, such as the formation of adaxial-adaxial polarity, lamina expansion, and floral organ development (Eckardtn, 2010; Ha et al., 2010; Tanaka et al., 2012). It has been reported that the YABBY plays different roles in lateral organ development, such as leaves (Eckardtn, 2010;

Ha et al., 2010), floral organs (Yamada et al., 2011; Tanaka et al., 2012; Fourquin et al., 2014) and fruit development (Han et al., 2015).

There are six YABBY members in *Arabidopsis*, YABBY1 (FIL, FILAMENTOUS FLOWER), YABBY2 (YAB2), YABBY3 (YAB3), YABBY4 (INO, INNER NO OUTER), YABBY5 (YAB5), and CRC (CRABS CLAW). *FIL*, *YAB2*, and *YAB3* are always expressed in the primordia of lateral organs, which determine the abaxial cell fates (Siegfried et al., 1999) and participate in floral organ formation and leaf development (Sawa et al., 1999; Eshed et al., 2004; Stahle et al., 2009), and they act redundantly to promote vegetative organ development (Bowman, 2000; Eckardtn, 2010). INO and CRC are expressed in specific tissues: INO participates in the development of ovule outer integument, while CRC have been reported to be associated with polarity of nectary and carpel (Bowman and Smyth, 1999; Zhang et al., 2013).

Six YABBY genes were identified in pomegranate. They were divided into five subfamilies (YAB1/3, YAB2, INO, CRC, and YAB5), based on protein sequence, motifs and similarity of exon-intron structure. *PgYABBYs* contained lots of hormone response and stress response elements. Subsequently, gene function prediction and protein-protein network analysis showed that *PgYABBYs* were associated with the development of apical meristem, flower, carpel, and ovule. Analysis of *PgYABBY* genes expression in various structures and organs suggested that *PgYABBYs* were highly activated in flower, leaf, and seed coat (Figure 1). Analysis of expression during flower development in pomegranate showed that *PgINO* might play a critical role in regulating the differentiation of flowers (Zhao et al., 2020). This study provided a theoretical basis for function research and utilization of YABBY genes in pomegranate.

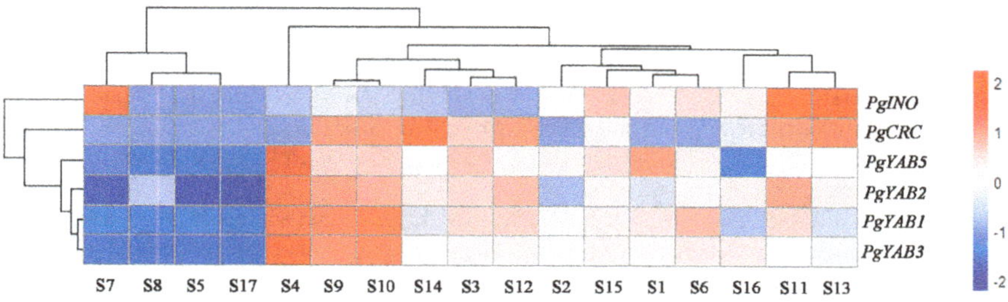

Fig. 1 Thermograph of YABBY family gene expression in different organs of pomegranate. ***Note*:** S1: 'Baiyushizi' endocarp; S2: 'Tunisiruanzi' endocarp; S3: flower; S4: leaf; S5: root; S6: endocarp; S7: exocarp; S8: pericarp; S9: hermaphrodite flower I; S10: functional male flower I; S11: hermaphrodite flower II; S12: functional male flower II; S13: hermaphrodite flower III; S14: functional male flower III; S15: 'Black127' mixture of flowers, leaves, fruits, and roots; S16: 'nana' mixture of flowers, leaves, fruits, and roots; S17: 'Wonderful' pericarp (S3–S8 were 'Dabenzi', S9–S14 were 'Tunisiruanzi').

3. Hormones Regulating Flower Organ Differentiation

Hormones, such as auxin and cytokinin, are involved in the complex molecular network that regulates the coordinated development of plant organs. There is evidence that many hormones play important functions in ovule primordia formation and female fertility (Bencivenga et al., 2012; Huang et al., 2013; Cai et al., 2020). Plants with reduced cytokinin production or perception show a drastic reduction in ovule numbers and female fertility (Hutchison et al., 2006; Riefler et al., 2006; Kinoshita-Tsujimura and Kakimoto, 2011). The transcription factors BEL1 and SPL/NZZ, previously described as key regulators of ovule development, are needed for the auxin and cytokinin signaling pathways for the correct patterning of the ovule (Bencivenga et al., 2012). BR regulates the expression level of genes related to ovule development, including *HLL*, *ANT* and *AP2* (Huang et al., 2013).

In the pomegranate genome, a total of 19 putative ARF candidate genes were identified (Huang et al., 2019). An NJ phylogenetic tree of ARF proteins was clustered ARFs into four classes (Class I, II, III, and IV). The roles of *PgARF* in the development of pomegranate leaves and flower tissues were inferred from the homologous genes *AtARF* in *A. thaliana*. The RNA-Seq expression profile suggested that most of the *PgARF* genes were highly expressed in root, leaf, flower, floral bud, fruit, inner seed coat, outer seed coat, and peel. *PgARF4b* was lowly expressed in the inner seed coat, and *PgARF5* highly expressed in the outer seed coat. These results suggested tissue-specific expression patterns of *ARF* family genes. Compared with the other *ARF* genes, *PgARF3a* had a very weak expression. This might be related to the pseudogenization of *PgARF3a* resulting in the gene inactivation. Conversely, distinct results were found in *PgARF24* which were highly expressed in all tissues. It inferred to be related with the neofunctionalization.

References

Aida, M., T. Ishida and M. Tasaka. 1999. Shoot apical meristem and cotyledon formation during *Arabidopsis* embryogenesis: Interaction among the *CUP-SHAPED COTYLEDON* and *SHOOT MERISTEMLESS* genes. *Development, 126*: 1563.

Alvarezbuylla, E.R., S. Pelaz, S.J. Liljegren, S.E. Gold, C. Burge, G.S. Ditta, L.R. De Pouplana, L.P. Martinezcastilla and M.F. Yanofsky. 2000. An ancestral MADS-box gene duplication occurred before the divergence of plants and animals. *Proc. Nati. Acad. Sci. USA, 97*: 5328–33.

Arnaud, N. and V.R. Pautot. 2014. Ring the BELL and tie the KNOX: Roles for TALEs in gynoecium development. *Front. Plant Sci., 5*: 93.

Arora, R., P. Agarwal, S. Ray, A.K. Singh, V.P. Singh, A.K. Tyagi and S. Kapoor. 2007. MADS-box gene family in rice: Genome-wide identification, organization, and expression profiling during reproductive development and stress. *BMC Genom., 8*: 242.

Belles-Boix, E., O. Hamant, S.M. Witiak, H. Morin, J. Traas and V. Pautot. 2006. KNAT6: An *Arabidopsis* homeobox gene involved in meristem activity and organ separation. *Plant Cell, 18*: 1900–1907.

Bencivenga, S., S. Simonini, Eva. Benková and L. Colombo. 2012. The transcription factors BEL1 and SPL are required for cytokinin and auxin signaling during ovule development in *Arabidopsis*. *The Plant Cell, 24*: 2886–97.

Bhargava, A., S.D. Mansfield, H. Hall, C.J. Douglas and B.E. Ellis. 2010. MYB75 functions in regulation of secondary cell wall formation in the *Arabidopsis* inflorescence stem. *Plant Physiol., 154*: 1428–38.

Bhatt, A.M., J.P. Etchells, C. Canales, A. Lagodienko and H.G. Dickinson. 2004. VAAMANA, a BEL1-like homeodomain protein, interacts with KNOX proteins BP and STM and regulates inflorescence stem growth in *Arabidopsis*. *Gene, 328*: 103–11.

Bowman, J.L. and D.R. Smyth. 1999. CRABS CLAW: A gene that regulates carpel and nectary development in *Arabidopsis*, encodes a novel protein with zinc finger and helix-loophelix domains. *Development, 126*(11): 2387–96.

Bowman, J.L. 2000. The YABBY gene family and abaxial cell fate. *Curr. Opin. Plant Biol., 3*(1): 17–22.

Brambilla, V., R. Battaglia, M. Colombo, S. Masiero, S. Bencivenga, M.M. Kater and L. Colombo. 2007. Genetic and molecular interactions between *BELL1* and MADS Box factors support ovule development in *Arabidopsis. Plant Cell, 19*: 2544–56.

Byrne, M.E., A. Groover, J.R. Fontana and R.A. Martienssen, 2003. Phyllotactic pattern and stem cell fate are determined by the Arabidopsis homeobox gene *BELLRINGER. Development, 130*: 3941–50.

Cai, H.Y., M.N. Chai, F.Q. Chen, Y.M. Huang, M. Zhang, Q. He, L.P. Liu, M.K. Yan and Y. Qin. 2020. HBI1 acts downstream of ERECTA and SWR1 in regulating inflorescence architecture through the activation of the brassinosteroid and auxin signaling pathways. *New Phytologist, 229*(1): 414–28.

Chandler, J.W. 2011. Founder cell specification. *Trends Plant Sci., 16*: 607–13.

Chandler, J.W., B. Jacobs, M. Cole, P. Comelli and W. Werr. 2011. *DORNRÖSCHEN-LIKE* expression marks *Arabidopsis* floral organ founder cells and precedes auxin response maxima. *Plant Mol. Biol., 76*: 171–85.

Chen, L.N., J. Zhang, H.X. Li, J. Niu, H. Xue, B.B. Liu, Q. Wang, X. Luo, F.H. Zhang, D.G. Zhao and S.Y. Cao. 2017. Transcriptomic analysis reveals candidate genes for female sterility in pomegranate flowers. *Front Plant Sci., 8*: 1430.

Cnops, G., P. Neyt, J. Raes, M. Petrarulo, H. Nelissen, N. Malenica, C. Luschnig, O. Tietz, F.A. Ditengou and K. Palme. 2006. The *TORNADO1* and *TORNADO2* genes function in several patterning processes during early leaf development in *Arabidopsis thaliana. Plant Cell, 18*: 852–66.

Coen, E.S. and E.M. Meyerowitz. 1991. The war of the whorls: Genetic interactions controlling flower development. *Nature, 353*: 31–37.

Díaz-Riquelme, J., D. Lijavetzky, J.M. Martinezzapater and M.J. Carmona. 2009. Genome-wide analysis of MIKCC-Type MADS-box genes in grapevine. *Plant Physiol., 149*, 354–69.

Eckardtn, N.A. 2010. YABBY genes and the development and origin of seed plant leaves. *Plant Cell, 22*(7): 2103.

Engin, H. and Z. Gokbayrak. 2017. Micromorphology of pollen grains from bisexual and functional male flowers of pomegranate. *Agrofor. International Journal, 2*: 2.

Eshed, Y., A. Izhaki, S.F. Baum, S.K. Floyd and J.L. Bowman. 2004. Asymmetric leaf development and blade expansion in *Arabidopsis* are mediated by KANADI and YABBY activities. *Development, 131*(12): 2997–3006.

Fourquin, C., A. Primo, I. Martinez-Fernandez, E. Huet-Trujillo and C. Ferrandiz. 2014. The *CRC* orthologue from Pisum sativum shows conserved functions in carpel morphogenesis and vascular development. *Ann. Bot., 114*(7): 1535–44.

Furumizu, C., J.P. Alvarez, K. Sakakibara and J.L. Bowman. 2015. Antagonistic roles for KNOX1 and KNOX2 genes in patterning the land plant body plan following an ancient gene duplication. *PLoS Genet., 11*.

Golz, J.F., M. Roccaro, R. Kuzoff and A. Hudson. 2004. GRAMINIFOLIZ promotes growth and polarity of Antirrhinum leaves. *Development, 131*(15): 3661–70.

Ha, C.M., J.H. Jun and J.C. Fletcher. 2010. Control of Arabidopsis leaf morphogenesis through regulation of the YABBY and KNOX families of transcription factors. *Genetics, 186*(1): 197–206.

Hackbusch, J., K. Richter, J. Muller, F. Salamini and J.F. Uhrig. 2005. A central role of *Arabidopsis thaliana* ovate family proteins in networking and subcellular localization of 3-aa loop extension homeodomain proteins. *Proc. Natl. Acad. Sci. USA, 102*: 4908–12.

Hamant, O. and V. Pautot. 2010. Plant development: A TALE story[J]. *C. R. Biol., 333*: 371–81.

Han, H.Q., Y. Liu, M.M. Jiang, H.Y. Ge and H.Y. Chen. 2015. Identification and expression analysis of YABBY family genes associated with fruit shape in tomato (*Solanum lycopersicum* L.). *Genet. Mol. Res., 14*(2): 7079–91.

Heijmans, K., P. Morel and M. Vandenbussche. 2012. MADS-box genes and floral development: The Dark Side. *J. Exp. Bot., 63*: 5397–5404.

Henschel, K., R. Kofuji, M. Hasebe, H. Saedler, T. Munster and G. Theissen. 2002. Two ancient classes of MIKC-type MADS-box genes are present in the Moss Physcomitrella Patens. *Mol. Biol. Evol.*, *19*: 801–14.
Huang, H.Y., W.B. Jiang, Y.W. Hu, P. Wu, J.Y. Zhu, W.Q. Liang, Z.Y. Wang and W.H. Lin. 2013. BR signal influences *Arabidopsis* ovule and seed number through regulating related genes expression by BZR1. *Molecular Plant*, *6*: 456–69.
Huang, X.B., T.K. Zhang, C.Y. Liu, Y.J. Zhao, H.M. Wei, J.Q. Zhou and Z.H. Yuan. 2019. Genome-wide identification and expression analysis of auxin response factor (*ARF*) gene family in *Punica granatum*. *Journal of Fruit Science*, *36*(1): 43–55.
Hutchison, C.E., J. Li, C. Argueso, M. Gonzalez, E. Lee, M.W. Lewis, B.B. Maxwell, T.D. Perdue, G.E. Schaller, J.M. Alonso, J.R. Ecker and J.J. Kieber. 2006. The *Arabidopsis* histidine phosphotransfer proteins are redundant positive regulators of cytokinin signaling. *Plant Cell*, *18*: 3073–87.
Irish, V. F. 2008. The *Arabidopsis* petal: A model for plant organogenesis. *Trends Plant Sci.*, *13*: 430–36.
Kim, D., Y. Cho, H. Ryu, Y. Kim, T. Kim and I. Hwang. 2013. BLH1 and KNAT3 modulate ABA responses during germination and early seedling development in *Arabidopsis*. *Plant J.*, *75*: 755–66.
Kinoshita-Tsujimura, K. and T. Kakimoto. 2011. Cytokinin receptors in sporophytes are essential for male and female functions in *Arabidopsis thaliana*. *Plant Signal. Behav.*, *6*: 66–71.
Kondhare, K.R., P.V. Vetal, H.S. Kalsi and A.K. Banerjee. 2019. BEL1-like protein (StBEL5) regulates *CYCLING DOF FACTOR1* (*StCDF1*) through tandem TGAC core motifs in potato. *J. Plant Physiol.*, *241*: 153014.
Li, E., S. Wang, Y. Liu, J. Chen and C.J. Douglas. 2011. OVATE FAMILY PROTEIN4 (OFP4) interaction with KNAT7 regulates secondary cell wall formation in *Arabidopsis thaliana*. *Plant J.*, *67*: 328–41.
Li, E., A. Bhargava, W. Qiang, M. Friedmann, N. Forneris, R. Savidge, L. Johnson, S. Mansfield, B. Ellis and C. Douglas 2012. The Class II KNOX gene KNAT7 negatively regulates secondary wall formation in *Arabidopsis* and is functionally conserved in Populus. *New Phytol.*, *194*: 102–15.
Lin, T., P. Sharma, D.H. Gonzalez, I.L. Viola and D.J. Hannapel. 2013. The impact of the long-distance transport of a BEL1-Like Messenger RNA on development. *Plant Physiol.*, *161*: 760–72.
Ma, Q., N. Wang, P. Hao, H. Sun and S. Yu. 2019. Genome-wide identification and characterization of TALE superfamily genes in cotton reveals their functions in regulating secondary cell wall biosynthesis. *BMC Plant Biol.*, *19*: 432.
Magnani, E. and S. Hake, 2008. *KNOX* lost the OX: The *Arabidopsis KNATM* gene defines a novel class of KNOX transcriptional regulators missing the homeodomain. *Plant Cell*, *20*, 875–87.
Mahajan, A., S. Bhogale, I.H. Kang, D.J. Hannapel and A.K. Banerjee. 2012. The mRNA of a Knotted1-like transcription factor of potato is phloem mobile. *Plant Mol. Biol.*, *79*, 595–608.
Meng, L., Z. Fan, Q. Zhang, C. Wang, Y. Gao, Y. Deng, B. Zhu, H. Zhu, J. Chen and W. Shan. 2018. *BEL1-LIKE HOMEODOMAIN 11* regulates chloroplast development and chlorophyll synthesis in tomato fruit. *Plant J.*, *94*: 1126–40.
Moyroud, E. and B.J. Glover. 2017. The evolution of diverse floral morphologies. *Curr. Biol.*, *27*: 941–51.
Riefler, M., O. Novak, M. Strnad and T. Schmülling. 2006. *Arabidopsis* cytokinin receptor mutants reveal functions in shoot growth, leaf senescence, seed size, germination, root development, and cytokinin metabolism. *Plant Cell*, *18*: 40–54.
Sakakibara, K., S. Ando, H.K. Yip, Y. Tamada, Y. Hiwatashi, T. Murata, H. Deguchi, M. Hasebe, J.L. Bowman. 2013. KNOX2 genes regulate the haploid-to-diploid morphological transition in land plants. *Science*, *339*: 1067–70.
Sauret-Güeto, S., K. Schiessl, A. Bangham, R. Sablowski and E. Coen. 2013. *JAGGED* controls *Arabidopsis* petal growth and shape by interacting with a divergent polarity field. *PLoS Biol.*, *11*: e1001550.
Sawa, S., K. Watanabe, K. Goto, E. Kanaya, E.H. Morita and K. Okada. 1999. FILAMENTOUS FLOWER, a meristem and organ identity gene of *Arabidopsis*, encodes a protein with a zinc finger and HMG-related domains. *Genes Dev.*, *13*(9): 1079–88.
Sieber, P., M. Petrascheck, A. Barberis and K. Schneitz. 2004. Organ polarity in *Arabidopsis NOZZLE* physically interacts with members of the YABBY family. *Plant Physiol.*, *135*(4): 2172–85.

Siegfried, K.R., Y. Eshed, S.F. Baum, D. Otsuga, G.N. Drews, J.L. Bowman. 1999. Members of the YABBY gene family specify abaxial cell fate in *Arabidopsis*. *Development, 126*(18): 4117–28.

Shan, H.Y., J. Cheng, R. Zhang, X. Yao and H.Z. Kong. 2019. Developmental mechanisms involved in the diversification of flowers. *Nature Plants, 5*: 917–23.

Shani, E., O. Yanai and N. Ori. 2006. The role of hormones in shoot apical meristem function. *Curr. Opin. Plant Biol., 9*: 484–89.

Smaczniak, C., R.G.H. Immink, G.C. Angenent and K. Kaufmann. 2012. Developmental and evolutionary diversity of plant MADS-domain factors: Insights from recent studies. *Development, 139*: 3081–98.

Smith, H.M., I. Boschke, S. Hake. 2002. Selective interaction of plant homeodomain proteins mediates high DNA-binding affinity. *Proc. Natl. Acad. Sci. USA, 99*: 9579–84.

Sommer, H., J.P. Beltran, P. Huijser, H. Pape, W. Lonnig, H. Saedler and Z.D. Schwarzsommer. 1990. A homeotic gene involved in the control of flower morphogenesis in *Antirrhinum majus*: The protein shows homology to transcription factors. *EMBO J., 9*: 605–13.

Stahle, M.I., J. Kuehlich, L. Staron, A.G. Arnim and J.F. Golz. 2009. YABBYs and the transcriptional corepressors LEUNIG and LEUNIG_HOMOLOG maintain leaf polarity and meristem activity in *Arabidopsis*. *Plant Cell, 21*(10): 3105–18.

Tanaka, W., Toriba, T., Ohmori, Y., Yoshida, A., Kawai, A., Mayama-Tsuchida, T., Ichikawa, H., Mitsuda, N., Ohme-Takagi, M., Hirano, H.Y. 2012. The YABBY gene TONGARI-BOUSHI1 is involved in lateral organ development and maintenance of meristem organization in the rice spikelet. *Plant Cell, 24*(1): 80–95.

Theisen, G., R. Melzer and F. Rumpler. 2016. MADS-domain transcription factors and the floral quartet model of flower development: Linking plant development and evolution. *Development, 143*: 3259–71.

Walcher-Chevillet, C.L. and E.M. Kramer. 2016. Breaking the mold: Understanding the evolution and development of lateral organs in diverse plant models. *Curr. Opin. Genet. Dev., 39*: 79–84.

Wang Y.Y., Y.J. Zhao, M. Yan, H.L. Zhao, X.H. Zhang and Z.H. Yuan. 2020. Genome-wide Identification and Expression Analysis of TALE Gene Family in Pomegranate (*Punica granatum* L.). *Agronomy, 10*(6): 829.

Wetzstein, H.Y., N. Ravid, E. Wilkins and A.P. Martinelli. 2011. A morphological and histological characterization of bisexual and male flower types in pomegranate. *J. Amer. Soc. Hort. Sci., 136*(2): 83–92.

Wetzstein, H.Y., W.G. Yi and J.A. Porter. 2013. Flower position and size impact ovule number per flower, fruitset, and fruit size in Pomegranate. *J. Amer. Soc. Hort. Sci., 138*(3): 159–66.

Xu, Z., Q. Zhang, L. Sun, D. Du, T. Cheng, H. Pan, W. Yang and J. Wang. 2014. Genome-wide identification, characterization and expression analysis of the MADS-box gene family in *Prunus mume*. *Mol. Genet. Genom., 289*: 903–20.

Yamada, T., S. Yokota, Y. Hirayama, R. Imaichi, M. Kato and C.S. Gasser. 2011. Ancestral expression patterns and evolutionary diversification of YABBY genes in angiosperms. *Plant J., 67*(1): 26–36.

Yanofsky, M.F., H. Ma, J.L. Bowman, G.N. Drews, K.A. Feldmann and E.M. Meyerowitz. 1990. The protein encoded by the *Arabidopsis* homeotic gene agamous resembles transcription factors. *Nature, 346*: 35–39.

Zhang, X.L., Z.P. Yang, J. Zhang and L.G. Zhang. 2013. Ectopic expression of *BraYAB1-702*, a member of YABBY gene family in Chinese cabbage causes leaf curling, inhibition of development of shoot apical meristem, and flowering stage delaying in *Arabidopsis thaliana*. *Int. J. Mol. Sci., 14*(7): 14872–91.

Zhao Y.J., C.Y. Liu, X.Q. Zhao, Y.Y. Wang, M. Yan and Z.H. Yuan. 2021. Cloning and Spatiotemporal Expression Analysis of *PgWUS* and *PgBEL1* in *Punica granatum*. *Acta Horticulturae Sinica*. (in Chinese)

Zhao, Y.J., H.L. Zhao, Y.Y. Wang, X.H. Zhang, X.Q Zhao and Z.H. Yuan. 2020. Genome-Wide Identification and Expression Analysis of MIKC-Type MADS-Box Gene Family in *Punica granatum* L. *Agronomy, 10*(8): 1197.

Zhao, Y.J., C.Y. Liu, D.P. Ge, M. Yan, Y. Ren, X.B. Huang and Z.H. Yuan. 2020. Genome-wide identification and expression of YABBY genes family during flower development in *Punica granatum* L. *Gene*, *752*: 144784. Elsevier.
Zhong, R., C. Lee, J. Zhou, R.L. Mccarthy and Z.H. Ye. 2008. A battery of transcription factors involved in the regulation of secondary cell wall biosynthesis in *Arabidopsis*. *Plant Cell*, *20*: 2763–82.

7

Seed Coat Development

Jiyu Li[1] and *Gaihua Qin*[1*]

Pomegranate (*Punica granatum* L.) is a fruit-bearing deciduous shrub or small tree cultivated worldwide. Pomegranate fruits possess rich nutrition, unique flavor, and important economic value, and it is favorite by people for its nutrients benefiting to human health. The seed coat which is composed of outer and inner seed coats, is the edible proportion of the pomegranate fruit. The soft and juicy outer seed coat is rich in nutrients, which is the main edible part. The inner seed coat is rich in lignin, cellulose and hemicellulose, which forms the seed hardness. Seed hardness is one of the important quality characteristics of fresh pomegranate, which determines the sensory quality of the fruit and consumer acceptance. The seed hardness formation is a process of lignification of inner seed coat cells during seed coat development. The seed hardness is depended on genotype, growing technology and environment, while the genetics of cultivars plays the decisive roles. Some candidate genes involving in seed coat development especially the seed hardness formation had been identified basing on transcriptome, proteomic, and other technologies, but their functional characteristics are unclear. It is of great importance to further investigate and clarify the molecular genetic mechanism of pomegranate seed coat development, and provided guidance for genetic improvement and breeding in pomegranate.

1. Introduction

Pomegranate (*Punica granatum* L.) belongs to the family Lythraceae, is favored by people for possessing rich nutrition, unique flavor, and economic value (Qin et al., 2017). As more attentions were paid on the nutrition and health protection

[1] Anhui Academy of Agricultural Sciences, Hefei 230001, China.
[*] Corresponding author: qghahstu@163.com

functions of pomegranate fruits, its market demand increased rapidly and the industry developed gradually. The edible proportion of the pomegranate fruit depends largely on the size of the seed coat, which is composed of outer and inner seed coats. The fleshy and juicy outer seed coat is rich in organic and inorganic nutrients, including sugars, organic acids, micronutrient, and water (Prakash et al., 2011), while the inner seed coat is rich in lignin, cellulose, and hemicellulose (Uçar et al., 2009; Zarei et al., 2016). The difference of lignin and cellulose accumulation in the inner seed coats implies the different seed hardness in cultivars. Based on the seed hardness, the pomegranate cultivars were classified in three categories: soft-seeded, semi-soft-seeded, and hard-seeded (Khadivi et al., 2015).

Seed hardness is an important quality character determining the sensory perception and consumer acceptance (Marmol et al., 2015; Qin et al., 2020). Excessive seed hardness will lead to decline of quality characteristics of pomegranate fruits, which in turn, influence eating quality and economic value of pomegranate fruits. For commercial purposes, seed hardness can also affect the processing of pomegranates for juice extraction or other food products. Pomegranate varieties with softer seeds may be more suitable for juicing, as they release their juice more readily, while harder seeds may require specialized equipment or processing techniques to extract the juice effectively. Seed hardness is mainly determined by the pomegranate genotype, meanwhile it is affected by planting technology and external environment. Therefore, knowledge of the mechanism and regulatory controls of soft seed development, will facilitate breeding and development of pomegranate cultivars with low seed hardness. This will in turn enhance market competitiveness of pomegranate and promote sustainable development of the pomegranate industry.

This chapter will cover current knowledge and advances in our understanding of the formation mechanism of pomegranate seed hardness and will offer insights into future research studies in this field.

2. Structure of Seed Coats and the Seed Hardness Development

Pomegranate seeds are composed of three parts: seed coat, endosperm, and embryo. The seed coat is divided into outer seed coat and inner seed coat. The outer seed coat is soft, juicy and rich in nutrients, which is the edible part of the fresh fruit. The inner seed coat is dense, and rich in lignin, hemicellulose, and cellulose, which is the basis for the formation of seed hardness. Anatomical structure of pomegranate seed coats revealed that the outer seed coat is composed of two cell layers including parenchyma cell layer and palisade cell layer, and the inner seed coat is composed of three cell layers including phellogen, sclerenchymatous cell layer, and parenchyma cell layer (Qin et al., 2018). Pomegranate inner and outer seed coat are developed from integuments; however, its developmental mechanism is different to plants, e.g., *Arabidopsis*. For *Arabidopsis,* the inner integument gradually degenerates during the development process, while the outer integument accumulates more proanthocyanidins, then develops into a hard seed coat (Coen et al., 2018). During plant evolution, hard seed coats can protect the embryo and avoid digestion by animals or insects disseminating seeds for fostering offspring. Therefore, the bright and juicy outer seed coat and the hard inner seed coat of pomegranate may be the result of evolutional selection. It was found that the marker gene of integument development INNER NO OUTER (INO) and its transcription regulator BELL1 (BEL1) were under positive selection during evolution and potentially contributed to the development of the fleshy outer seed coat (Qin et al., 2017). Zhao et al. (2023) found that PgCRC and PgINO interact with PgBEL1 to regulate ovule and seed development in pomegranate (Zhao et al., 2023).

It was found that the seed coat of pomegranate began to develop rapidly 40 days after pollination (DAP), the outer seed coat expanded rapidly and the inner seed coat began to harden (Xue et al., 2016). The formation of seed hardness is a process of lignification of the inner seed coat cells, in which the lignin of the inner seed coat cells continuously accumulates and the secondary cell wall gradually thickens. Studies showed that the cell wall thickness of the seed coat from a hard-seed pomegranate is significantly greater than that from a soft-seed pomegranate (Xie et al., 2017; Qin et al. (2018). Anatomy of inner seed coat showed that only part of it was lignified for soft seed pomegranate, while the entire inner seed coat, except for the cells around the vascular bundles, appeared lignified in a hard seed pomegranate (Pujari et al., 2015).

The main components of the cell wall, lignin, cellulose, and hemicellulose are connected to each other to form a network structure, which constitutes the skeleton of plant cells. The content of lignin and cellulose accounted for 21.44% and 18.71% of the dry matter of pomegranate seeds, respectively (Dalimov et al., 2003). Positive correlation of seed hardness to the lignin content in inner seed coat was detected (Zarei et al., 2016; Qin et al., 2018). Lignin is composed of the primary monolignols *p*-coumaryl alcohol, coniferyl alcohol, and sinapyl alcohol, and so on, which are formed predominantly by oxidative polymerization of the three major monolignols to generate the hydroxyphenyl(H), guaiacyl (G), and syringyl (S) lignin subunits (Boerjan et al., 2003). Widely-target metabolic profiling of pomegranate seed coats revealed that coniferyl alcohol and sinapyl alcohol were the main monolignols accumulated in seed coats, while little *p*-coumaryl alcohol was detected (Qin et al., 2020). That means, S lignin and G lignin are dominant lignin subunits in pomegranate seed coats. In addition, the accumulation amounts of cellulose and hemicellulose and their precursors in seed coat was significantly different between hard-seed and soft-seed pomegranates, which implied that the different accumulation of cellulose and hemicellulose contributing to the formation of pomegranate seed hardness (Zarei et al., 2016).

3. Evaluation of Seed Hardness

The evaluation of pomegranate seed hardness is usually based on the edible sense, determination of cellulose and lignin content, and hardness determination (Jalikop et al., 1988; Zhang et al., 2015, 2017). The edible sense of seed hardness is a process of subjective evolution, which was used widely. The relationship between lignin and hemicellulose content and seed hardness has been confirmed, so the contents of lignin and hemicellulose were used to evaluate the seed hardness. Hardness determination can be used to evaluate the seed hardness directly, that which had high repeatability and reliability. Fruit hardness tester or grain hardness tester were used widely to determine the pomegranate seed hardness (Lu et al., 2006; Wang et al., 2010). Subsequently, texture analyzer was widely used to determine the pomegranate seed hardness, to get the further reliable results. As we know, texture analyzer can simulate the chewing movement of the oral cavity, select the appropriate probe according to the characteristics of the sample, and edit the special running program to automatically analyze the sample, so the result is reasonable stable. Qin et al. (2020) found that the seed hardness of pomegranate has continuous distribution in germplasm (Qin et al., 2020). There were studies that showed that seed hardness was distributed in the range of 1.7~10 kg•cm^{-2} among pomegranate varieties (Yang et al., 2015; Zhang et al., 2015; Zhao et al., 2016; Xie et al., 2017; Qin et al., 2018). Jalikop et al. (1988) considered that those having seed hardness < 4 kg•cm^{-2} are soft-seeded cultivars, those having seed hardness = 4 ~ 10 kg•cm^{-2} are semi-soft-seeded cultivars, and those having seed hardness > 10 kg•cm^{-2} are hard-seeded cultivars (Jalikop et al., 1988). Lu et al. (2006) considered that those having seed hardness < 3.67 kg•cm^{-2} are soft-seeded cultivars, those having seed hardness = 3.67 ~ 4.20 kg•cm^{-2} are semi-soft-seeded cultivars, and those having seed hardness > 4.2 kg•cm^{-2} are hard-seeded cultivars (Lu et al., 2006). Besides, there were arguments on pomegranate seed hardness, and it was considered that pomegranate varieties should be classified into finer grades, such as hard seed, semi-hard seed, semi-soft seed, soft seed, and super soft seed. The standards of classification are reasonable, while there is limitation for further study. Therefore, a standard accepted by population is need.

4. Factors Affecting Seed Hardness Formation

It is well known that seed hardness of pomegranate is mainly determined by its genetype, though environmental situations and cultivation measures have impacts on the seed hardness formation. Generally, fully mature pomegranate seeds are harder than immature ones because secondary cell wall thickening and lignin accumulation occurs in the inner seed coat at the later stage of fruit development, which leads to the formation of seed hardness. Testa and fruit-cavity xenia is widespread in fruit trees, which is also present in pomegranates (Xue et al., 2016; Hong et al., 2020). In the hybrid offspring with the soft-seeded cultivar 'Tunisia' as the female parent, the seed hardness differs depending on the male parent, and the seed hardness is the smallest in the selfed progeny, indicating that xenia has a direct effect on the seed hardness of pomegranate (Pujari et al., 2015). The correlation of seed hardness to the tree age showed that seed hardness decreased with the increase of tree age (Xue, et al., 2016). The location of the fruits on the tree also has an effect on seed hardness. Studies showed that the seeds of the fruits on the north and inside of the tree canopy were softer, while those on the south and southwest of the tree canopy were harder. The difference may be the result of different nutritional conditions and growth microenvironment of fruits in the tree (Lu et al., 2006).

Seed hardness is different according to the soils in which pomegranate planted and the fertilizer used. For soft-seed pomegranate 'Tunisia', the seed hardness was the highest for growing in sandy soil, followed by loam soil and red loam soil (Xue et al., 2017). The application of phosphate fertilizer can reduce the seed hardness, while urea, compound fertilizer, and potassium fertilizer have no effect on the seed hardness (Si et al., 2017). Gibberellin, 2,4-dichlorophenoxyacetic acid (2,4-D) and naphthalene acetic acid are widely used as plant growth regulators. Duction of seed hardness was detected in pomegranate that was sprayed with the growth regulators during flowering (Chen et al., 2020). Although the effects of plant growth regulators on seed hardness formation has been confirmed, the application technology of plant growth regulator on pomegranate seed softening is still lacking.

5. Molecular Mechanism of Seed Hardness Development

As an important material basis for the lignification of pomegranate inner seed coat, the synthesis, metabolism, and accumulation of polyphenol polymer lignin, polysaccharide polymer cellulose, and hemicellulose are the key to studying the formation mechanism of pomegranate seed hardness. Phenylalanine or tyrosine is catalyzed by a series of enzymes to form monolignols, which are then polymerized into lignin under the action of polymerase (Li et al., 2007; Boerjan et al., 2003; Humphreys et al., 2002). The formation and polymerization of monolignols have been studied clearly in model plants, but it is not clear in pomegranates. Qin et al. (2017) annotated the genes related to lignin, cellulose, and hemicellulose in pomegranate metabolism at the whole-genome scale, then surveyed the expression profiles of these genes in soft- and hard-seeded cultivars during fruit development and found that the formation of seed hardness is related to the specific expression of genes involved in lignin, cellulose, hemicellulose synthesis, and degradation (Qin et al., 2017). This inference was further verified on proteomic level (Niu et al., 2018). Since the genes involved in the synthesis and degradation of lignin, cellulose, and hemicellulose are polygene families, it is necessary to focus on candidate genes.

Several genes and transcriptional regulators involving in lignin, cellulose, hemicellulose synthesis, and degradation were identified and considered as the candidate genes involved in seed hardness formation based on the transcriptome and omics research (Zhang et al., 2015; Dong et al., 2016; Huang et al., 2017; Xue et al., 2017; Dong et al., 2018; Qin et al., 2020). *Pgr008023*, *Pgr008163*, *Pgr009460*, and *Pgr025550* encoding ABCG transporters, *Pgr011171.1* encoding POD, *Pgr013634.1* encoding F5H, *Pgr022328.1*, *Pgr006310.1*, *Pgr006318.1*, and *Pgr006319.1* encoding CAD, and *Pgr011171.1* encoding POD had lots of transcript accumulation in inner seed coat, accompanied with the sinapyl alcohol and coniferyl alcohol accumulation were considered as candidate genes involved in seed hardness formation (Qin et al., 2020). Ectopic expression of *PgABCG14* in *Arabidopsis* promoted plant growth and increased lignin accumulation in stems (Yu et al., 2022). It was reported that an SUT sucrose transporter encoded by *PgL0145810.1* may play a negative regulatory role in seed hardness formation (Poudel et al., 2020). Besides, a number of selective sites related to seed hardness composed of *SUC8*, *SUC6*, *FOXO*, and *MAPK* genes were considered as the candidate genes based on the whole genome resequencing (Luo et al., 2019). Furthermore, *Pgr000815.1* encoding AUX-IAA might be involved in the lignin metabolism, *Pgr011491.1* encoding AUX-IAA2, and *Pgr017424.1* encoding NAC66 might be involved in the metabolism of cellulose and hemicellulose based on the results of co-expression network analysis of lignin, cellulose, and hemicellulose metabolism-related genes and the transcription factors for inner seed coats at different developmental stages of hard-seed cultivar and soft-seed cultivar (Qin et al., 2020). Besides, it was found that *PgL0137670* encoding NAC transcript factor has an SNP site (T-C) at 166 bp, and the allelic variant T site of *PgL0137670* was associated with the characteristics of soft-seed, according to the comparative transcriptome analysis of soft-seeded cultivar and hard-seeded cultivar (Xia et al., 2018). UDP glycosyltransferases (UGTs) play an indispensable role in regulating lignin biosynthesis (Lin et al., 2016). In pomegranate, the relative expression levels of *PgUGTE10* and *PgUGTL11* in 'Tunisia' were higher than in 'Dabenzi', and the expressions of these two genes were highly correlated with the expression of genes involved in lignin biosynthesis pathways, indicating that *PgUGTE10* and *PgUGTL11* are potential candidate genes involved in seed hardness development by catalyzing the glycosylation of specific substrates (Li et al., 2023).

Cellulose synthase (CesA) usually acts as part of cellulose synthase complexes (CSCs) that include several catalytic subunits known as CesA proteins. CesA 4, 7, and 8 are subunits that assemble into CSCs and biosynthesize cellulose in plant secondary cell walls (Taylor et al., 2003). *Pgr007560*, *Pgr012627*, and *Pgr027371* encode CesA 4, 7, and 8; they had high transcripts accumulations in inner seed coats, which were considered as candidate genes contributing to cellulose biosynthesis in the inner seed coats. Studies on microRNA in the inner seed coat with different seed hardness showed that the expression of transcription factors MYB, NAC, and WRKY involved in the regulation of lignin synthesis may be regulated by miR164e and miR172b (Luo et al., 2018). The transcription factor *PgSND1-like* is confirmed to be involved in lignin biosynthesis by heterologous transformation (Xia et al., 2018). In summary, to better understand the genetic mechanism of seed hardness formation, functions and its regulatory network of candidate genes involved in seed hardness formation are needed, based on the progress on the candidate genes' identification.

6. Limitation and Prospect of Seed Coat Development

Great progress on the metabolic and genetic mechanism of seed hardness formation has been made with the development of genomic, transcriptic, and metablic technology. Many genes involving seed hardness formation were identified.

However, the function of candidate genes need to be further studied. Due to the multiple metabolic pathways involved in the formation of seed hardness, the genetic mechanism is complex, and the genetic transformation system of pomegranate has not been established as yet; this leads to difficulties in verifying the gene function and its genetic regulation. Therefore, it is of great importance to further study and investigate, and clarify the formation mechanism of pomegranate seed hardness, and provide guidance for genetic improvement and breeding in pomegranate. The following are some suggestions for further study:

(a) Strengthening the collection, evaluation, and utilization of pomegranate germplasms to enrich the selection of soft-seed pomegranate.
(b) Identify the genetic locus of soft-seed traits by constructing segregated populations, using resequencing technology, and combining phenotypic data. Isolate key genes and molecular markers to accelerate the process of molecular marker-assisted breeding of soft-seeded pomegranate.
(c) Combine genomics, transcriptomics, metabolomics, and other omics technologies to identify the novel genes involved in seed hardness formation. Building the transgenic technology of pomegranate to study the gene function and its genetic mechanism of seed hardness formation is important.

References

Alcaraz-Marmol, F., Calin-Sanchez, A., Nuncio-Jauregui, N., Carbonell-Barrachina, A. A., Hernandez, F. and Jose Martinez, J. 2015. Classification of pomegranate cultivars according to their seed hardness and wood perception. *Journal of Texture Studies, 46*(6): 467–74.

Boerjan, W., Ralph, J. and Baucher, M. 2003. Lignin biosynthesis. *Annual Review of Plant Biology, 54*: 519–46.

Chen, L.N., Niu, J., Liu, B.B., Jing, D., Luo, X., Li, H.X., Xia, X.C., Yang, X.W., Zhang, F.H., Cao, D., Wang, Q. and Cao, S.Y. 2020. Effects of foliar application of plant growth regulators at different flowering stages on fruit setting and quality in pomegranate. *Journal of Fruit Science, 2*: 244–53.

Coen O. and Magnani, E. 2018. Seed coat thickness in the evolution of angiosperms. *Cell Mol. Life Sci., 75*: 2509–18.

Dalimov, D.N., Dalimov, G.N. and Bhatt, M. 2003. Chemical composition and lignins of tomato and pomegranate seeds. *Chemistry of Natural Compounds, 39*(1): 37–40.

Dong, L.L., Gong, L.Y., Chen, L., Tu, J.L. and Zhang, S.M. 2016. Cloning and expression analysis of PgCCR from pomegranate. *Journal of Nanjing Agricultural University, 39*(05): 747–53.

Dong, L.L., Dong, L.L., Xiong, F., Liu, N., Wang, Q. and Zhang, S.M. 2018. Molecular cloning and expression analysis of PgLAC in pomegranate. *American Journal of Molecular Biology, 8*(3): 145-55.

Hong, J., Huang, R., Huang, C., Wang, J. and Li, Y. 2020. Research progress and prospects of xenia. *Plant Physiology Journal, 56*(02): 151–62.

Huang, R., Xiong, F., Chen, L., Zhang, S.M.Z., Dong, L.L. 2017. Cloning and expression analysis of lignin biosynthesis-related gene PgMYB308 in Pomegranate. *Acta Botanica Boreali-OccidentaliaSinica, 37*(12): 2357–62.

Humphreys, J.M. and Chapple, C. 2002. Rewriting the lignin roadmap. *Current Opinion in Plant Biology, 5*(3): 224–29.

Jalikop, S.H. and Kumar, P.S. 1998. Use of soft, semi-soft, and hard-seeded types of pomegranate (*Punice granatum*) for improvement of fruit attributes. *Indian Journal of Agricultural Sciences, 68*(2): 87–91.

Khadivi-Khub, A., Kameli, M., Moshfeghi, N. and Aziz Ebrahimi. 2015. Phenotypic characterization and relatedness among some Iranian pomegranate (*Punica granatum* L.) accessions. *Trees, 293*(3): 893–901.

Li, G.X., Li, J.Y., Qin, GH., Liu, C.Y., Liu, X., Cao, Z., Jia, B.T. and Zhang, H.P. 2023. Characterization and Expression Analysis of the UDPG glycosyltransferase Family in Pomegranate (*Punica granatum* L.). *Horticulturae, 9*: 119.

Li, L.B., Liu, L., He, C.F., Dong, Y.M. and Peng, Z.H. 2007. Research progresses on the genes encoding the key enzymes in biosynthetic pathway of lignin. *Molecular Plant Breeding, S1*: 45–51.

Lin, J.S., Huang, X.X., Li, Q., Cao, Y., Bao, Y., Meng, X.F., Li, Y.J., Fu, C. and Hou, B.K. 2016. UDP-glycosyltransferase 72B1 catalyzes the glucose conjugation of monolignols and is essential for the normal cell wall lignification in *Arabidopsis thaliana. Plant J., 88*(1): 26–42.

Lu, L.J., Gong, X.M. and Zhu, L.W. 2006. Study on seed hardness of pomegranate cultivars in China. *Journal of Anhui Agricultural University, 33*(3): 356–59.

Luo X., Cao D., Zhang, J.F., Chen, L., Xia, X.C., Li, H.X., Zhao, D.G., Zhang, F.H., Xue, H., Chen, L.N., Li, Y.Z. and Cao, S.Y. 2018. Integrated microRNA and mRNA expression profiling reveals a complex network regulating pomegranate (*Punica granatum* L.) seed hardness. *Scientific Reports, 8*(1): 9292.

Luo, X., Li, H.X., Wu, Z.K., Yao, W., Zhao, P., Cao, D., Yu, H.Y., Li, K.D., Poudel, K., Zhao, D.G., Zhang, F.H., Xia, X.C., Chen, L.N., Wang Q., Jing, D. and Cao, S.Y. 2019. The pomegranate (*Punica granatum* L.) draft genome dissects genetic divergence between soft- and hard-seeded cultivars. *Plant Biotechnology Journal,18*(4): 955–68.

Niu, J., Cao, D., Li, HX., Xue, H., Chen, L., Liu, B.B. and Cao, S.Y. 2018. Quantitative proteomics of pomegranate varieties with contrasting seed hardness during seed development stages. *Tree Genet. Genomes, 14*(1): 14.

Poudel, K., Luo, X., Chen, L.N., Jing, D., Xia, X.C., Tang, L.Y., Li, H.X. and Cao, S.Y. 2020. Identification of the SUT Gene Family in Pomegranate (*Punica granatum* L.) and Functional Analysis of PgL0145810.1. *International Journal of Molecular Sciences, 21*(18): 6608.

Prakash, C.V.S. and Prakash, I. 2011. Bioactive chemical constituents from pomegranate (*Punica granatum* L.) juice, seed, and peel. A review. *Int. J. Res. Chem. Environ., 1*: 1–18.

Pujari, K.H. and Rane, D.A. 2015. Concept of seed hardness in pomegranate: I) anatomical studies in soft and hard seeds of 'Muskat' pomegranate. *Acta Horticulturae, 1098*: 97–104.

Qin, GH., Xu, C.Y., Ming, R., Tang, H.B., Guyot, R., Kramer, E.M., Hu, Y.D., Yi, X.K., Qi, Y.J., Xu, X.Y., Gao, Z.H., Pan, H.F., Jian, J.B., Tian, Y.P., Yue, Z. and Xu, Y.L. 2017. The pomegranate (*Punica granatum* L.) genome and the genomics of punicalagin biosynthesis. *Plant J., 91*: 1108–28.

Qin, G.H., Liu, C.Y., Xu, Y.L., Li, Y.L., Qi, Y.J., Pan, H.F., Yi, X.K. and Gao, Z.H. 2018. Microstructure observation of seed coats and its development of pomegranate (*Punica granatum* L). *Chin. J. Trop. Crops, 39*, 489–93.

Qin, G.H., Liu, C.Y., Li, J.Y., Qi, Y.J., Gao, Z.H., Zhang, X.L., Yi, X.K., Pan, H.F., Ming, R. and Xu, Y.L. 2020. Diversity of metabolite accumulation patterns in inner and outer seed coats of pomegranate: Exploring their relationship with genetic mechanisms of seed coat development. *Horticulture Research, 7*(6).

Si, S.X., Niu, J., Chen, L.N. and Cao, S.Y. 2017. Effect of different reagents on the proportion of fertile flower and fruit yield and quality of pomegranate. *Journal of Northwest Forestry University, 32*(004): 111–116.

Taylor, N.G., Howells, R.M., Httly, A.K., Vickers, K. and Turner, S.R. 2003. Interactions among three distinct CesA proteins essential for cellulose synthesis. *Proceedings of the National Academy of Sciences of the United States of America, 100*(3): 1450–55.

Uçar, S. and Karagöz, S. 2009. The slow pyrolysis of pomegranate seeds: The effect of temperature on the product yields and bio-oil properties. *J. Anal. Appl. Pyrol., 84*, 151–56.

Wang, X.F., Zhou, Y.F., Zhao, C.H., Sun, L. and Wu, Q.G. 2010. Determination of fruit quality of 36 pomegranate cultivars. *Chinese Journal of Tropical Crops, 31*(01): 136–40.

Xia, X.C., Li, H.X., Cao, D., Luo, X., Yang, X.W., Chen, L.N., Liu, B.B., Wang, Q., Jing. D. and Cao, S.Y. 2018. Characterization of an NAC transcription factor involved in the regulation of pomegranate seed hardness (*Punica granatum* L.). *Plant Physiology and Biochemistry*, *139*: 379–88.

Xie, X.B., Huang, Y., Tian, S.P., Li, G.L. and Cao, S.Y. (2017) Relationship of Seed Hardness Development and Microstructure of Seed Coat Cell in Soft Seed Pomegranate. *Acta Horticulturae Sinica*, *44*(06): 1174–80.

Xue, H. 2017. *Factors Affecting the Seed Hardness and the Transcriptome Analysis of Soft-seed Pomegranate*. Chinese Academy of Agricultural Sciences Dissertation.

Xue, H., Cao, S.Y., Chen, L.N., Li, H.X., Niu, J., Zhang, F.H. and Zhao, D.G. 2016. Comparison of seed structure and hardness between 'Yudazi' and 'Tunisia'. *Journal of Fruit Science*, *33*(5): 563–69.

Xue, H., Cao, S.Y., Niu, J., Li, H.X., Zhang, F.H. and Zhao, D.G. 2016. Effects of xenia on fruit setting and quality in 'Tunisia' pomegranate. *Journal of Fruit Science*, *33*(02): 196–201.

Xue, H., Xue, H., Cao, S.Y., Li, H.X., Zhang, J., Niu, J., Chen, L.N., Zhang, F.H. and Zhao, D.G. (2017). *De novo* transcriptome assembly and quantification reveal differentially expressed genes between soft-seed and hard-seed pomegranate (*Punica granatum* L.). *PloS One*, *12*(6): e0178809.

Xue, H., Xue, H., Cao, S.Y., Liu, B.B., Li, H.X., Niu, J. and Zhang, J. 2017. Effect of different soil types and fertilizer on seed hardness of Tunisian soft-seed pomegranate. *Acta Agriculturae Jiangxi*, *29*(01): 43–46.

Yang, Z.J., An, G.C., Shi, Y.P., Qiao, R.J., Li, J.H., Zhang, Y.G., Guo, K. and Yang, H.T. 2015. Eating quality comparison of six pomegranate cultivars in Zaozhuang Shandong. *Food Science and Technology*, *40*(01): 45–50.

Yu, Q., Li, JY., Qin, GH., Liu, CY., Cao, Z., Jia, BT., Xu, YL., Li, GX., Yang, Y., Su, Y. and Zhang, H. 2022. Characterization of the ABC Transporter G Subfamily in Pomegranate and Function Analysis of PgrABCG14. *Int. J. Mol. Sci.*, *23*(19): 11661.

Zarei, A., Zamani, Z., Fatahi, R., Mousavi, A., Salami, S.A., Avila C. M. Cánovas F. 2016. Differential expression of cell-wall-related genes in the seeds of soft- and hard-seeded pomegranate genotypes. *Sci. Hort.*, *205*: 7–16.

Zhang, L.H., Zhu, X.M., Xu, J.X. and Lü, H. 2017. Related factors affecting hardness of pomegranate seed. *Northern Horticulture*, *23*: 47–51.

Zhang, S.M., Gong, L.Y., Cao, D.Q., Zhang, Y.J. and Yang, J. 2015. Total lignin content in pomegranate seed coat and cloning and expression analysis of PgCOMT gene. *Journal of Tropical and Subtropical Botany*, *23*: 65–73.

Zhao, D.C., Jia, M., Tang, G.M., Liang, Y., Wang, X.F., Shu, X.G., Liang, J., Qi, Y.K., Sun L. and Qu, L.J. 2016. Establishment of the detecting method on the fruit texture of pomegranate by puncture test. *Chinese Journal of Tropical Crops*, *37*(07): 1419–23.

Zhao, Y.J., Wang, Y.Y., Yan, M., Liu, C.Y., Yuan, Z. 2023. BELL1 interacts with CRABS, CLAW, and INNER NO OUTER to regulate ovule and seed development in pomegranate. *Plant Physiol.*, *191*(2): 1066–83.

Uncovering the Molecular Basis for Pomegranate Fruit Coloration

Xueqing Zhao[1] and *Zhaohe Yuan*[1*]

Anthocyanins, in combination with carotenoids or chlorophylls, are responsible for almost all fruit coloration. The accumulation of anthocyanin pigments in fruit is an important determinant of ripeness and quality, providing essential cultivar differentiation and implication in the health attributes of foods. A high variability in fruit color observed among different pomegranate accessions arises from the differentially accumulation of anthocyanins. Though various kinds of anthocyanins have been identified in pomegranate fruit, limited information is available contributing to the molecular mechanisms underlying fruit color intensity variation in the fruit. In this chapter, an updated summary of published reports on identified individual anthocyanins from pomegranate species is presented, with specific emphasis on revealing molecular mechanisms for controlling coloration traits. The review of the current status of mechanism research of genetic regulation of the anthocyanin biosynthesis will facilitate the management, breeding, and engineering of anthocyanin content or production of red color in pomegranate.

1. Introduction

Pomegranates (*Punica granatum*) have been a popular fruit throughout human history and are experiencing a surge in popularity at present due to the health benefits associated with their juice. Among pomegranate fruit characters, the red coloration of the fruit largely determines the consumer appeal and impacts significantly on the market value of the produce. The red color of pomegranate is primarily associated

1 College of Forestry, Nanjing Forestry University, Nanjing 210037, China.

* Corresponding author: zhyuan88@hotmail.com

with anthocyanin pigment. Anthocyanin reddening in fruit has been a subject of genetic studies, and the molecular genetic control of anthocyanins biosynthesis is now one of the best understood of all secondary metabolic pathways. Cultivated pomegranate show substantial diversity in fruit color, including red, pink, yellow, white. The majority of pomegranate cultivars possess anthocyanin pigmentation in the aril of the fruit as well as skin. The wide range of color phenotype of pomegranate fruits makes it an excellent model plant for elucidating anthocyanin regulation mechanism.

Up to now, more than one hundred anthocyanin components have been identified in various plant tissues, including fruit skins, aril juices, flowers. Most of anthocyanins structural genes and several regulatory genes involved in biosynthetic pathway have been isolated. The different expression patterns of anthocyanin genes in various pomegranate cultivars have also been monitored. Furthermore, the molecular mechanism of anthocyanins absence in 'white' and 'black' pomegranate has already been revealed. With the release of pomegranate genome sequence, it gives greater access to intensive exploration of controlled genetic expression and gene discovery involved in fruit anthocyanin biosynthesis. However, the genomic analysis and transcriptomic approaches are far from being exploited to undercover the anthocyanin biosynthesis and regulation network in pomegranate.

This chapter covers the recent advances in understanding the regulation of anthocyanins biosynthesis in pomegranate fruit especially in skin. The molecular knowledge will provide basic information to the breeders and researchers, contributing to development of pomegranate industry with sustainable productions and increase of pomegranate consumption as new profitable crops.

2. Diversity for Anthocyanin Composition in Pomegranate Germplasm

2.1 The Chemical and Category of Anthocyanins

Plant compounds that are perceived by humans to have color are generally referred to as 'pigments'. Anthocyanins, a class of flavonoids derived ultimately from phenylalanine, are water-soluble, synthesized in the cytosol, and localized in vacuoles. They provide a wide range of colors ranging from orange, red to violet, blue for fruits and vegetables. Besides as defense and protection against light stress, anthocyanins also play an important reproductive role as attractants in plant and animal interactions. In recent years, there has been an unprecedented expansion of knowledge about anthocyanins pigments, not only for their coloring traits, but for their potential health-promoting properties. Understanding the chemical structure of the pigment is helpful in assessing the bioavailability and pharmacological activity. There have also been substantial improvements in analytical technology that have led to the discovery of novel anthocyanin compounds, but only few of their molecules have been studied intensively.

Evidences from spectrum, molecule, and functional genome have revealed that all anthocyanins have an identical flavonoid C6-C3-C6 skeleton structure, which is also referred as 2-phenylchromenylium (flavylium) (Anderson and Jordheim, 2006). Chemically, anthocyanins are glycosides whose aglycones are so-called sugar-free anthocyanidins. The combination of sugar group in aglycones could confer stability and water solubility of the molecules, therefore, anthocyanidins rarely occur naturally in the form of aglycone. The most common sugar moieties, which are mainly bond to the molecule at the C3-position of the C-ring or the C5- or C7-position of the A-ring, are glucose, galactose, arabinose, rhamnose, and xylose. More than 90% of sugar substituent is glucose (Castaneda-Ovando et al., 2009). Depending on the number and position of the hydroxyl, sugar and methoxyl groups, as well as the extent of sugar acylation present at different positions of the basic structure, different anthocyanins have been described. It is considered that more than 8,000 different anthocyanins in land plants (Reis et al., 2016). To date, chemical studies have documented over 700 anthocyanins in diverse plant species. However, only 27 individual anthocyanidins have been found in nature (Krga and Milenkovic, 2019), and the commonly occurring six of them are cyanidin (Cy), delphinidin (Dp), pelargonidin (Pg), petunidin (Pt), peonidin (Pn) and malvidin (Mv) (Table 1). In fact, approximately 90% of anthocyanins are based on Cy, Dp, Pg, and their methylated derivatives (Anderson and Jordheim, 2006). The most widely distributed anthocyanin species in fruit and vegetables is cyanidin-3-O-glucoside (Cy3G) (Khoo et al., 2019).

Table 1 Six common anthocyanidins in the plant kingdom.

Anthocyanidin	*R1*	*R2*	*Molecular Weight*
delphinidin (Dp)	OH	OH	303
cyanidin (Cy)	OH	H	287
pelargonidin (Pg)	H	H	271
peonidin (Pn)	OCH3	H	301
petunidin (Pt)	OH	OCH3	317
malvidin (Mv)	OCH3	OCH3	331

2.2 *The Identified Anthocyanins in Pomegranate*

Pomegranate is among one of the first fruit crops to be domesticated by ancient times. Over the course of time successive selections created most of the pomegranate cultivars grown today all over the world. At present, more than 500 cultivars are known around the world (Jaakola and Laura, 2013). Studies have demonstrated a high color genetic diversity of pomegranate germplasm. This diversity is displayed mostly in the fruit skin and arils. Pomegranate cultivars display a wide array of color phenotypic differences in fruit skin, ranging from white-yellow, green, pink, and red to dark purple. The fruit aril color varies from white to deep red (Figure 1).

Fig. 1 The color variation of pomegranate fruits.

The pomegranate fruit color is determined largely by the presence of anthocyanin pigments. The color variability in different cultivars derived from different contents of anthocyanin and different relative amounts of its derivatives. Presently, large amounts of different anthocyanins have been identified in various parts of the pomegranate trees, including fruits, leaves, and flowers. Six major anthocyanin compounds have been identified in pomegranate fruit, including 3-glucosides and 3,5-diglucosides of cyanidin, delphinidin, and pelargonidin (Fellah et al., 2018; Russo et al., 2018; Zhao et al., 2013). It has been reported the presence of colored anthocyanin-flavanol and flavanol-anthocyanin adducts in juice, even in minor concentration (Sentandreu et al., 2013; Sentandreu et al., 2010, 2012). In fresh fruit juice, the concentration of anthocyanin-flavanol was much lower even when compared with those of flavanol-anthocyanin adducts (Sentandreu et al., 2012). The obtained results demonstrated that the fruit was a rich source of anthocyanins, especially the peels (exocarp) (Bar-Ya'akov et al., 2019). The individual anthocyanidin and anthocyanin compounds presented in different parts of pomegranate are listed in Table 2.

3. Inheritance of the Color Traits in Pomegranate

Though the preferences of skin and aril color of pomegranate are listed as the breeding objective in traditional and modern breeding programs, inheritance mode of fruit color in pomegranate is poorly understood. Because of high heterozygosity, trait inheritance analysis (including fruit color) is difficult in pomegranate (Holland and Bar-Ya'akov, 2018). Studies on the inheritance of fruit color are only a few cases. By using Daru and Ganesh F2 crosses progenies, Jalikop et al. (2005) revealed that red and pink aril color was always dominant to white. The same result was also obtained by Ben-Simhon et al. (2015), who revealed that white fruit skin was recessive to red color, and the white phenotype had a single-gene Mendelian type of inheritance. The results of a Spanish breeding program also indicated that red aril color is dominant over pale aril color (Bartual et al., 2015). Similarly, the 'black' pomegranate phenotype was controlled mainly by a single recessive gene showing a Mendelian inheritance ration of one gene (Trainin et al., 2021). From these results, it can be concluded that monogenic traits determine the color of fruit and red and pink colors are dominant to yellow. However, according to the Jalikop (2011), fruit peel color was a polygenic trait and was affected by sunlight.

Table 2 Identified anthocyanins in pomegranate.

No.	*Anthocyanins/Anthocyanidins*	*References*
1	cyanidin	Fellah et al. (2018)
2	cyanidin-3-glucoside	Fischer et al. (2011); Mena et al. (2011); PalaToklucu (2011); Gomez-Caravaca et al. (2013); Zhao et al. (2013); Lantzouraki et al. (2014); Akhavan et al. (2015); Abid et al. (2017); Ambigaipalan et al. (2017); Fellah et al. (2018); Paul et al. (2018); Russo et al. (2018); Singh et al. (2019); Kostka et al. (2020)
3	cyanidin-3,5-diglucoside	Fischer et al. (2011); Mena et al. (2011); PalaToklucu (2011); Gomez-Caravaca et al. (2013); Sentandreu et al. (2013); Zhao et al. (2013); Lantzouraki et al. (2014); Akhavan et al. (2015); Ambigaipalan et al. (2017); Fellah et al. (2018); Paul et al. (2018); Russo et al. (2018); Singh et al. (2019); Kostka et al. (2020)
4	cyanidin-3-rutinoside	Fischer et al. (2011); Gomez-Caravaca et al. (2013); Sentandreu et al. (2013); Abid et al. (2017); Ambigaipalan et al. (2017); Fellah et al. (2018)
5	cyanidin-pentoside	Fischer et al. (2011); Sentandreu et al. (2013); Lantzouraki et al. (2014); Akhavan et al. (2015); Abid et al. (2017); Ambigaipalan et al. (2017)
6	cyanidin-3,5-pentoside-hexoside	Fischer et al. (2011); Gomez-Caravaca et al. (2013); Lantzouraki et al. (2014); Akhavan et al. (2015); Ambigaipalan et al. (2017)
7	cyanidin-trihexoside	Sentandreu et al. (2013)
8	cyanidin-3,5-caffeoyl-hexoside	Sentandreu et al. (2013)
9	cyanidin-3-hexoside	Sentandreu et al. (2013); Lantzouraki et al. (2014)
10	cyanidin-caffeoyl	Sentandreu et al. (2013); Lantzouraki et al. (2014)
11	cyanidin-3-(p-coumaroyl) hexoside	Sentandreu et al. (2013)
12	cyanidin-3-(p-coumaroyl) glucoside/ cyanidin-3-O-D-Rhamnosyl-(1,6)-D-glucoside	Singh et al. (2019)
13	cyanidin-3-O-(6"-p-coumaroyl-glucoside)	Fellah et al. (2018); Singh et al. (2019)
14	cyanidin-xyloside	Fellah et al. (2018)
15	cyanidin-3-O-sophoroside	Fellah et al. (2018)
16	cyanidin-3-O-arabinoside	Fellah et al. (2018)
17	cyanidin-3-O-galactoside	Fellah et al. (2018)

Contd.

Table 2 *Contd.*

No.	*Anthocyanins/Anthocyanidins*	*References*
18	cyanidin-3-O-(6"-acetyl-glucoside)	Fellah et al. (2018)
19	cyanidin-3-O-(6"-acetyl-galactoside)	Fellah et al. (2018)
20	cyanidin-3-O-(6"-caffeoyl-glucoside)	Fellah et al. (2018)
21	cyanidin-3-O-caffeoylglucoside)	Lantzouraki et al. (2014)
22	cyanidin-3-O-(6"-succinyl-glucoside)	Fellah et al. (2018)
23	cyanidin-3-hexoside-(epi) gallocatechin	Sentandreu et al. (2012); Sentandreu et al. (2013)
24	cyanidin-3-hexoside-(epi) catechin	Sentandreu et al. (2012); Sentandreu et al. (2013)
25	cyanidin-3-hexoside-(epi) afzelechin	Sentandreu et al. (2012); Sentandreu et al. (2013)
26	delphinidin	WuTian (2017)
27	delphinidin-3-glucoside	Fischer et al. (2011); Mena et al. (2011); PalaToklucu (2011); Sentandreu et al. (2013); Zhao et al. (2013); Lantzouraki et al. (2014); Akhavan et al. (2015); Ambigaipalan et al. (2017); Fellah et al. (2018); Russo et al. (2018); Singh et al. (2019); Kostka et al. (2020)
28	delphinidin-3,5-diglucoside	Fischer et al. (2011); Mena et al. (2011); PalaToklucu (2011); Gomez-Caravaca et al. (2013); Sentandreu et al. (2013); Zhao et al. (2013); Lantzouraki et al. (2014); Akhavan et al. (2015); Ambigaipalan et al. (2017); Fellah et al. (2018); Russo et al. (2018); Singh et al. (2019); Kostka et al. (2020)
29	delphinidin-3,5-diglucoside-6"-O-4, 6'''-O-1-cyclic-malyl diester	Singh et al. (2019)
30	delphinidin-trihexoside	Sentandreu et al. (2013)
31	delphinidin-3,7,3'-triglucoside	Singh et al. (2019)
32	delphinidin-3-[6-(2-acetylrhamnosyl) glucoside]	Singh et al. (2019)
33	delphinidin 3-(2-xylosylgalactoside)-5-glucoside	Singh et al. (2019)
34	delphinidin-3-(2G-xylosylrutinoside)	Singh et al. (2019)

Contd.

Table 2 *Contd.*

No.	*Anthocyanins/Anthocyanidins*	*References*
35	delphinidin-3-fructoside	Singh et al. (2019)
36	delphinidin-3-lathyroside	Singh et al. (2019)
37	delphinidin-3,5-dihexoside	Zhao et al. (2013); Lantzouraki et al. (2014)
38	delphinidin-3,5-pentoside-hexoside	Sentandreu et al. (2013); Lantzouraki et al. (2014)
39	delphinidin-3-pentoside	Sentandreu et al. (2013); Lantzouraki et al. (2014); Akhavan et al. (2015)
40	delphinidin-3-rutinoside	Sentandreu et al. (2013); Fellah et al. (2018)
41	delphinidin-3-arabinoside	Fellah et al. (2018)
42	delphinidin-3-sambubioside	Fellah et al. (2018)
43	delphinidin-3-xyloside	Fellah et al. (2018)
44	delphinidin-3-O-caffeoylglucoside	Lantzouraki et al. (2014)
45	delphinidin-3,5-caffeoyl-hexoside	Sentandreu et al. (2013); Lantzouraki et al. (2014)
46	delphinidin-pentoside	Sentandreu et al. (2013); Akhavan et al. (2015)
47	delphinidin-3-(p-coumaroyl) hexoside	Sentandreu et al. (2013)
48	delphinidin-caffeoyl	Sentandreu et al. (2013)
49	Delphinidin-caffeoyl (epi) Catechin-cyanidin-3,5-dihexoside	Lantzouraki et al. (2014)
50	delphinidin-3-O-glucosyl-glucoside	Fellah et al. (2018)
51	delphinidin-3-O-galactoside	Fellah et al. (2018)
52	delphinidin-trihexoside	Sentandreu et al. (2013)
53	delphinidin-3-O-(6"-acetyl-glucoside)	Fellah et al. (2018)
54	delphinidin-3-O-(6"-acetyl-galactoside)	Fellah et al. (2018)
55	delphinidin-3-O-(6"-p-coumaroyl-glucoside)	Fellah et al. (2018)
56	delphinidin-3-hexoside-(epi) gallocatechi	Sentandreu et al. (2012)
57	delphinidin-3-hexoside-(epi) catechin	Sentandreu et al. (2012)
58	delphinidin-3-hexoside-(epi) afzelechin	Sentandreu et al. (2012)

Contd.

Table 2 *Contd.*

No.	*Anthocyanins/Anthocyanidins*	*References*
59	pelargonidin	Fellah et al. (2018)
60	pelargonidin-3-O-glucoside	Fischer et al. (2011); Mena et al. (2011); PalaToklucu (2011); Gomez-Caravaca et al. (2013); Lantzouraki et al. (2014); Akhavan et al. (2015); Ambigaipalan et al. (2017); Fellah et al. (2018); Russo et al. (2018); Singh et al. (2019); Kostka et al. (2020)
61	pelargonidin-3,5-diglucoside	Fischer et al. (2011); Mena et al. (2011); PalaToklucu (2011); Gomez-Caravaca et al. (2013); Zhao et al. (2013); Lantzouraki et al. (2014); Akhavan et al. (2015); Abid et al. (2017); Fellah et al. (2018); Russo et al. (2018)
62	pelargonidin-pentoside	Sentandreu et al. (2013); Abid et al. (2017)
63	pelargonidin-3,5-pentoside-hexoside	Sentandreu et al. (2013)
64	pelargonidin-3,5-caffeoyl-hexoside	Sentandreu et al. (2013); Lantzouraki et al. (2014)
65	pelargonidin-3-O-rutinoside	Fellah et al. (2018); Singh et al. (2019)
66	pelargonidin-3-O-arabinoside	Fellah et al. (2018)
67	pelargonidin-3-O-sambubioside	Fellah et al. (2018)
68	pelargonidin-3-O-sophoroside	Fellah et al. (2018)
69	pelargonidin-3-O-galactoside	Fellah et al. (2018)
70	pelargonidin-3-O-(6''-malonyl-glucoside)	Fellah et al. (2018)
71	pelargonidin-3-O-(6''-succinyl-glucoside)	Fellah et al. (2018)
72	Pelargonidin-3-O-caffeoylglucoside	Lantzouraki et al. (2014)
73	pelargonidin -3-O-(6-O-caffeoyl-beta-D-glucoside) 5-O-beta-D-glucoside	Singh et al. (2019)
74	pelargonidin-3-hexoside-(epi) gallocatechin	Sentandreu et al. (2012)
75	pelargonidin-7-glucoside	Singh et al. (2019)
76	peonidin	Fellah et al. (2018)
77	peonidin-hexoside	Zhao et al. (2013)
78	peonidin-3-O-glucoside	Fellah et al. (2018)
79	peonidin-3-O-rutinoside	Fellah et al. (2018)
80	peonidin-3-O-arabinoside	Fellah et al. (2018)

Contd.

Table 2 *Contd.*

No.	*Anthocyanins/Anthocyanidins*	*References*
81	peonidin-3-O-galactoside	Fellah et al. (2018)
82	peonidin-3-O-(6"-acetyl-galactoside)	Fellah et al. (2018)
83	peonidin-3-O-(6"-acetyl-glucoside)	Fellah et al. (2018)
84	petunidin-3-O-glucoside	Fellah et al. (2018); Singh et al. (2019)
85	petunidin-3-O-rutinoside	Fellah et al. (2018)
86	petunidin-3-O-rhamnoside	Fellah et al. (2018)
87	petunidin-3-O-arabinoside	Fellah et al. (2018)
88	petunidin-3-O-galactoside	Fellah et al. (2018)
89	petunidin-3-O-(6"-acetyl-glucoside)	Fellah et al. (2018)
90	petunidin-3-O-(6"-acetyl-galactoside)	Fellah et al. (2018)
91	petunidin-3-O-(6"-p-coumaroyl-glucoside)	Fellah et al. (2018)
92	petunidin- 3-glucoside-5-(6"-acetylglucoside)	Singh et al. (2019)
93	malvidin-3-O-glucoside	Fellah et al. (2018)
94	malvidin-3-O-arabinoside	Fellah et al. (2018)
95	malvidin-3-O-galactoside	Fellah et al. (2018)
96	malvidin-3,5-O-diglucoside	Fellah et al. (2018)
97	malvidin-3-O-(6"-acetyl-galactoside)	Fellah et al. (2018)
98	malvidin-3-O-(6"-acetyl-glucoside)	Fellah et al. (2018)
99	vitisin A	Fellah et al. (2018)
100	pinotin A	Fellah et al. (2018)
101	(epi)gallocatechin-cyanidin-3-hexoside	Sentandreu et al. (2010); Gomez-Caravaca et al. (2013); Sentandreu et al. (2013); Ambigaipalan et al. (2017)
102	(epi)gallocatechin-cyanidin -3,5-dihexoside	Sentandreu et al. (2010); Sentandreu et al. (2013); Lantzouraki et al. (2014)
103	(epi)gallocatechin-delphinidin-3-hexoside	Sentandreu et al. (2010); Sentandreu et al. (2013)
104	(epi)gallocatechin-delphinidin-3,5 -dihexoside	Sentandreu et al. (2010); Sentandreu et al. (2013); Lantzouraki et al. (2014)

Contd.

Table 2 *Contd.*

No.	*Anthocyanins/Anthocyanidins*	*References*
105	(epi)gallocatechin-pelargonidin-3-hexoside	Sentandreu et al. (2010); Gomez-Caravaca et al. (2013); Sentandreu et al. (2013); Ambigaipalan et al. (2017)
106	(epi)gallocatechin-pelargonidin -3,5-dihexoside	Sentandreu et al. (2010)
107	(epi) catechin-cyanidin-3-hexoside	Sentandreu et al. (2010); Gomez-Caravaca et al. (2013); Sentandreu et al. (2013)
108	(epi) catechin-cyanidin-3,5-dihexoside	Sentandreu et al. (2010); Sentandreu et al. (2013); Lantzouraki et al. (2014)
109	(epi) catechin-delphinidin-3-hexoside	Sentandreu et al. (2010); Sentandreu et al. (2013)
110	(epi) catechin-delphinidin-3,5-dihexoside	Sentandreu et al. (2010); Sentandreu et al. (2013)
111	(epi) catechin-pelargonidin-3-hexoside	Sentandreu et al. (2010)
112	(epi) catechin-pelargonidin-3,5-dihexoside	Sentandreu et al. (2010)
113	(epi) afzelechin-cyanidin-3-hexoside	Sentandreu et al. (2010); Sentandreu et al. (2013)
114	(epi) afzelechin-cyanidin-3,5-dihexoside	Sentandreu et al. (2010); Sentandreu et al. (2013)
115	(epi) afzelechin-delphinidin-3-hexoside	Sentandreu et al. (2010); Gomez-Caravaca et al. (2013); Sentandreu et al. (2013); Ambigaipalan et al. (2017)
116	(epi) afzelechin-delphinidin-3,5-dihexoside	Sentandreu et al. (2010)
117	(epi) afzelechin-pelargonidin-3-hexoside	Sentandreu et al. (2010); Sentandreu et al. (2013)
118	(epi) afzelechin-pelargonidin -3,5-dihexoside	Sentandreu et al. (2010)
119	cyanidin-3-hexoside-(epi) gallocatechin	Sentandreu et al. (2012)
120	cyanidin-3-hexoside-(epi) catechin	Sentandreu et al. (2012)
121	cyanidin-3-hexoside-(epi) afzelechin	Sentandreu et al. (2012)
122	delphinidin-3-hexoside-(epi) gallocatechin	Sentandreu et al. (2012)

Contd.

Table 2 *Contd.*

No.	*Anthocyanins/Anthocyanidins*	*References*
123	delphinidin-3-hexoside-(epi) catechin	Sentandreu et al. (2012)
124	delphinidin-3-hexoside-(epi) afzelechin	Sentandreu et al. (2012)
125	pelargonidin-3-hexoside-(epi) gallocatechin	Sentandreu et al. (2012)

Zaouay and Mars (2014) estimated the partitioning of variance and genetic parameters for fruit quality including skin and juice color in 28 accessions of Tunisian pomegranate. The results revealed that the variety contributed most to the variance in skin and juice color. Among the 28 pomegranate clones, estimates of PCV (phenotypic coefficient of variation) and GCV (genotypic coefficient of variation) were relatively high for skin and juice color, meaning that fruit quality improvement in pomegranate could be achieved even if selection was based on phenotype. Also, PCV value were higher than GCV value for skin and juice color, which reflected the influence of environment on the expression of traits.

In pomegranate, skin and juice color had a high broad-sense heritability values indicating considerable genetic variation. These relatively high heritability values demonstrated that selection for these traits through breeding is feasible. Also, the high heritability values of skin color and juice color accompanied with high GA (genetic advance) suggested an important additive gene effects.

4. Transcriptional Regulation of Anthocyanin Biosynthesis

4.1 The Anthocyanins Biosynthetic Pathway

The anthocyanin biosynthesis pathway has been extensively studied in numerous model plants and fruit crops at biochemical, genetic, and molecular levels, and is now one of the best understood of all secondary metabolic pathways in plants.

The biosynthesis of anthocyanins is derived from the flavonoid branch of the phenylpropanoid metabolic pathway, that is, anthocyanins share common biosynthetic precursors with flavonoids and proanthocyanidins. The anthocyanin biosynthetic pathway starts with the condensation of 4-coumaroyl-CoA with three molecules of malonyl-CoA, leading to the formation of naringenin chalcone by CHS (chalcone synthase). Next, the naringenin chalcone is isomerized to naringenin through the action of CHI (chalcone isomerase). After hydroxylation at the three positions of naringenin, dihydroflavonols are formed and will be further catalyzed to yield leucoanthocyanidins by DFR (dihydroflavonol 4-reductase). Subsequently, the action of ANS (anthocyanin synthase)/LDOX (leucoanthocyanidin oxidase) generates the anthocyanidins which are then converted to glucosylated anthocyanins by UFGT (UDP-glucose: flavonoid 3-*O*-glucosyltransferase) (Tanaka, 2008).

Anthocyanin biosynthesis takes place in cytosol with the help of enzymes CHS, CHI, DFR, F3H (flavanone 3-hydroxylase), F3′H (flavonoid 3'-hydroxylase), F3′5′H (flavonoid 3'5'-hydroxylase), and ANS, after that anthocyanins are transported to vacuoles. Recent studies have demonstrated that nearly all the enzymes involved in anthocyanidin biosynthesis have been isolated and functionally characterized (Pervaiz et al., 2017), but the sequential modification of anthocyanidins metabolism, such as glycosylation, methylation and acylation, remains relatively unexplored.

Anthocyanin accumulation can also be regulated on parallel pathways that utilize the same substrate. The proanthocyanidins (condensed tannins) pathway splits from the anthocyanins pathway via enzymes including LAR (leucoanthocyanidin reductase) and ANR (anthocyanidin reductase). LAR converts leucoanthocyanidin into catechin, while ANR competes with the UFGT, converting anthocyanidins into condensed tannins (Bogs et al., 2005).

4.1.1 The structural genes involved in the anthocyanin biosynthetic pathway

Anthocyanin accumulation is determined by the activity of structural genes, which are divided into general phenylpropanoid metabolism genes, early (EBGs, including *CHS*, *CHI*, *FLS*, *F3'H*, and *F3H*) and late biosynthesis genes (LBGs, including *DFR*, *ANS/LDOX*, *UFGT*). EBGs are involved in the production of common precursors of multiple flavonoids, whereas LBGs are more specific to the anthocyanins.

4.1.2 The transcription factors involved in the anthocyanin biosynthetic pathway

The factors activating the expression of the structural genes and controlling the synthesis of certain pigments at a particular time in certain part of the plant are referred to as transcription factors (TFs). Enzymes/proteins involved in anthocyanin biosynthesis regulation are members of protein families containing R2R3-MYB protein, a basic helix-loop-helix (bHLH, MYC) protein, and WD40-type co-regulators (Laitinen et al., 2008; Feng et al., 2010), each of these regulatory factors, separately or together regulate expression of structural genes. Studies of anthocyanin biosynthesis have revealed that the expression of late genes in the anthocyanin biosynthetic pathway depends predominantly on the activity of MYB-bHLH-WD40 transcriptional complexes (also named MBW complex) (Gonzalez et al., 2008; Xu et al., 2015).

MYB receives intense research interest as it is the main family of the transcription factors that are implicated in promotion or suppression of gene transcription in the plant kingdom (Chen et al., 2019; Nanavi et al., 2020). In anthocyanin biosynthesis, MYB TFs have been identified to be the major determinant regulators in the MBW complex. They are characterized by a structurally conserved DNA-binding domain consisting of single or multiple imperfect repeats; those associated with the anthocyanin pathway are of the two-repeat (R2R3) class (Allan et al., 2008).

A small group of single MYB (R3) repeat proteins have only one MYB domain, lack a transcriptional activation domain and function as negative regulators of transcription (Lloyd et al., 2017).

In plants, bHLH proteins comprise the second largest transcription factor class. They are characterized by a DNA binding basic Helix-Loop-Helix domain (Allan et al., 2008). The bHLH TF family has been implicated in a range of functions in plants, frequently in combination with MYBs to regulate many cellular key processes.

Unlike the MYB and bHLH domains, it is difficult to define a distinct family of proteins based solely on the presence of a WD repeat motif (Lloyd et al., 2017). The WD was defined as an approximately 40 amino acid structural repeat ending with tryptophan-aspartic acid (W-D). Due to large variations in the position and length of structural elements composing WD repeat motifs found in different proteins, it is uncertain to define and identify WD repeats.

4.2 The Anthocyanin Biosynthesis in the Red Peel of Pomegranate

4.2.1 Anthocyanin structural genes

In pomegranate, most structural genes that encode enzymes involved in anthocyanin biosynthesis pathway have been cloned (Arlotta et al., 2020; Ben-Simhon et al., 2015; Zhao et al., 2015). In 2011, Ono et al. (2011) first reported transcriptome set of the pomegranate fruit peel using next-generation technologies. All candidate genes for anthocyanin biosynthesis were identified using RNA-Seq on the Illumina Genome Analyzer platform. Pomegranate genome which was released in 2018 (Yuan et al., 2018), curated annotation identified 26 potentially structural anthocyanins genes responsible for peel and aril red coloration. These genes included 2 *CHS*, 1 *CHI*, 5 *F3H*, 2 *F3'H*, 1 *F3'5'H*, 2 *DFR*, 1 *ANS/LDOX*, 5 *UFGT*, 7 *AOMT* (anthocyanin O-methyltransferase).

Expressions of several key genes in the anthocyanin biosynthetic pathway were studied in different cultivars and during fruit development. EBGs of the anthocyanin pathway, such as *PAL*, *CHI*, *CHS*, and *F3'H* were expressed in a low level in the peel of pink cultivar, but was dramatically increasing in the red cultivar as the fruit matures (Harel-Beja et al., 2019), suggesting that the transcriptional abundance of these genes decide the density of red decoration for fruit skin. On contrast, Ben-Simhon et al. (2011) reported that the expression level of *CHS* was not correlated with the skin color and the timing of red color appearance while *DFR* and *ANS* were coordinately regulated in cyanidin biosynthesis in pomegranate fruit skin. The high expression level of *DFR* gene positively correlated with skin color phenotype were also documented in different Iranian pomegranate genotypes (Rouholamin et al., 2015).

UFGT is the final gene in the anthocyanin biosynthetic pathway that catalyzes the transfer of glucose moiety from UDP-glucose to the 3-OH group of anthocyanidin, which could enhance the stability and solubility to various anthocyanidins. In the wild pomegranate, the expression of *PgUFGT* gene was the highest in red fruits as compared to leaves, flowers, and green fruits. Meanwhile, there was a positive correlation between the expression level of *PgUFGT* and total anthocyanin content (Kaur et al., 2019). During fruit development, two transcription abundance peaks of *PgUFGT* were recorded in red pomegranate, while only one expression peaked in white pomegranate, indicating *PgUFGT* played different roles in two cultivars (Zhao et al., 2017).

The obtained results suggest that EBGs and LBGs coordinately regulate anthocyanin biosynthesis, as recorded in many other horticultural crops. In fact, because of competition for the same substrate with other polyphenols, as flavonols and proanthocyanins, the structural genes involved in the anthocyanin biosynthesis may play more sophisticated functions in transcription regulation. A case in point is the study by Zhao et al. (2015), who detected the expression level of six anthocyanin-related structural genes (*PgCHS, PgCHI, PgDFR, PgANS, PgUFGT*) in red and white pomegranate during fruit development phases, and found that almost all of the genes (except *PgANS*), especially LSGs, displayed unanimously expression patterns in both cultivars, strongly suggested that they coordinately control flavonoid biosynthesis in pomegranate.

Just after formation of relatively unstable anthocyanidins, they could be modified by series of modification enzymes, resulting in specific decoration patterns that greatly influence their function and stability. Cytosolic modification reactions include glycosylation, methylation, and acylation, with glycosylation by glycosyltransferases being the most frequently occurring (Yonekura-Sakakibara et al., 2009). For pomegranate, knowledge of the biochemistry and molecular biology of these modification reactions is unavailable, except for the only molecular cloning and transcriptional expression analyses of anthocyanin glycosyltransferases *UGT/UFGT* mentioned above. Recently, based on three published pomegranate genome sequences, Zhang et al. (2021) screened seven anthocyanin O-methyltransferases (AOMTs) genes, and determined their expression levels in peel and aril during fruit different developmental stages. The results demonstrated that *PgOMT01* and *PgOMT03* only expressed in aril, *PgOMT04*, *PgOMT07*, and *PgOMT10* were more highly expressed in aril than in the peel, indicating *PgAOMT* genes had aril/peel-specific expression patterns in pomegranate (Zhang et al., 2021).

Therefore, expression analysis demonstrated the expression profile of the anthocyanin pathway-related genes in different genotypes and developing phases, suggesting the coordinated and sophisticated transcription of EBGs and LBGs in red decoration of the pomegranate fruit.

4.2.2 Anthocyanin regulatory genes

In pomegranate, the full or partial length of nucleotide sequence of MYB (Arlotta et al., 2020; Ben-Simhon et al., 2011; Khaksar et al., 2015; Zhao et al., 2015), bHLH and WD40 (Ben-Simhon et al., 2011) have been cloned by RACE-PCR method. By genomic and RNA-Seq analysis, Yuan et al. (2018) identified 7 *R2R3-MYB*, 9 *bHLH*, and 13 *WD40* genes which displayed high expression in both peel and aril, suggesting that their roles in regulating anthocyanin production in pomegranate fruit was prevalent.

Of the regulatory genes, MYB is the most focused in elucidation of anthocyanins biosynthesis. A serial of results indicated a strong association between skin color in mature pomegranate fruits with the *MYB* transcripts. Khaksar et al. (2015) found a higher expression level of *PgMYB* (accession number KF631414) gene in dark purple cultivar "Poost Siyah Yazd" than in a red skin accession "Bozi Isfahan" when the fruit became over-ripened. A higher transcription abundance of *AN2* (MYB, JF747151) in black skin genotype than in red skin genotype was also recorded (Rouholamin et al., 2015). But according to the results of Rouholamin et al. (2015), there were no consistent differences for the expression of *AN2* (MYB) among the four different colored genotypes, because the higher transcriptional levels were also found in white and green colored genotypes than in the red pomegranate.

During fruit developing stages, the transcript of *MYB* was gradually increased as the fruit matured both in fruit peel (Ben-Simhon et al., 2011; Khaksar et al., 2015) and in aril (Arlotta et al., 2020), revealing a strong association between fruit color with the *MYB* transcripts. Nevertheless, as reported by Zhao et al. (2015), who demonstrated an early expression peak of *PgMYB* (KF841621) in red and white pomegranate cultivar, MYB may act as multiple regulatory functions, not only in anthocyanins biosynthesis, but in other metabolism process.

The expression of MYB exhibits a tissue-specific expression manner. For *PgAn2* and *PAP-like* (MT495438), the transcriptional abundance was relatively high in flowers (Arlotta et al., 2020; Ben-Simhon et al., 2011). However, in three Iran pomegranate accessions, *PgMYB* (KF631414) transcript was detected at low levels in flowers (Khaksar et al., 2015). The different expression patterns of *MYB* may arise from genotype differences, or different members of MYB family. The identity of these two *MYB* genes (KF631414 vs. MT495438) was only 41% at the nucleotide acid level, though both may be involved in the biosynthesis of anthocyanins in pomegranate.

Ben-Simhon et al. (2011) described three regulatory genes: *PgWD40*, *PgAn1*(bHLH), and *PgAn2* (MYB), and studied the activity of *PgWD40* in an Arabidopsis *ttg1* mutant and was shown to be dependent on MYB function. In addition, high positive correlations were found between the steady state mRNA levels of the regulatory genes *PgWD40* and *PgAn2* (MYB) and the structural genes *PgDFR* and *PgLDOX* in both accessions. Also, the expression of *PgWD40* and *PgAn2* (MYB) genes were highly correlated with cyanidin derivatives (red pigment) accumulation in developing pomegranate fruit skin of two accessions with different timing of color appearance, indicating that *PgWD40* and *PgAn2* (MYB) were responsible for specific control of cyanidin synthesis through action on *DFR* and *LDOX* (Ben-Simhon et al., 2011). By using four different color genotypes, Rouholamin et al. (2015) also demonstrated that the expression of *WD40* gene was higher in red and black skin genotypes in comparison to green and white skin genotypes. The expression of both *WD40* and *DFR* genes followed the same trend, indicating *DFR* may be the targeted gene of *WD40* in the anthocyanin biosynthesis of pomegranate skin color, though the correlation with other structural genes were not compared.

As Rouholamin et al. (2015) described, though both *PgAn2* (MYB) and *PgAn1*(bHLH) positively correlated with cyanidin biosynthesis in pomegranate, much lower correlation was found with the *PgAn1*. White and green pomegranate skin had the highest *AN1* (bHLH) expression, whereas bright red and black skin genotypes accumulated the lowest expression level of *AN1* (Rouholamin et al., 2015). These results suggested that bHLH may promote the peel coloration in a different manner. More studies are required to elucidate the role of *PgAn1* (bHLH) in control of anthocyanin biosynthesis in pomegranate.

4.3 The Anthocyanin Biosynthesis in the Red Aril of Pomegranate

The pomegranate fruit inside is full of seeds or grains of prismatic shape. Each grain is the edible portion of the fruit, and it constitutes the pomegranate seed (Melgarejo et al., 2020). Most of authors name the edible part which is fleshy, juicy coating surrounding each seed as the 'aril'. In fact, the whole pomegranate seed cannot be called aril, because it is exariled (Melgarejo et al., 2020). To be consistent with descriptions in the publications, 'aril' is still used to represent the outer seed coat (the edible part) of the pomegranate fruit.

The aril color varies largely among cultivars, ranging from white, pink to red and dark red, which is as diverse as skin color. The attractive red color of pomegranate juice is one of the parameters that are evaluated for the commercial classification of the product in relation to its quality, which influences consumer behavior. During fruit ripening, a rapid increase in anthocyanin pigment concentration of aril juice was reported in red pomegranate accessions (Gil et al., 1995; Kulkarni and Aradhya, 2005; Shwartz et al., 2009), but the aril color was not always in parallel with the peel color. Some researchers assumed that the fruit's outer skin color was not correlated significantly with the color of the arils (Dafny-Yalin et al., 2010; Holland et al., 2009). Shwartz et al. (2009) revealed that the peel color index was only positively correlated with ascorbic acid of aril juice, but not correlated with other aril juice parameters, such as sugar (glucose and fructose) contents, pH and total soluble solids. In 'black' pomegranate, high contents of anthocyanins accumulated in the fruit peel but not in arils (Trainin et al., 2021). All these findings indicated that there may be a different anthocyanin biosynthetic mechanism in aril.

The studies on structural genes involved in anthocyanin accumulation in pomegranate aril are limited, compared with that in fruit skin. Attanayake et al. (2019) studied major anthocyanin biosynthesis gene expression in pomegranate arils. Using a cultivar 'Nimali' with yellowish peel and pink arils, the authors found that there was no clear change in the expression of *CHI* and it was at a very low level during fruit development. Expression of all other genes *F3H*, *DFR*, and *ANS* were upregulated from 120 DAI (days after initiation) when the aril color turned from white to pink, and unchanged until ripening. Additionally, a high and positive correlation was found between total cyanidins content and expression of major genes in the flavonoid biosynthetic pathway, specifically *F3H*, *DFR*, and *ANS*, suggesting that these genes decide the overall color of the arils. The association pattern observed that the up-regulation of the structural genes, especially the LBGs, corresponded to the accumulation of anthocyanins in the outer seed coat, was also confirmed by Qin et al. (2017) based on the genome data of 'Dabenzi'.

Recently, Arlotta et al. (2020) identified a series of flavonoid-pathway TFs in pomegranate arils, including *PgPAP-like*, *PgMYB5-like*, *PgMYB4-like*, and unraveled the interaction mode of *PgMYB5-like+PgbHLH*. The authors suggested a possible function of *PgPAP-like* MYB as a stimulator of anthocyanin biosynthesis in pomegranate, based on its corresponding expression pattern with the *PgCHS* and *PgF3'5'H* and the content of most of the anthocyanins. *PgMYB4* may be a repressor of flavonoid biosynthesis, for the lack of correlation between its expression profile and anthocyanin accumulation. Functional analysis revealed that *PgMYB5-like* strongly interacted with *PgbHLH*, probably activating *F3H*, *F3'H*, and *F3'5'H* resulting in the production of intermediates of the flavonoids pathway, especially dihydroflavonols, while anthocyanins were not produced (Arlotta et al., 2020). Differently from TFs, the expression profiles for most of the selected flavonoid structural genes were highly correlated with anthocyanins. *PgCHS* was highly up regulated in ripe arils of 'Wonderful', while its expression could not be detected in 'Valenciana' pomegranate arils. This result was not in accordance with that in fruit peel (Ben-Simhon et al., 2011; Zhao et al., 2015). The positive correlation of transcription level of *PgCHS*, *PgF3'5'H*, *PgDFR*, and *PgUFGT* with the selected structural genes and with TFs evidenced that these structure genes were directly involved in aril flavonoid biosynthesis pathway, especially the anthocyanin class.

Anthocyanin synthesis depends primarily on light intensity with temperature, carbohydrates, and hormones all playing a lesser role (Thomson et al., 2018). By fruit bagging experiments, it is confirmed that the biosynthesis of anthocyanins is light-responsive in numerous fruit trees, such as apple (Feng et al., 2014), litchi (Wei et al., 2011), grape (Azuma et al., 2012). For red aril, color accumulation is not affected by light deprivation because of the thick pericarp, meaning that the anthocyanin biosynthesis in aril is light-independent.

On the other hand, temperature of the fruit surface can strongly influence the process of anthocyanin biosynthesis. For plants in general, low temperatures tend to promote skin coloration while higher temperatures do not (Steyn et al., 2010; Thomson et al., 2018). For instance, the optimum temperatures for anthocyanin accumulation were 12°C in unripe apples and 16–24°C in ripe apples (Faragher, 1983). Indeed, the intensity of the red color of fruit aril was inversely related to the sum of heat units accumulated during fruit development and ripening (Borochov-Neori et al., 2009). Also, the red color intensity of the aril was enhanced when shade nets were employed to reduce air temperature in fruit vicinity during ripening in preliminary experiments (Borochov-Neori et al., 2009). A previous study showed that, contrary to the effect on skin anthocyanins, arils of fruit from the outer canopy were less pigmented than in fruit from inner canopy (Gil et al., 1995). Di Stefano et al. (2020) observed that pomegranate juice with South and North exposure showed higher anthocyanins content, thus, they speculated that fruit exposure to sun radiation in East and West positions might reduce the accumulation of anthocyanins due to elevated temperature, which causes a reduction in anthocyanin synthesis or even partial degradation. Hence, the lower temperature may be favorable to the anthocyanin biosynthesis for pomegranate arils.

Taken together, temperature affects the peels and arils differently (Schwartz et al., 2009). It is speculated that synthesis of anthocyanins in aril is not light-dependent and favorable of lower temperature. The genetic components that control aril color appearance require further elucidation in pomegranate.

5. The Genetic Nature of Fruit Color Mutation

Different cultivars of pomegranate provide groundwork for identifying genes responsible for variations in color pigmentation. Much progress has been made regarding the white fruit mutation at the molecular level. Also, there are increasing interest in the color mutation mechanism of 'black' pomegranate.

5.1 The White Fruit

Phenotypes mutation with beneficial color traits contribute to our understanding of the genetic and molecular mechanism controlling anthocyanin accumulation. The skin color of the 'White' pomegranate at different developmental stages of the fruit varies from green to yellowish, and the aril color during developmental stages is white. The phenotypic characteristics of the 'white' pomegranate strongly suggested that it lacks the typical pomegranate color rendered by anthocyanins in all tissues of the plant, especially in flowers and fruits (Ben-Simhon et al., 2015; Zhao et al., 2015).

Gene-expression analysis indicated that all of the structural and regulatory genes involved in anthocyanin biosynthesis in pomegranate, except for the *ANS* gene, were expressed in all of the examined tissue of the 'white' pomegranate (Ben-Simhon et al., 2015; Zhao et al., 2015). However, the expression of *PgAN2*(MYB) seemed to be tissue-specific and not related to the 'white' pomegranate (Ben-Simhon et al., 2015). Further HPLC study showed that other flavonoids (kaempferol, quercetin, catechin and catechin derivatives) corresponding to the upstream stages of LDOX function accumulated in the "white pomegranate" skin (Ben-Simhon et al., 2015; Zhao et al., 2014). These data indicated that the 'white' phenotype may result from null expression of the *PgLDOX* gene. Subsequently, by comparison of genomic sequences of *PgLDOX* in 'white' and colored pomegranate accessions, Ben-Simhon et al. (2015) found that an SNP (the 'white' pomegranate contained a T/T combination and the colored accessions C/C) was located 1,008 bp downstream of ATG but does not change the amino acid sequence, suggesting that the SNP was not responsible for the 'white' phenotype. Further analysis revealed the presence of an insertion in the coding region of *PgLDOX* (located between position 90-91 downstream of the ATG initiation codon) of 'white' pomegranate and the F2 progeny which showed the 'white' phenotype, while colored accessions did not have this insertion. It was speculated that this insertion sequence was the direct cause of the failure of anthocyanin synthesis in white-peel pomegranate.

Although the position of the putative insertion has been identified, Ben-Simhon et al. (2015) failed to ascertain the identity of the inserted sequence. Afterwards, Jeong et al. (2018) showed that the *ANS* gene in the white pomegranate was inactivated by a chromosomal translocation, not a large-sized insertion. Analysis revealed that the 27,786 bp block containing 88-bp 5' partial sequences of the *ANS* gene, might be translocated into an approximately 22-kb upstream region in an inverted orientation (Jeong et al., 2018). It would be impossible to connect 5' and 3' region of the *ANS* gene in the while accessions of pomegranate due to the inverted orientation of the translocated block.

Overall, studies show that due to a mutation of the *ANS* gene of the anthocyanin biosynthetic pathway, and subsequently the lack of expression of *PgANS* resulted in the inability to produce anthocyanin in the "white pomegranate".

5.2 The Black Fruit

'Black' pomegranates are characterized by purple to black peel color. Unique accumulation of delphinidin together with the high levels of cyanidin derivatives are responsible for the purple color of the fruit (Trainin et al., 2021; Zhao et al., 2013). Some work has been done to identify the genetic components that are responsible for the 'black' trait (Rouholamin et al., 2015; Trainin et al., 2021).

Comparative expression analysis found that though *AN2* (MYB) was constantly expressed in four different skin genotypes, the expression level of *WD40* in Black skin genotype was the highest, while *AN1* (bHLH) showed the lowest transcription (Rouholamin et al., 2015). Similar results were recorded by Trainin et al. (2021), who observed the higher MYB and lower bHLH rate of transcription in black compared with red pomegranate. As far as structural genes, *DFR* gene transcripts showed 3 and 5 times more than red and green skin genotypes (Rouholamin et al., 2015). However, according to the opinion of Trainin et al. (2021), except for the *F3'5'H* gene, other structural genes, including *CHS*, *CHI*, *F3H*, *DFR*, *F3'H*, *LDOX*, *ANR*, and *LAR*, were expressed at a much higher level in the peel of red pomegranate as compared to 'Black' genotype (Trainin et al., 2021), indicating that these genes were not responsible for the Black phenotype of the pomegranate.

By mapping the Black trait in the F2 population and comparative expression analysis, Trainin et al. (2021) strongly suggested that a single base deletion of *ANR* gene created a nonsense mutation, shortening the deduced protein by 25 aa. This mutation potentially inactivated or reduce the activity of the ANR protein, reducing competition over anthocyanidins formation and resulting in overproduction of anthocyanin (Trainin et al., 2021). The 'Black' phenotype in pomegranate suggested a novel regulatory pathway that controls color production in fruit by partially or completely blocking a competing pathway. Nonetheless, more work needs to be done to verify if the same *ANR* mutation occurred in other Black pomegranate cultivars.

Different from other red commercial cultivars, color accumulation in peel of 'black' pomegranate was not affected by light deprivation (Trainin et al., 2021). The mechanism behind the Black color formation in pomegranate probable be novel and needs to be further elucidated comprehensively.

6. Scopes and Future Perspectives

Anthocyanins are the major natural pigments in pomegranate known for their antioxidative properties. Variations in composition and concentration of individual anthocyanins produce the diversity of fruit color phenotypes. Though some achievements have been made in fruit color evolution of the pomegranate, there are some key considerations to take into account in the near future.

6.1 Mapping and Analysis QTL Controlling Color Traits

The fine mapping of quantitative trait loci (QTL) is the important basis for QTL cloning and marker-assisted selection (MAS). Genetic linkage maps based on molecular markers provides a powerful tool for understanding of anthocyanin phenotypes based on genetics. The very first linkage map in pomegranate was constructed based on AFLP markers by Sarkhosh et al. (2012), using an F1 mapping population and the two-way pseudo-testcross strategy. This AFLP-based genetic map covered 435.5, 344.2, and 335.5cM of the haploid genomes of female parent, male parent, and the consensus map. However, there were no phenotypic data collected for the progeny in the segregating population, which could not lead to marker assisted breeding. Harel-Beja et al. (2015) reported a high-resolution genetic map based on 1,092 SNP markers combined with phenotypic traits. The map was based on an F2 population of pomegranate accessions which represent evolutionary distant pomegranate accessions. This map is the initial step towards the usage of molecular markers for breeding as it contains several traits including color which are important for breeding. However, the only linkage map was based on the temporarily separated population, thus affecting the accuracy of subsequent gene mapping. The advancement of high throughput sequencing technology and release of pomegranate genome data set made it possible for screening of NGS-based molecular markers, construction of highly saturated linkage maps, and fine mapping of color-related genes.

6.2 The Transport Mechanism from Endoplasmic Reticulum (ER) to the Vacuole

In plant cells, anthocyanins are synthesized on the cytoplasmic surface of the endoplasmic reticulum (ER) and then are transported and accumulated in the central vacuole of cells. Though the biosynthesis pathway and the regulation of anthocyanidins and anthocyanins have been investigated thoroughly in various plants, little is known about the possible mechanisms of its final stage: anthocyanins traffic between the ER and the vacuole. Transporter-mediated and vesicle-mediated transport are two major hypotheses proposed for the transport of anthocyanins to the vacuole (Shi and Xie, 2014). So far, four different models have been proposed to explain the transport of anthocyanin from biosynthetic sites to the central vacuole, and four types of transporters have been associated with the transport of anthocyanins: Glutathione s-transferase (GST), multidrug resistance-associated protein, multidrug and toxic compound extrusion, bilitranslocase-homologue (Zhao and Dixon, 2010), The functions of these proteins and related genes, especially GST, have also been studied in apple (Jiang et al., 2019), kiwifruit (Liu et al., 2019), grape (Ricardo et al., 2016), strawberry (Gao et al., 2019; Lin et al., 2020). Although different models have been proposed, cellular and subcellular information is still lacking for reconciliation of different lines of evidence in various anthocyanin sequestration studies. For pomegranate, no studies have been reported regarding anthocyanins transporting.

6.3 The Transcriptional Regulation Mechanism of Anthocyanins in Fruit

Transcriptional regulation often involves the combinatorial action of multiple TFs from the same, or different, families that modulate the expression of the same target genes (Hassani et al., 2020). It is known that combinations of the R2R3–MYB, bHLH, and WDR proteins and their interactions determine a set of genes involved in anthocyanin pathway. Generally, both bHLH and MYB bind to specific DNA sequences within the promoter regions of their target genes to enhance or repress the gene's expression (Espley et al., 2009; Fickett and Wasserman, 2000). As reviewed by Xu et al. (2015), anthocyanin biosynthesis is regulated by different sets of MBW complexes in specific plants. For instance, the apple MdTTG1 (a WD40 protein) interacts with bHLH but not MYB proteins to form a complex regulate anthocyanin accumulation (An et al., 2012). While in Chinese Bayberry (*Myrica rubra*), MrWD40-1 physically interacted with both MrMYB and MrbHLH to regulate anthocyanin biosynthesis (Liu et al., 2013). Though the function of pomegranate *PgWD40* is dependent on functionally known MYB protein (Ben-Simhon et al., 2011), the specific interaction with MYB/bHLH and the target genes are not well-defined. It is also worth noting that several other TFs have also been reported to regulate anthocyanin biosynthesis via interaction with the MBW complex, such as COP1 (Maier et al., 2013), JAZ (Qi et al., 2011), NAC (Zhou et al., 2015), SPL (Gou et al., 2011), bZIP (An et al., 2018) and WRKY (Verweij et al., 2016). However, it is limited for pomegranate in understanding how MBW members as well as other TFs functionally activate the biosynthesis of anthocyanins at different developmental stages and environmental conditions.

Furthermore, the anthocyanin pathway in the arils and peels may be controlled differently. Though a series of experiments have demonstrated the transcriptional regulation of red color formation in fruit outer skin, the anthocyanin biosynthesis pathway and related functional genes in arils still remain to be confirmed. Owing to lack of comparison analyses, it is unclear whether the anthocyanin biosynthesis is separately controlled or coordinately regulated in inner arils and in outer peels.

6.4 Posttranscriptional and Posttranslational Regulation of Anthocyanin Biosynthesis

Plant cells can regulate anthocyanin biosynthesis by acting both transcriptionally and post-translationally. Substantial investigations have demonstrated the biosynthesis regulation of anthocyanins at the transcriptional level as reviewed (Gu et al., 2019; Nanavi et al., 2020; Shi and Xie, 2014). With the advancement of molecular biology and omics technology, studies on the mechanisms at post-transcriptional and post-translational levels are now on their way (Saminathan et al., 2016).

Current research has identified some post-transcriptional modulators mostly microRNAs (miRNAs) to be involved in anthocyanin biosynthesis (He et al., 2019; Liu et al., 2016; Varsha et al., 2019). In pomegranate, miRNAs play important roles in female sterility (Chen et al., 2020), and seed hardness (Luo et al., 2018). However, knowledge and roles of miRNAs in pomegranate anthocyanin biosynthesis have not been explored.

The molecular basis involved in the potential post-translational and regulatory processes has been narrowly described. Post-translational modification includes processes such as phosphorylation, S-nitrosylation, ubiquitination, and Small Ubiquitin-like Modifier (SUMO)-ylation (Nanavi et al., 2020). So far, post-translational modification mechanisms regulating transcription factor activity seem to attract researchers' attention, but the clear mechanisms at post-translational level remain to be discovered in plants including pomegranate.

6.5 The External Factors Influenced the Anthocyanin Accumulation

Anthocyanin biosynthesis can be induced by various abiotic factors such as light, temperature, sugars, and inner hormones. Numerous significant advances have made in elucidating the molecular mechanisms of anthocyanin biosynthesis in response to these factors (Gu et al., 2019; Shi and Xie, 2014). Nevertheless, pertinent researches are still limited in pomegranate.

Bar-Ya'akov et al. (2019) made a summary of the influences of environmental factors particularly climate and geographic conditions on color accumulation. In brief, heat (Borochov-Neori et al., 2009), season (Borochov-Neori et al., 2011), temperature (Schwartz et al., 2009), salt stress (Borochov-Neori et al., 2014) all exert substantial influences on anthocyanin accumulation and composition. The significance and underlying mechanisms of how do anthocyanins respond to the environmental changes are not yet understood. For instance, the biosynthesis of anthocyanins in maturing red fruit is a light-dependent process, while color accumulation in peel of 'black' pomegranate is not affected by light deprivation (Trainin et al., 2021). The genetic components that control sensitivity to light need to be elucidated in pomegranate especially in 'black' fruit.

7. Concluding Remarks

Colored pomegranate is one of the breeding objectives around the world. For breeding programs to achieve excellent exterior quality and improve value-added fruit containing optimized anthocyanin content, better knowledge of the regulatory mechanism of anthocyanin biosynthesis is required. Though a wealth of knowledge is now available on the molecular biology of anthocyanin biosynthesis, studies on pomegranate are at the beginning.

Most of the current studies have been focused on cloning and expression analysis of the pathway genes. Despite the prominent progress in pomegranate genomics, functional verification and mapping of genes involved in fruit and aril color have been achieved in only very few cases in pomegranate. The next focus is the better understanding of the color diversity of pomegranate fruits in the occurrence of genes and regulators from various regulatory levels, as well as the interactions of co-pigments particularly those share the same precursors of the pathway, including flavonoids and proanthocyanins. A combination of genomic, transcriptomic, proteomic, and metabolomic approaches will enable further enhancing our knowledge of anthocyanin biosynthesis mechanisms in pomegranate, and this knowledge may be used for supporting breeding efforts towards developing new pomegranate cultivars with tailored anthocyanin composition.

References

Abid, M., Yaich, H., Cheikhrouhou, S., Khemakhem, I., Bouaziz, M., Attia, H. and Ayadi, M.A. 2017. Antioxidant properties and phenolic profile characterization by LC-MS/MS of selected Tunisian pomegranate peels. *Journal of Food Science and Technology-Mysore*, *54*: 2890–01. https://doi.org/10.1007/s13197-017-2727-0.

Akhavan, H., Barzegar, M., Weidlich, H. and Zimmermann, B.F. 2015. Phenolic compounds and antioxidant activity of juices from ten Iranian pomegranate cultivars depend on extraction. *Journal of Chemistry*, *2015*: 1–7. https://doi.org/10.1155/2015/907101.

Allan, A.C., Hellens, R.P. and Laing, W.A. 2008. MYB transcription factors that color our fruit. *Trends in Plant Science*, *13*: 99–102. https://doi.org/10.1016/j.tplants.2007.11.012.

Ambigaipalan, P., Camargo, A.D. and Shahidi, F. 2017. Identification of phenolic antioxidants and bioactives of pomegranate seeds following juice extraction using HPLC-DAD-ESI-MS^n. *Food Chemistry*, *221*: 1883–94. https://doi.org/10.1016/j.foodchem.2016.10.058.

An, J., Yao, J., Xu, R., You, C., Wang, X. and Hao, Y. 2018. Apple *bZIP* transcription factor MdbZIP44 regulates abscisic acid-promoted anthocyanin accumulation. *Plant, Cell, and Environment*:, *41*. https://doi.org/10.1111/pce.13393.

An, X., Tian, Y., Chen, K., Wang, X. and Hao, Y. 2012. The apple WD40 protein MdTTG1 interacts with bHLH but not MYB proteins to regulate anthocyanin accumulation. *Journal of Plant Physiology*, *169*: 710–17. https://doi.org/10.1016/j.jplph.2012.01.015.

Anderson, O.M. and Jordheim, M. (2006). The anthocyanins. *In*: O. M. Anderson & K. R. Markham (Eds.), *Flavonoids: Chemistry, Biochemistry, and Applications*, 472–551). Boca Raton, FL, USA: CRC Press/Taylor&Francis Group.

Arlotta, C., Puglia, G.D., Genovese, C., Toscano, V. and Raccuia, S.A. 2020. MYB5-like and bHLH influence flavonoid composition in pomegranate. *Plant Science*, *298*: 110563. https://doi.org/10.1016/j.plantsci.2020.110563.

Attanayake, R., Rajapaksha, R., Weerakkody, P. and Bandaranayake, P.C.G. 2019. The effect of maturity status on biochemical composition, antioxidant activity, and anthocyanin biosynthesis gene expression in a pomegranate (*Punica granatum* L.) cultivar with red flowers, yellow peel, and pinkish arils. *Journal of Plant Growth Regulation, 38*: 992–1006. https://doi.org/10.1007/s00344-018-09909-2.

Azuma, A., Yakushiji, H., Koshita, Y. and Kobayashi, S. 2012. Flavonoid biosynthesis-related genes in grape skin are differentially regulated by temperature and light conditions. *Planta, 236*: 1067–80. https://doi.org/10.1007/s00425-012-1650-x.

Bar-Ya'akov, I., Tian, L., Amir, R. and Holland, D. 2019. Primary metabolites, anthocyanins, and hydrolyzable tannins in the pomegranate fruit. *Frontiers in Plant Science, 10*: 620. https://doi.org/10.3389/fpls.2019.00620.

Bartual, J., Palou, L. and Perez-Gago, M.B. 2015. Characterization of fruit traits from 'Mollar de Elche' pomegranate progenies. *Acta Horticulturae, 1106*: 25–30. https://doi.org/10.17660/ActaHortic.2015.1106.5.

Ben-Simhon, Z., Judeinstein, S., Nadler-Hassar, T., Trainin, T., Bar-Ya'akov, I., Borochov-Neori H. and Holland, D. 2011. A pomegranate (*Punica granatum* L.) WD40-repeat gene is a functional homologue of Arabidopsis *TTG1* and is involved in the regulation of anthocyanin biosynthesis during pomegranate fruit development. *Planta, 234*: 865–81. https://doi.org/10.1007/s00425-011-1438-4.

Ben-Simhon, Z., Judeinstein, S., Trainin, T., Harel-Beja, R., Bar-Ya'akov, I., Borochov-Neori, H. and Holland, D. 2015. A 'White' anthocyanin-less pomegranate (*Punica granatum* L.) caused by an insertion in the coding region of the leucoanthocyanidin dioxygenase (LDOX; ANS) gene. *PloS One, 10*: e0142777. https://doi.org/10.1371/journal.pone.0142777.

Bogs, J., Downey, M., Harvey, J.S., Ashton, A.R., Tanner, G.J. and Robinson, P.S. 2005. Proanthocyanidin synthesis and expression of genes encoding leucoanthocyanidin reductase and anthocyanidin reductase in developing grape berries and grapevine leaves. *Plant Physiology, 139*: 652–663. https://doi.org/10.1104/pp.105.064238.

Borochov-Neori, H., Judeinstein, S., Harari, M., Bar-Ya'akov, I., Patil, B.S., Lurie, S. and Holland, D. 2011. Climate effects on anthocyanin accumulation and composition in the pomegranate (*Punica granatum* L.) fruit arils. *Journal of Agricultural and Food Chemistry, 59*: 5325–34. https://doi.org/10.1021/jf2003688.

Borochov-Neori, H., Judeinstein, S., Tripler, E., Harari, M., Greenberg, A., Shomer, I. and Holland, D. 2009. Seasonal and cultivar variations in antioxidant and sensory quality of pomegranate (*Punica granatum* L.) fruit. *Journal of Food Composition and Analysis, 22*: 189–95. https://doi.org/10.1016/j.jfca.2008.10.011.

Borochov-Neori, H., Judeinstein, S., Tripler, E., Holland, D. and Lazarovitch, N. 2014. Salinity effects on color and health traits in the pomegranate (*Punica granatum* L.) fruit peel. *International Journal of Postharvest Technology and Innovation, 4*: 54. https://doi.org/10.1504/IJPTI.2014.064145.

Castaneda-Ovando, A., de Lourdes, Pacheco-Hernandez M., Elena, Paez-Hernandez M., Rodriguez, J.A. and Andres Galan-Vidal, C. 2009. Chemical studies of anthocyanins: A review. *Food Chemistry, 113*: 859–71. https://doi.org/10.1016/j.foodchem.2008.09.001.

Chen L., Hu B., Qin Y., Hu G. and Zhao, J. 2019. Advance of the negative regulation of anthocyanin biosynthesis by MYB transcription factors. *Plant Physiology and Biochemistry, 136*: 178–87. https://doi.org/10.1016/j.plaphy.2019.01.024.

Chen, L., Luo, X., Yang, X., Jing, D., Xia, X., Li, H., Poudel, K. and Cao, S. 2020. Small RNA and mRNA sequencing reveal the roles of microRNAs involved in pomegranate female sterility. *International Journal of Molecular Sciences, 21*: 558. https://doi.org/10.3390/ijms21020558.

Dafny-Yalin, M., Glazer, I., Bar-Ilan, I., Kerem, Z., Holland, D. and Amir, R. 2010. Color, sugars, and organic acids composition in aril juices and peel homogenates prepared from different pomegranate accessions. *Journal of Agricultural and Food Chemistry, 58*: 4342–52. https://doi.org/10.1021/jf904337t.

de Pascual-Teresa, S., Moreno, D.A. and Garcia-Viguera, C. 2010. Flavanols and anthocyanins in cardiovascular health: A review of current evidence. International *Journal of Molecular Sciences, 11*: 1679–1703. https://doi.org/10.3390/ijms11041679.

Di Stefano, V., Scandurra, S., Pagliaro, A., Martino, V.D. and Melilli, M.G. 2020. Effect of sunlight exposure on anthocyanin and non-anthocyanin phenolic levels in pomegranate juices by high resolution mass spectrometry approach. *Foods*, *9*: 1161. https://doi.org/10.3390/foods9091161.

Espley, R.V., Brendolise, C., Chagne, D., Kutty-Amma, S., Green, S., Volz, R., Putterill, J., Schouten, H.J., Gardiner, S.E. and Hellens, R.P. 2009. Multiple repeats of a promoter segment causes transcription factor autoregulation in red apples. *Plant Cell*, *21*: 168–83. https://doi.org/10.1105/tpc.108.059329.

Faragher, J.D. 1983. Temperature regulation of anthocyanin accumulation in apple skin. *Journal of Experimental Botany*, *34*: 10. https://doi.org/10.1093/jxb/34.10.1291.

Fellah, B., Bannour, M., Rocchetti, G., Lucini, L. and Ferchichi, A. 2018. Phenolic profiling and antioxidant capacity in flowers, leaves, and peels of Tunisian cultivars of *Punica granatum* L. *Journal of Food Science and Technology-Mysore*, *55*: 3606–15. https://doi.org/10.1007/s13197-018-3286-8.

Feng, F., Li, M., Ma, F. and Cheng, L. 2014. The effects of bagging and debagging on external fruit quality, metabolites, and the expression of anthocyanin biosynthetic genes in 'Jonagold' apple (*Malus domestica* Borkh.). *Scientia Horticulturae*, *165*: 123–31. https://doi.org/10.1016/j.scienta.2013.11.008.

Fickett, J.W. and Wasserman, W.W. 2000. Discovery and modeling of transcriptional regulatory regions. *Current Opinion in Biotechnology*, *11*: 19–24. https://doi.org/10.1016/s0958-1669(99)00049-x.

Fischer, U.A., Carle, R. and Kammerer, D.R. 2011. Identification and quantification of phenolic compounds from pomegranate (*Punica granatum* L.) peel, mesocarp, aril, and differently produced juices by HPLC-DAD-ESI/MS^n. *Food Chemistry*, *127*: 807–21. https://doi.org/10.1016/j.foodchem.2010.12.156.

Gao, Q., Luo, H., Li, Y., Liu, Z. and Kang, C. 2019. Genetic modulation of RAP alters fruit coloration in both wild and cultivated strawberry. *Plant Biotechnology Journal*, *18*. https://doi.org/10.1111/pbi.13317.

Gil, M.I., García-Viguera, C., Artés, F. and Tomás-Barberán, F.A. 1995. Changes in pomegranate juice pigmentation during ripening. *Journal of the Science of Food & Agriculture*, *68*: 77–81. https://doi.org/10.1002/jsfa.2740680113.

Gomez-Caravaca, A.M., Verardo, V., Toselli, M., Segura-Carretero, A., Fernandez-Gutierrez, A. and Caboni, M.F. 2013. Determination of the major phenolic compounds in pomegranate juices by HPLC-DAD-ESI-MS. *Journal of Agricultural and Food Chemistry*, *61*: 5328–37. https://doi.org/10.1021/jf400684n.

Gonzalez, A., Zhao, M., Leavitt, J.M. and Lloyd, A.M. 2008. Regulation of the anthocyanin biosynthetic pathway by the TTG1/bHLH/Myb transcriptional complex in Arabidopsis seedlings. *Plant Journal*, *53*: 814–27. https://doi.org/10.1111/j.1365-313X.2007.03373.x.

Gou, J., Felippes, F., Liu, C., Weigel, D. and Wang, J. 2011. Negative regulation of anthocyanin biosynthesis in Arabidopsis by a miR156-targeted SPL transcription factor. *Plant Cell*, *23*: 1512–22. https://doi.org/10.1105/tpc.111.084525.

Gu, K., Wang, C., Hu, D. and Hao, Y. 2019. How do anthocyanins paint our horticultural products? *Scientia Horticulturae*, *249*: 257–62. https://doi.org/10.1016/j.scienta.2019.01.034.

Harel-Beja, R., Sherman, A., Rubinstein, M., Eshed, R., Bar-Ya'akov, I., Trainin, T., Ophir, R. and Holland, D. 2015. A novel genetic map of pomegranate based on transcript markers enriched with QTLs for fruit quality traits. *Tree Genetics and Genomes*, *11*. https://doi.org/10.1007/s11295-015-0936-0.

Harel-Beja, R., Tian, L., Freilich, S., Habashi, R., Borochov-Neori, H., Lahav, T., Trainin, T., Doron-Faigenboim, A., Ophir, R. and Bar-Ya'Akov, I. 2019. Gene expression and metabolite profiling analyses of developing pomegranate fruit peel reveal interactions between anthocyanin and punicalagin production. *Tree Genetics and Genomes*, *15*: 22.21–22.13. https://doi.org/10.1007/s11295-019-1329-6.

Hassani, D., Fu, X., Shen, Q., Khalid, M. and Tang, K. 2020. Parallel transcriptional regulation of artemisinin and flavonoid biosynthesis. *Trends in Plant Science*, *25*. https://doi.org/10.1016/j.tplants.2020.01.001.

He, L., Tang, R., Shi, X., Wang, W. and Jia, X. 2019. Uncovering anthocyanin biosynthesis related microRNAs and their target genes by small RNA and degradome sequencing in tuberous roots of sweetpotato. *BMC Plant Biology*, *19*: 232. https://doi.org/10.1186/s12870-019-1790-2.

Holland, D. and Bar-Ya'akov, I. 2018. Pomegranate (*Punica granatum* L.) breeding. *In*: J.M. Al-Khayri, S.M. Jain and D.V. Johnson (Eds.), *Advances in Plant Breeding Strategies: Fruits*, 601–47. Gewerbestrasse 11, 6330 Cham, Switzerland: Springer International Publishing AG.

Holland, D., Hatib, K. and Bar-Ya'akov, I. 2009. *Pomegranate: Botany, Horticulture, Breeding* (Vol. 35). John Wiley & Sons, Inc.

Jaakola and Laura. 2013. New insights into the regulation of anthocyanin biosynthesis in fruits. *Trends in Plant Science*, *18*: 477–83. https://doi.org/10.1016/j.tplants.2013.06.003.

Jalikop, S.H. 2011. Breeding of pomegranate and annonaceous fruits. *Acta Horticulturae*, *890*: 191–97. https://doi.org/10.17660/ActaHortic.2011.890.25.

Jalikop, S.H., Rawal, R.D. and Kumar, R. 2005. Exploitation of sub-temperate pomegranate Daru in breeding tropical varieties. *Acta Horticulturae*, *696*: 107–12. https://doi.org/10.17660/ActaHortic.2005.696.18.

Jeong, H., Park, M. and Kim, S. 2018. Identification of chromosomal translocation causing inactivation of the gene encoding anthocyanidin synthase in white pomegranate (*Punica granatum* L.), and development of a molecular marker for genotypic selection of fruit colors. *Horticulture Environment and Biotechnology*, *59*: 857–64. https://doi.org/10.1007/s13580-018-0082-3.

Jiang, S., Chen, M., He, N., Chen, X., Wang, N., Sun, Q., Zhang, T., Xu, H., Fang, H., Wang, Y., Zhang, Z., Wu, S. and Chen, X. 2019. MdGSTF6, activated by MdMYB1, plays an essential role in anthocyanin accumulation in apple. *Horticulture Research*, *6*: 40. https://doi.org/10.1038/s41438-019-0118-6.

Kaur, R., Kapoor, N., Aslam, L. and Mahajan, R. 2019. Molecular characterization of PgUFGT gene and R2R3-PgMYB transcription factor involved in flavonoid biosynthesis in four tissues of wild pomegranate (*Punica granatum* L.). *Journal of Genetics*, *98*: 94. https://doi.org/10.1007/s12041-019-1141-y.

Khaksar, G., Tabatabaei, B.E.S., Arzani, A., Ghobadi, C. and Ebrahimie, E. 2015. Functional analysis of a pomegranate (*Punica granatum* L.) MYB transcription factor involved in the regulation of anthocyanin biosynthesis. *Iranian Journal of Biotechnology*, *13*: 17–25. https://doi.org/10.15171/ijb.1045.

Khoo, H.E., Azlan, A. and Lim, S.M. 2019. Evidence-based therapeutic effects of anthocyanins from foods. *Pakistan Journal of Nutrition*, *18*: 1–11. https://doi.org/10.3923/pjn.2019.1.11.

Kostka, T., Ostberg-Potthoff, J.J., Briviba, K., Matsugo, S., Winterhalter, P. and Esatbeyoglu, T. 2020. Pomegranate (*Punica granatum* L.) extract and its anthocyanin and copigment fractions-free radical scavenging activity and influence on cellular oxidative stress. *Foods*, *9*: 1617. https://doi.org/10.3390/foods9111617.

Krga, I. and Milenkovic, D. 2019. Anthocyanins: From sources and bioavailability to cardiovascular-health benefits and molecular mechanisms of action. *Journal of Agricultural and Food Chemistry*, *67*: 1771–83. https://doi.org/10.1021/acs.jafc.8b06737.

Kulkarni, A.P. and Aradhya, S.M. 2005. Chemical changes and antioxidant activity in pomegranate arils during fruit development. *Food Chemistry*, *93*: 319–24. https://doi.org/10.1016/j.foodchem.2004.09.029.

Lantzouraki, D.Z., Sinanoglou, V.J., Zoumpoulakis, P.G., Glamočlija, J., Ćirić, A., Soković, M., Heropoulos, G. and Proestos, C. 2014. Antiradical–antimicrobial activity and phenolic profile of pomegranate (*Punica granatum* L.) juices from different cultivars: A comparative study. *RSC Advances*, *5*: 2602–14. https://doi.org/10.1039/C4RA11795F.

Lin, Y., Zhang, L., Zhang, J.H., Zhang, Y. and Tang, H. 2020. Identification of anthocyanins-related glutathione S-transferase (GST) genes in the genome of cultivated strawberry (*Fragaria ×*

ananassa). *International Journal of Molecular Sciences*, *21*: 8708. https://doi.org/10.3390/ijms21228708.

Liu, R., Lai, B., Hu, B., Qin, Y., Hu, G. and Zhao, J. 2016. Identification of microRNAs and their target genes related to the accumulation of anthocyanins in *Litchi chinensis* by high-throughput sequencing and degradome analysis. *Frontiers in Plant Science*, *7*. https://doi.org/10.3389/fpls.2016.02059.

Liu, X., Chao, F., Zhang, M. and Yin, X. 2013. The MrWD40-1 gene of Chinese bayberry (*Myrica rubra*) interacts with MYB and bHLH to enhance anthocyanin accumulation. *Plant Molecular Biology Reporter*, *31*: 1474–84. https://doi.org/10.1007/s11105-013-0621-0.

Liu, Y., Qi, Y., Zhang, A., Wu, H., Liu, Z. and Ren, X. 2019. Molecular cloning and functional characterization of AcGST1, an anthocyanin-related glutathione S-transferase gene in kiwifruit (*Actinidia chinensis*). *Plant Molecular Biology*, *100*: 1–15. https://doi.org/10.1007/s11103-019-00870-6.

Lloyd, A., Brockman, A., Aguirre, L., Campbell, A., Bean, A., Cantero, A. and Gonzalez, A. 2017. Advances in the MYB-bHLH-WD repeat (MBW) pigment regulatory model: Addition of a WRKY factor and co-option of an anthocyanin MYB for betalain regulation. *Plant and Cell Physiology*, *9*. https://doi.org/10.1093/pcp/pcx075.

Luo, X., Cao, D., Zhang, J., Li, C., Xia, X., Li, H., Zhao, D., Zhang, F., Hui, X. and Chen, L. 2018. Integrated microRNA and mRNA expression profiling reveals a complex network regulating pomegranate (*Punica granatum* L.) seed hardness. *Scientific Reports*, *8*: 9292. https://doi.org/10.1038/s41598-018-27664-y.

Maier, A., Schrader, A., Kokkelink, L., Falke, C. and Hoecker, U. 2013. Light and the E3 ubiquitin ligase COP1/SPA control the protein stability of the MYB transcription factors PAP1 and PAP2 involved in anthocyanin accumulation in Arabidopsis. *The Plant Journal*, *74*: 638–51. https://doi.org/10.1111/tpj.12153.

Melgarejo, P., Nunez-Gomez, D., Legua, P., Martinez-Nicolas, J.J. and Almansa, S.M. 2020. Pomegranate (*Punica granatum* L.), a dry pericarp fruit with fleshy seeds. *Trends in Food Science and Technology*, *102*: 232–36. https://doi.org/10.1016/j.tifs.2020.02.014.

Mena, P., Garcia-Viguera, C., Navarro-Rico, J., Moreno, D.A., Bartual, J., Saura, D. and Marti, N. 2011. Phytochemical characterization for industrial use of pomegranate (*Punica granatum* L.) cultivars grown in Spain. *Journal of the Science of Food and Agriculture*, *91*: 1893–1906. https://doi.org/10.1002/jsfa.4411.

Nanavi, S.M., Samec, D., Tomczyk, M., Milella, L., Russo, D., Habtemariam, S., Suntar, I., Rastrelli, L., Daglia, M., Xiao, J., Giampieri F., Battino, M., Sobarzo-Sanchez, E., Nabavi, S.F., Yousefi, B., Jeandet, P., Xu, S. and Shirooie, S. 2020. Flavonoid biosynthetic pathways in plants: versatile targets for metabolic engineering. *Biotechnology Advances*, *38*: 107316. https://doi.org/10.1016/j.biotechadv.2018.11.005.

Ono, N.N., Britton, M.T., Fass, J.N., Nicolet, C.M., Lin, D. and Tian, L. 2011. Exploring the transcriptome landscape of pomegranate fruit peel for natural product biosynthetic gene and SSR marker discovery. *Journal of Integrative Plant Biology*, *53*: 800–13. https://doi.org/10.1111/j.1744-7909.2011.01073.x.

Pala, C.U. and Toklucu, A.K. 2011. Effect of UV-C light on anthocyanin content and other quality parameters of pomegranate juice. *Journal of Food Composition and Analysis*, *24*: 790–95. https://doi.org/10.1016/j.jfca.2011.01.003.

Paul, A., Banerjee, K., Goon, A. and Saha, S. 2018. Chemo-profiling of anthocyanins and fatty acids present in pomegranate aril and seed grown in Indian condition and its bioaccessibility study. *Journal of Food Science and Technology-Mysore*, *55*: 2488–96. https://doi.org/10.1007/s13197-018-3166-2.

Pervaiz, T., Songtao, J., Faghihi, F., Haider, M.S. and Fang, J. 2017. Naturally occurring anthocyanin, structure, functions, and biosynthetic pathway in fruit plants. *Journal of Plant Biochemistry and Physiology*, *05*: 1000187. https://doi.org/10.4172/2329-9029.1000187.

Qi, T., Song, S., Ren, Q., Wu, D., Huang, H., Chen, Y., Fan, M., Peng, W., Ren, C. and Xie, D. 2011. The Jasmonate-ZIM-Domain proteins interact with the WD-Repeat/bHLH/MYB complexes to regulate jasmonate-mediated anthocyanin accumulation and trichome initiation in *Arabidopsis thaliana*. *The Plant Cell*, *23*: 1795–1814. https://doi.org/10.1105/tpc.111.083261.

Qin, G., Xu, C., Ming, R., Tang, H., Guyot, R., Kramer, E.M., Hu, Y., Yi, X., Qi, Y., Xu X., Gao, Z., Pan, H., Jian, J., Tain, Y., Yue, Z. and Xu, Y. 2017. The pomegranate (*Punica granatum* L.) genome and the genomics of punicalagin biosynthesis. *The Plant Journal*, *91*: 1108–28. https://doi.org/10.1111/tpj.13625.

Reis, J.F., Silva Monteiro, V.V., Gomes, R.d.S., do Carmo, M.M., da Costa, G.V., Ribera, P.C. and Monteiro, M.C. 2016. Action mechanism and cardiovascular effect of anthocyanins: A systematic review of animal and human studies. *Journal of Translational Medicine*, *14*. https://doi.org/10.1186/s12967-016-1076-5.

Ricardo, P.D., José, M.E., Josselyn, S.C., Enrique, G.V. and Simón, R.L. 2016. Differential roles for VviGST1, VviGST3, and VviGST4 in proanthocyanidin and anthocyanin transport in *Vitis vinífera*. *Frontiers in Plant Science*, *7*: 1166. https://doi.org/10.3389/fpls.2016.01166.

Rouholamin, S., Zahedi, B., Nazarian-Firouzabadi, F. and Saei, A. 2015. Expression analysis of anthocyanin biosynthesis key regulatory genes involved in pomegranate (*Punica granatum* L.). *Scientia Horticulturae*, *186*: 84–88. https://doi.org/10.1016/j.scienta.2015.02.017.

Russo, M., Fanali, C., Tripodo, G., Dugo, P., Muleo, R., Dugo, L., De Gara, L. and Mondello, L. 2018. Analysis of phenolic compounds in different parts of pomegranate (*Punica granatum*) fruit by HPLC-PDA-ESI/MS and evaluation of their antioxidant activity: Application to different Italian varieties. *Analytical and Bioanalytical Chemistry*, *410*: 3507–20. https://doi.org/10.1007/s00216-018-0854-8.

Saminathan, T., Bodunrin, A., Singh, N.V., Devarajan, R., Nimmakayala, P., Jeff, M., Aradhya, M. and Reddy, U.K. 2016. Genome-wide identification of microRNAs in pomegranate (*Punica granatum* L.) by high-throughput sequencing. *BMC Plant Biology*, *16*: 122. https://doi.org/10.1186/s12870-016-0807-3.

Sarkhosh, A., Zamani, Z., Fatahi, R., Wiedow, C., Chagne, D. and Gardiner, S.E. 2012. A pomegranate (*Punica granatum* L.) linkage map based on AFLP markers. *Journal of Horticultural Science and Biotechnology*, *87*: 1–6. https://doi.org/10.1080/14620316.2012.11512821.

Schwartz, E., Tzulker, R. and Glazer, I. 2009. Environmental conditions affect the color, taste, and antioxidant capacity of 11 pomegranate accessions' fruits. *Journal of Agricultural and Food Chemistry*, *57*: 9197–9209. https://doi.org/10.1021/jf901466c.

Sentandreu, E., Cerdan-Calero, M. and Sendra, J.M. 2013. Phenolic profile characterization of pomegranate (*Punica granatum*) juice by high-performance liquid chromatography with diode array detection coupled to an electrospray ion trap mass analyzer. *Journal of Food Composition and Analysis*, *30*: 32–40. https://doi.org/10.1016/j.jfca.2013.01.003.

Sentandreu, E., Navarro, J.L. and Sendra, J.M. 2010. LC-DAD-ESI/MS^n determination of direct condensation flavanol-anthocyanin adducts in pressure extracted pomegranate (*Punica granatum* L.) juice. *Journal of Agricultural and Food Chemistry*, *58*: 10560–67. https://doi.org/10.1021/jf101978z.

Sentandreu, E., Navarro, J.L. and Sendra, J.M. 2012. Identification of new colored anthocyanin-flavanol adducts in pressure-extracted pomegranate (*Punica granatum* L.) juice by High-performance Liquid Chromatography/Electrospray Ionization Mass Spectrometry. *Food Analytical Methods*, *5*: 702–09. https://doi.org/10.1007/s12161-011-9301-6.

Shi, M. and Xie, D. 2014. Biosynthesis and metabolic engineering of anthocyanins in *Arabidopsis thaliana*. *Recent Patents on Biotechnology*, *8*: 47–60. https://doi.org/10.2174/1872208307666131218123538.

Shwartz, E., Glazer, I., Bar-Ya'akov, I., Matityahu, I., Bar-Ilan, I., Holland, D. and Amir, R. 2009. Changes in chemical constituents during the maturation and ripening of two commercially important pomegranate accessions. *Food Chemistry*, *115*: 965–73.

Singh, S.P., Pal, R.K., Saini, M.K., Singh, J., Gaikwad, N., Parashuram, S. and Kaur, C. 2019. Targeted metabolite profiling to gain chemometric insight into Indian pomegranate cultivars and elite germplasm. *Journal of the Science of Food and Agriculture*, *99*: 5073–82. https://doi.org/10.1002/jsfa.9751.

Steyn, W.J., Wand, S.J.E., Jacobs, G., Rosecrance, R.C. and Roberts, S.C. 2010. Evidence for a photoprotective function of low-temperature-induced anthocyanin accumulation in apple and pear peel. *Physiologia Plantarum*, *136*: 461–72. https://doi.org/10.1111/j.1399-3054.2009.01246.x.

Tanaka, Y. 2008. Plant pigments for coloration: Anthocyanins, betalains, and carotenoids. *Plant Journal*, *54*: 733–49. https://doi.org/10.1111/j.1365-313X.2008.03447.x.

Thomson, G.E., Turpin, S. and Goodwin, I. 2018. A review of preharvest anthocyanin development in full red and blush cultivars of European pear. *New Zealand Journal of Crop and Horticultural Science*, *46*: 81–100. https://doi.org/10.1080/01140671.2017.1351378.

Trainin, T., Harel-Beja, R., Bar-Ya'akov, I., Ben-Simhon, Z., Yahalomi, R., Borochov-Neori, H., Ophir, R., Sherman, A., Doron-Faigenboim, A. and Holland, D. 2021. Fine mapping of the "black" peel color in pomegranate (*Punica granatum* L.) strongly suggests that a mutation in the *Anthocyanidin Reductase* (ANR) gene is responsible for the traint. *Frontiers in Plant Science*, *12*: 642019. https://doi.org/10.3389/fpls.2021.642019.

Varsha, T., Chenna, S., Ashwin, N., Awadhesh, P. and Shivaprasad, P.V. 2019. miR828 and miR858 regulate VvMYB114 to promote anthocyanin and flavonol accumulation in grapes. *Journal of Experimental Botany*, *18*. https://doi.org/10.1093/jxb/erz264.

Verweij, W., Spelt, C.E. and Bliek, M. 2016. Functionally similar WRKY proteins regulate vacuolar acidification in Petunia and hair development in Arabidopsis. *The Plant Cell*, *28*. https://doi.org/10.1105/tpc.15.00608.

Wei, Y., Hu, F., Hu, G., Li, X. and Huang, X. 2011. Differential expression of anthocyanin biosynthetic genes in relation to anthocyanin accumulation in the pericarp of *Litchi Chinensis* Sonn. *PLoS One*. https://doi.org/10.1371/journal.pone.0019455.

Wu, S. and Tian, L. 2017. Diverse phytochemicals and bioactivities in the ancient fruit and modern functional food pomegranate (*Punica granatum*). *Molecules*, *22*. https://doi.org/10.3390/molecules22101606.

Wu, Y., Han, T., Lyu, L., Li, W. and Wu, W. 2023. Research progess in understanding the biosynthesis and regualtion of plant anthocyanins. Scientia Horticulturae, 321: 112374. https://doi.org/10.1016/j.scienta.2023.112374.

Xu, W., Dubos, C. and Lepiniec, L.C. 2015. Transcriptional control of flavonoid biosynthesis by MYB–bHLH–WDR complexes. *Trends in Plant Science*, *20*: 176 –85. https://doi.org/10.1016/j.tplants.2014.12.001.

Yonekura-Sakakibara, K., Nakayama, T., Yamazaki, M. and Saito, K. (2009). Modification and Stabilization of Anthocyanins. *In*: C. Winefield, K. Davies and K. Gould (Eds.), *Anthocyanins: Biosynthesis, Functions, and Applications*, 169–90. New York, NY: Springer New York.

Yuan, Z., Fang, Y., Zhang, T., Fei, Z., Han, F., Liu, C., Liu, M., Xiao, W., Zhang, W., Wu, S., Zhang, M., Ju, Y., Xu, H., Dai, H., Liu, Y., Chen, Y., Wang, L., Zhou, J., Guan, D., Yan, M., Xia, Y., Huang, X., Liu, D., Wei, H. and Zheng, H. 2018. The pomegranate (*Punica granatum* L.) genome provides insights into fruit quality and ovule developmental biology. *Plant Biotechnology Journal*, *16*: 1363–74. https://doi.org/10.1111/pbi.12875.

Zaouay, F. and Mars, M. 2014. Phenotypic variation and estimation of genetic parameters to improve fruit quality in Tunisian pomegranate (*Punica granatum* L.) accessions. *The Journal of Horticultural Science and Biotechnology*, *89*: 221–28. https://doi.org/10.1080/14620316.2014.11513072.

Zhang, X., Yuan, W., Zhao, Y., Ren, Y., Zhao, X. and Yuan, Z. 2021. Genome-wide identification and evolutionary analysis of AOMT gene family in pomegranate (*Punica granatum*). *Agronomy*, *11*: 318.https://doi.org/10.3390/agronomy11020318.

Zhao, J. and Dixon, R.A. 2010. The 'ins' and 'outs' of flavonoid transport. *Trends in Plant Science*, *15*: 72–80. https://doi.org/10.1016/j.tplants.2009.11.006.

Zhao, X., Li, B. and Yuan, Z. 2017. Cloning and expression analysis of *PgUFGT* gene from *Punica granatum* L. *Acta Botanica Boreali-Occidentalia Sinica, 37*: 646–53. https://doi.org/10.7606/j.issn.1000-4025.2017.04.0646.

Zhao, X., Yuan, Z., Fang, Y., Yin, Y. and Feng, L. 2013. Characterization and evaluation of major anthocyanins in pomegranate (*Punica granatum* L.) peel of different cultivars and their development phases. *European Food Research and Technology, 236*: 109–17. https://doi.org/10.1007/s00217-012-1869-6.

Zhao, X., Yuan, Z., Fang, Y., Yin, Y. and Feng, L. 2014. Flavonols and flavones changes in pomegranate (*Punica granatum* L.) fruit peel during fruit development. *Journal of Agricultural Science and Technology, 16*: 1649–59.

Zhao, X., Yuan, Z., Feng, L. and Fang, Y. 2015. Cloning and expression of anthocyanin biosynthetic genes in red and white pomegranate. *Journal of Plant Research, 128*: 687–96. https://doi.org/10.1007/s10265-015-0717-8.

Zhou, H., Wang, K., Wang, H., Gu, C., Dare, A.P., Espley, R.V., He, H., Allan, A.C. and Han, Y. 2015. Molecular genetics of blood-fleshed peach reveals activation of anthocyanin biosynthesis by NAC transcription factors. *The Plant Journal, 82*: 105–121. https://doi.org/10.1111/tpj.12792.

Mechanism of Pomegranate Fruit Cracking

Yuying Wang[1] and *Zhaohe Yuan*[1*]

Pomegranate (*Punica granatum* L.) is an ancient fruit that has become an emerging functional fruit due to its peel and aril, which are rich in medicinally valuable ellagitannin-based compounds with antioxidant, antibacterial and anti-inflammatory properties. Generally, when the fruit is in the ripening stage and rainfall is encountered, fruit cracking is more pronounced and pomegranate cracking limits the quality and marketability of the fruit, causing great economic losses. Fruit cracking is affected by genetic, physiological, environmental, and postharvest storage factors.

1. Introduction

In 2003, the APGII classification method classified the pomegranate family as Lythraceae, and *Punica granatum* is a deciduous shrub or small tree of the genus Punica, and is an ancient fruit tree species (Yuan et al., 2018; Kandylis and Kokkinomagoulos, 2020). Scientists rank it in the top five of the oldest cultivated fruits along with olives, grapes, dates and figs (Aslanova and Magerramov, 2012), while references to pomegranates exist in the *Koran* and the *Bible* (Teixeira et al., 2013). In many religions and cultures, the pomegranate is considered an auspicious symbol, mainly a symbol of life, luck, and fertility (Jadon et al., 2012). Originally from Iran, Afghanistan and the Caucasus and other Central Asian regions, pomegranate has been cultivated for more than 2000 years, and was introduced to China by Zhang Qian on his mission to the West in 138–125 B.C. It was one of the first fruit tree species introduced in China (Qin et al., 2017). Due to the complex and variable climate and geographical conditions, as well as the long

[1] College of Forestry, Nanjing Forestry University, Nanjing 210037, China.
* Corresponding author: zhyuan88@hotmail.com

period of natural selection and artificial domestication, famous cultivation regions and cultivation groups have been formed in China, such as Kashgar in Xinjiang, Lintong in Shaanxi, Kaifeng in Henan, Huaiyuan in Anhui, Yicheng in Shandong, Huili in Sichuan, and Mengzi in Yunnan (Long et al., 2020). Pomegranate is a valuable source of nutrients, it contains hydrolyzed tannins, concentrated tannins, flavonols, anthocyanins, phenolic acids, and organic acids. It is also characterized by a low pH (usually <4.0), high acidity (even juice citric acid up to 20 g/L) and sugar content (mainly fructose and glucose) reaching 70–180 g/L (Kandylis and Kokkinomagoulos, 2020). The peel is a source of phenolics, minerals, and complex polysaccharides, and besides water (85%), the aril also contains sugar, pectin, organic acids, phenolics, and flavonoids, mainly anthocyanins.

Pomegranate is an important economic crop in tropical and subtropical regions of the world. The past few decades have seen tremendous growth in area, production and exports around the world. But ripe fruit cracking is an important physiological disease, and the yield loss caused by this undesirable phenomenon sometimes reaches more than 65%. In addition, cracked fruit is not suitable for transportation and is prone to rot (Malhotra et al., 1983). Fruit cracking varies by variety, but in harsh climatic conditions, the problem is more severe than in pomegranates in arid areas, with losses of up to 65–75% (Prasad et al., 2003; Chandra et al., 2011). Economic losses due to fruit splitting vary from 10–40% and sometimes up to 70% (Pal et al., 2017). Fruit cracking is one of the most important physiological disorders, as it leads to a decrease in fruit marketing value of about 50% (Holland and Bar-Ya'akov, 2018). The causes associated with cracking may be improper irrigation, environmental factors, and nutritional deficiencies, especially boron, calcium, and potassium salts (Singh et al., 2020). In addition, it has been reported to be associated with high evapotranspiration, low humidity, water imbalance, and drastic diurnal temperature fluctuations during fruit growth and development. Cracking is more pronounced when the fruits experience rainfall during ripening.

The peel plays a crucial role in crack resistance, transportability, storability, and shelf-life quality. Fruit cracking is a common physiological disease that occurs when the fruit surface cracks in response to a mismatch between internal growth and the external environment. Li et al. (2018) found that the wax layer of the cuticular membrane (CM) on the fruit surface of *Ziziphus jujuba* 'Hupingzao' was cracked, and some of the cracks extended to the cuticular layer and formed larger cracks after white ripening. The fruit cracking can seriously affect the fruit appearance and quality, and the cracked fruits are susceptible to different degrees of storage diseases and fungal and bacterial infections, which can shorten the storage and shelf-life of the fruits, thus causing significant economic losses (Thomidis and Exadaktylou, 2013). For example, cherry (Correia et al., 2018), apple (Ginzberg and Stern, 2019), grape (Zhang et al., 2020), pomegranate (Hosein-Beigi et al., 2019; Singh et al., 2020), lychee (Wang et al., 2019), citrus (Cronjé et al., 2013), pear (Kwon et al., 2016), tomato (Jiang et al., 2019a; Xue et al., 2020), and watermelon (Jiang et al., 2020b) are often subject to fruit cracking. In general, fruit cracking is more pronounced when rainfall is encountered when the fruit is at the ripening stage (Thomidis and Exadaktylou, 2013; Khalil and Aly, 2013).

2. Factors Involved in Fruit Cracking

Fruit cracking is influenced by genetic, physiological, environmental, and postharvest storage factors (Khadivi-Khub, 2014), such as the genotype of the fruit (Yazici and Ercisli, 2017), the cuticle and its wax (Balbontin et al., 2014), the nature, composition, and endogenous hormones of the cell wall (Jiang et al., 2019a), the characteristics of the fruit itself (fruit size and hardness (Saei et al., 2014)), anatomical structure of the pericarp (Konarska, 2013) and biomechanics (Saei et al., 2014)), moisture (Galindo et al., 2014), and nutrients (Knoche et al., 2011), etc.

2.1 Genetic Factors

In 1981, Cuartero et al. (1981) found that fruit cracking was determined by genetic characteristics. The sensitivity of fruit cracking was different in different cultivars (Beyer et al., 2002a). Fruit cracking was not controlled by a single gene, but by multiple genes (Vaidyanathan and Harrigan, 2005). The fruit cracking traits can act on offspring through inheritance, and the degree of cracking varies greatly between different varieties (Correia et al., 2018). Polat et al. (2012) observed that the pomegranate fruit cracking rate in an early seedless variety is lower than two late varieties Hicaz and Katırbaş. This result is consistent with that of Yazici and Ercişli (2017). Yamaguchi et al. (2002) investigated 37 sweet cherry varieties for fruit cracking and found that fruit cracking varied significantly among varieties and was positively correlated with fruit quality and flesh hardness, and Simon et al. (2004) identified that sweet cherries with different fruit quality had different rates of fruit cracking. Greco et al. (2008) analyzed crack susceptibility of 30 sweet cherry varieties in Italy and confirmed that crack susceptibility was genotype dependent.

Quantitative trait loci (QTL) methods play an important role in elucidating the genetic basis of fruit dehiscence. In Capel et al. (2017), QTL controlling fruit dehiscence were identified by constructing a genetic map of a tomato recombinant inbred line (RIL) population. Zhang et al. (2020) localized 40 QTL for fruit quality traits using a new linkage map based on enzymatic amplified polymorphic sequences, and QTL for fruit quality traits were mainly concentrated near LG6 and LG9, and two QTL were found to be associated with exocarp thickness, peel hardness, and dehiscence (ET7.1, ET11.1), one (EPFI3.1), three (FCR4.1, FCR6.1, FCR9.1).

The cuticle consists of keratin and wax, CM covers the stems, leaves, flowers and fruits of plants and is a natural waterproof barrier that protects plant tissues from stress injury, invasive pathogens and insect herbivores (Dominguez et al., 2011; Lewandowska et al., 2020; Qiao et al., 2020), CM plays a major role in fruit cracking (Peschel et al., 2007). The main components of keratin are mainly oxidized fatty acids rich in carbon chains, 16 or 18 (C16 or C18) in length, and polyesters of propylene glycol, while the wax components are mainly very long-chain saturated fatty acids, derivatives (alkanes, aldehydes, alcohols, esters, etc.), flavonoids and triterpenes (Fich et al., 2016; Samuels et al., 2008). It was found that waxes and keratin are deposited in the stratum corneum during growth and "fix" the strain in the stratum corneum (Knoche and Lang, 2017). Several genes related to cuticle formation and regulation have been identified in Arabidopsis, tomato, apple, and cherry (Lee and Suh, 2015; Shi et al., 2013; Lashbrooke et al., 2015; Alkio et al., 2012). Related studies found that some proteins (BODYGUARD (BDG)), enzymes (Glycerol-3-Phosphate Acyltransferase, GPAT6), etc. play an important role in cuticle synthesis (Jakobson et al., 2016; Petit et al., 2016), in addition to fruit cuticles are regulated during development by their composition to polysaccharide ratio, etc. (Espana et al., 2014). Alkio et al. (2014) sequenced and annotated the exocarp of cherries and annotated cuticle and wax synthesis in biological processes. Rios et al. (2015) used gas chromatography-mass spectrometry (GC-MS) and found that varieties with higher concentrations of C29 alkanes (Kordia, Regina, and Lapins) were less prone to cracking and that a higher proportion of C29 alkanes in cherry epidermal waxes played an important role in fruit cracking resistance. Konarska et al. (2013) found that apples with fewer crystalline waxes were prone to cracking. Zoffoli et al. (2008) found by metabolomics and transcriptomics that the Havana pepper genotype PI 257145 had much higher keratin content than PI 224448.

SHINE branch transcription factors act on epidermal patterning and fruit cuticle formation (Shi et al., 2011). Zhu et al. (2020) performed peel transcriptome sequencing analysis of the grape variety 'Xiangfei', which is prone to dehiscence, and found that genes related to cell wall metabolism and cuticle biosynthesis play an important role in dehiscence.

2.2 Physiological Factors

Fruit characteristics such as fruit shape, size, and hardness can affect the occurrence of fruit cracking. The physiological metabolism inside the fruit is closely related to the growth and development of the fruit, and studies have shown that fruit soluble sugar, soluble solids, pectin, enzymes and endogenous hormones have a close relationship with fruit cracking (Chen et al., 2019; Khadivi-Khub, 2014). Sugar is an important component of fruit and the material basis of its growth and development. Li et al. (2020) found that the crack resistance of varieties was related to the content of reducing sugar, soluble total sugar and cellulose, but not to the starch content. The reducing sugar content and cellulose content of the crack-resistant varieties ('Muzao' and 'Xiangzao') were higher than those of the crack-susceptible varieties ('Goutouzao', 'Junzao', and 'Tuanzao'). Total soluble sugars of crack-resistant varieties were lower than those of crack-susceptible varieties (Li et al., 2020).

2.3 Environmental Factors

Water is the main environmental factor causing fruit cracking, and both soil water and water on the fruit surfaces are important factors inducing fruit cracking (Beyer et al., 2002b). Exposing the fruit surface to liquid water or high-water vapor will lead to the formation of cutin membrane microcracks. Studies have shown that pomegranate fruits are sensitive to water deficit at the end of the fruit growth and ripening period. When rainfall affects previously water-stressed pomegranates, an asymmetric increase in pericarp swelling pressure occurs because aril swelling increases much more than pericarp swelling. The increase in aril pressure stresses the peel and makes it susceptible to rupture (Galindo et al., 2014; Peña et al., 2013). In addition, improper irrigation during fruit ripening, the occurrence of sunburn, and mineral deficiencies can also cause fruit cracking, and Yazici and Özgüven (2009) showed that sunburned fruits have a higher probability of fruit cracking. Excessive temperature differences can lead to the accumulation of carbohydrates, reducing the plants osmotic potential and allowing them to absorb more water, grow faster, and crack more easily (Wang and Camp, 2000).

2.4 Postharvest Storage Factors

During postharvest storage, the structure of lenticels, microcracks, wax patterns, and pericarp tissue components changed in fruits, which affected fruit quality (Lufu et al., 2021). Therefore, it is particularly important to carry out postharvest treatments on fruits. In the U.S. Pacific Northwest (PNW), sweet cherries are hydrocooled as soon as possible after harvest and shipped in cold flume water during packing to reduce respiration rates and extend storage/shipping life (Wang et al., 2015). Wang et al. (2015) revealed that hydrocooling cherry fruit in appropriate $CaCl_2$ solutions (i.e., 0.2-0.5%) for 5 min and then passing the fruit in cold flume water for 15 min increased fruit firmness, peel color, and reduced cracking and decay following 4 weeks of cold storage.

3. Mechanism of Pomegranate Fruit Cracking

Saei et al. (2014) compared the sensitive cultivar 'Malas-e-Saveh' with the 'Yusef-Khani' cultivar with strong crack resistance, and the peel thickness was significantly different between the two varieties, but not significantly different from cracking fruit relation. Fruit dehiscence was significantly positively correlated with fruit volume, fruit shape, and seed weight ratio. The failure strain of 'Yusef-Khani' was significantly higher than that of 'Malas-e-Saveh'. 'Malas-e Saveh' has higher secant modulus and toughness than 'Yousef-Khani'. A biological yield point was observed in 'Malas-e-Saveh', which is a major difference in biomechanical behavior between resistant and susceptible varieties. Galindo et al. (2014) found that pomegranate fruit is sensitive to water deficit at the end of fruit growth and ripening period. During these phenological stages, water may enter the fruit through the phloem rather than the xylem. Although leaf swelling was maintained under very severe water stress, fruit swelling was lost, resulting in reduced fruit swelling. When rainfall affected previously water-stressed pomegranate plants, an asymmetric increase in pericarp expansion pressure occurred because aril expansion was much greater than pericarp expansion. The increased pressure on the aril can stress the peel and make it prone to rupture. Subsequently, Hosein-Beigi et al. (2019) investigated the effect of calcium (Ca) (0, 0.75% and 1.5%), boron (B) (0 and 3000 ppm) and gibberellin (GA3) (0, 75 and 150 ppm) on the individual and combined effects of 'Malase-Torshe-Saveh' pomegranate fruit; all treatments improved the physical properties of pomegranate fruit, but GA3 treatment had the best effect, with the highest fruit length, fruit diameter, fruit weight per 100 seeds, and yield per plant. In addition, application of boron not only improved fruit size, weight, and yield, but also reduced fruit splitting mainly by increasing the water content of the peel. Ca treatment was the most effective in improving fruit quality traits including TSS and reducing fruit dehiscence. The combined spraying of Ca, B, and GA_3 was more effective than spraying alone, with the highest fruit yield and quality, and the least fruit cracking and sunburn symptoms, indicating that there was a certain synergistic effect between these three compounds.

In addition, the incidence of fruit cracking was 7.26% under the bagging treatment, the control group was 38.11%, indicating that bagging considerably diminished the fruit cracking rate. LncRNA libraries for fruit cracking (FC), fruit non-cracking (FNC), and fruit non-cracking under bagging (FB) in pomegranate were performed and analyzed via high-throughput transcriptome sequencing. A total of 3,194 lncRNAs were obtained with a total length of 4,898,846 nt and an average length of 1,533.77 nt in pomegranate. We identified 42 differentially expressed lncRNAs (DELs) and 137 differentially expressed mRNAs (DEGs) in FC vs FNC and 35 DELs and 160 DEGs in FB vs FC that formed co-expression networks respectively, suggesting that they are involved in phytohormone signaling pathway, lignin catabolic process, lipid transport/binding, cutin biosynthetic process, and cell wall organization. We also found that 18 *cis*-acting DELs regulated 18 target genes, and 10 *trans*-acting DELs regulated 24 target genes in FC vs FNC, 23 DELs regulate 23 target genes for the cis-acting lncRNAs and 12 DELs regulated 36 target genes in FB vs FC, which provides an understanding for the regulation of the fruit cracking. Gene Ontology (GO) and Kyoto Encyclopedia of Genes and Genomes (KEGG) analysis results demonstrated that DELs participated in calcium ion binding, glycerophospholipid metabolism, flavonoid biosynthetic process, cell wall biogenesis, xyloglucan metabolic process, hormone signal transduction, and starch and sucrose metabolism (Wang et al., 2022). Those findings provide new insights into the roles of lncRNAs in regulating the fruit cracking and lay the foundation for further improvement of pomegranate quality.

References

Alkio, M., Jonas, U., Declercq, M., Van Nocker, S. and Knoche, M. 2014. Transcriptional dynamics of the developing sweet cherry (*Prunus avium* L.) fruit: Sequencing, annotation, and expression profiling of exocarp-associated genes. *Hortic. Res.*, *1*: 11.

Alkio, M., Jonas, U., Sprink, T., van Nocker, S. and Knoche, M. 2012. Identification of putative candidate genes involved in cuticle formation in *Prunus avium* (sweet cherry) fruit. *Ann. Bot.*, *110*(1): 101–12.

Aslanova, M.S. and Magerramov, M.A. 2012. Physicochemical parameters and amino acid composition of new pomological sorts of pomegranate fruits. *Chem. Plant Raw. Mater.*, *1*: 165–69.

Balbontin, C., Ayala, H., Rubilar, J., Cote, J. and Figueroa, C.R. 2014. Transcriptional analysis of cell wall and cuticle related genes during fruit development of two sweet cherry cultivars with contrasting levels of cracking tolerance. *Chil. J. Agr. Res.*, *74*(2): 162–69.

Beyer, M., Hahn, R., Peschel, S., Harz, M. and Knoche, M. 2002a. Analyzing fruit shape in sweet cherry (*Prunus avium* L.). *Sci. Hortic.*, *96*: 139–50.

Beyer, M., Peschel, S., Knoche, M. and Knrgen, M. 2002b. Studies on water transport through the sweet cherry fruit surface, IV. Regions of preferential uptake. *HortScience*, *37*: 637–41.

Capel, C., Yuste-Lisbona, F., Lopez-Casado, G., Angosto, T., Cuartero, J., Lozano, R. and Capel, J. 2017. Multi-environment QTL mapping reveals genetic architecture of fruit cracking in a tomato RIL Solanum lycopersicum x S-pimpinellifolium population. *Theor. Appl. Genet.*, *130*(1): 213–22.

Chandra, R.S., Suroshe, J., Sharma, R.A. and Meshram, D.T. 2011. *Pomegranate Growing Manual*, 1–58. (Run on)NRC on Pomegranate, Solapur.

Chen, J., Duan, Y., Hu, Y., Li, W., Sun, D., Hu, H. and Xie, J. 2019. Transcriptome analysis of atemoya pericarp elucidates the role of polysaccharide metabolism in fruit ripening and cracking after harvest. *BMC Plant Biol.*, *19*: 219.

Correia, S., Schouten, R., Silva, A.P. and Gonçalves, B. 2018. Sweet cherry fruit cracking mechanisms and prevention strategies: A review. *Sci. Hortic.*, *240*: 369–77.

Cronjé, P.J.R., Stander, O.P.J. and Theron, K.I. 2013. Fruit splitting in Citrus. *Horticultural Reviews*, *41*: 177–200.

Cuartero, J., Palomares, G., Balasch, S. and Nuez, F. 1981. Tomato fruit cracking under plastic-house and in the open air. II. General and specific combining abilities. *BMC Infect. Dis.*, *15*: 1–13.

Dominguez, E., Heredia-Guerrero, J.A. and Heredia, A. 2011. The biophysical design of plant cuticles: An overview. *New Phytol.*, *189*(4): 938–49.

Espana, L., Heredia-Guerrero, J.A., Segado, P., Benitez, J.J., Heredia, A. and Dominguez, E. 2014. Biomechanical properties of the tomato (*Solanum lycopersicum*) fruit cuticle during development are modulated by changes in the relative amounts of its components. *New Phytol.*, *202*(3): 790–802.

Fich, E.A., Segerson, N.A. and Rose, J.K. 2016. The Plant polyester cutin: Biosynthesis, structure, and biological roles. *Annu. Rev. Plant Biol.*, *67*: 207–33.

Galindo, A., Rodríguez, P., Collado-González, J., Cruz, Z.N., Torrecillas, E., Ondoño, S., Corell, M., Moriana, A. and Torrecillas, A. 2014. Rainfall intensifies fruit peel cracking in water stressed pomegranate trees. *Agric. Forest Meteorol.*, *194*: 29–35.

Ginzberg I. and Stern, R.A. 2019. Control of fruit cracking by shaping skin traits: Apple as a model. *Crit. Rev. Plant Sci.*, *38*(5): 401–10.

Greco, P., Palasciano, M., Mariani, R., Pacifico, A. and Godini, A. 2008. Susceptibility to cracking of thirty sweet cherry cultivars. *Acta Hortic.*, 379–82.

Holland, D. and Bar-Ya'akov, I. 2018. Pomegranate (*Punica Granatum* L.) Breeding. *In*: *Advances in Plant Breeding Strategies: Fruits*, 601–47. Springer International Publishing.

Hosein-Beigi, M., Zarei, A., Rostaminia, M. and Erfani-Moghadam, J. 2019. Positive effects of foliar application of Ca, B and GA_3 on the qualitative and quantitative traits of pomegranate (*Punica granatum* L.) cv. "Malase-Torshe-Saveh". *Sci. Hortic.*, *254*: 40–47.

Jadon, G., Nainwani, R., Singh, D., Soni, P.K. and Diwaker, A.K. 2012. Antioxidant activity of various parts of *Punica Granatum*: A review. *Journal of Drug Delivery and Therapeutics*, *2*(6): 138–41.

Jakobson, L., Lindgren, L.O., Verdier, G., Laanemets, K., Brosche, M., Beisson, F. and Kollist, H. 2016. BODYGUARD is required for the biosynthesis of cutin in Arabidopsis. *New Phytol.*, *211*(2): 614–26.

Jiang, F., Lopez, A., Jeon, S., de Freitas, S.T., Yu, Q., Wu, Z., Labavitch, J.M., Tian, S., Powell, A.L.T. and Mitcham, E. 2019a. Disassembly of the fruit cell wall by the ripening-associated polygalacturonase and expansin influences tomato cracking. *Hortic. Res.*, *6*: 17.

Jiang, H.K., Tian, H.M., Yan, C.S., Jia, L., Wang, Y., Wang, M.X., Jiang, C.J., Li, Y.Y., Jiang, J.Y. and Fang, L. 2019b. RNA-seq analysis of watermelon (*Citrullus lanatus*) to identify genes involved in fruit cracking. *Sci. Hortic.*, *248*: 248–55.

Kandylis, P. and Kokkinomagoulos, E. 2020. Food applications and potential health benefits of pomegranate and its derivatives. *Foods*, *9*(2): 122.

Khadivi-Khub, A. 2014. Physiological and genetic factors influencing fruit cracking. *Acta Physiol. Plant*, *37*(1): 1718.

Khalil, H. and Aly, S. 2013. Cracking and fruit quality of pomegranate (*Punica granatum* L.) as affected by pre-harvest sprays of some growth regulators and mineral nutrients. *J. Hortic. Sci. Ornam. Plants*, *5*: 71–76.

Knoch,e M., Khanal, B.P. and Stopar, M. 2011. Russeting and microcracking of "Golden Delicious" apple fruit concomitantly decline due to gibberellin A4+7 application. *Journal of the American Society for Horticultural Science*, *136*(3): 159–64.

Knoche, M. and Lang, A. 2017. Ongoing growth challenges fruit skin integrity. *Crit. Rev. Plant Sci.*, *36*(3): 190–215.

Konarska, A. 2013. The structure of the fruit peel in two varieties of *Malus domestica* Borkh. (Rosaceae) before and after storage. *Protoplasma*, *250*(3): 701–14.

Kwon, Y., Han, H.H. and Park, H.S. 2016. The characteristics of cork and hypodermis tissues and cracking in Asian pear (*Pyrus pyrifolia* cv. Mansoo). *Sci. Hortic.*, *199*: 224–728.

Lashbrooke, J., Aharoni, A. and Costa, F. 2015. Genome investigation suggests *MdSHN3*, an APETALA2-domain transcription factor gene, to be a positive regulator of apple fruit cuticle formation and an inhibitor of russet development. *J. Exp. Bot.*, *66*(21): 6579–89.

Lee, S.B. and Suh, M.C. 2015. Advances in the understanding of cuticular waxes in Arabidopsis thaliana and crop species. *Plant Cell Rep.*, *34*(4): 557–72.

Lewandowska, M., Keyl, A. and Feussner, I. 2020. Wax biosynthesis in response to danger: its regulation upon abiotic and biotic stress. *New Phytol.*, *227*(3): 698–13.

Li, N., Song, Y.Q., Li, J., Chen, Y.Y., Xue, X.F. and Li, L.L. 2018. Development of the cuticular membrane and biomechanical properties in Hupingzao (*Ziziphus jujuba* Mill. 'Hupingzao'). *Sci. Hortic.*, *229*: 25–32.

Li, X., Chen, X., Wen, X., Zhao, Y. and Ma, H. 2020. Correlation analysis between the sugar components and fruit cracking in easily cracked and resistant Jujube. *Mol. Plant Breed.*, *18*: 6180–86.

Long, X., Feng, D., Yuan, Z., Lin, S. and Yan, C. 2020. *Modern Fruit Tree Cultivation in China*. Beijing.

Lufu, R., Ambaw, A. and Opara, U.L. 2021. Functional characterization of lenticels, micro-cracks, wax patterns, peel tissue fractions, and water loss of pomegranate fruit (cv. 'Wonderful') during storage. *Postharvest Biol. Technol.*, *178*: 111539.

Malhotra, V.K., Khajuria, H.N. and Jawanda, J.S. 1983. Studies on physico-chemical characteristics. 1-Physical characteristics. *Punjab Hort.* J., *23*: 153–57.

Pal, R.K., Singh, N.V. and Maity A. 2017. Pomegranate fruit cracking in dryland farming. *Curr. Sci.*, *112*: 896–97.

Peña, M.E., Artés-Hernández, F., Aguayo, E., Martínez-Hernández, G.B. and Gómez, P.A. 2013. Effect of sustained deficit irrigation on physicochemical properties, bioactive compounds and postharvest life of pomegranate fruit (cv. "Mollar de Elche"). *Postharvest Biol. Technol.*, *86*: 171–80.

Peschel, S., Franke, R., Schreiber, L. and Knoche, M. 2007. Composition of the cuticle of developing sweet cherry fruit. *Phytochemistry*, *68*(7): 1017–25.

Petit, J., Bres, C., Mauxion, J.P., Tai, F.W., Martin, L.B., Fich, E.A., Joubes, J., Rose, J.K., Domergue, F. and Rothan, C. 2016. The Glycerol-3-Phosphate Acyltransferase GPAT6 from tomato plays a central role in fruit cutin biosynthesis. *Plant Physiol.*, *171*(2): 894–913.

Polat, A.A., Caliskan, O. and Kamiloglu, O. 2012. Determination of pomological characteristics of some pomegranate cultivars in Drtyol (Turkey) conditions. *Acta Hortic.*, 401–505.

Prasad, R.N., Bankar, G.J. and Vashishtha, B.B. 2003. Effect of drip irrigation on growth, yield and quality of pomegranate in arid region. *Indian J. Hort.*, *60*: 140–42.

Qiao, P., Bourgault, R., Mohammadi, M., Matschi, S., Philippe, G., Smith, L.G., Gore, M.A., Molina, I. and Scanlon, M.J. 2020. Transcriptomic network analyses shed light on the regulation of cuticle development in maize leaves. *Proc. Natl. Acad. Sci. U.S.A.*, *117*(22): 12464–71.

Qin, G., Xu, C., Ming, R., Tang, H., Guyot, R., Kramer, E.M., Hu, Y., Yi, X., Qi, Y. and Xu, X. 2017. The pomegranate (*Punica granatum* L.) genome and the genomics of punicalagin biosynthesis. *Plant J.*, *91*(6): 1108–28.

Rios, J.C., Robledo, F., Schreiber, L., Zeisler, V., Lang, E., Carrasco, B. and Silva, H. 2015. Association between the concentration of n-alkanes and tolerance to cracking in commercial varieties of sweet cherry fruits. *Sci. Hortic.*, *197*: 57–65.

Saei, H., Sharifani, M.M., Dehghani, A., Seifi, E. and Akbarpour, V. 2014. Description of biomechanical forces and physiological parameters of fruit cracking in pomegranate. *Sci. Hortic.*, *178*: 224–30.

Samuels, A., Kunst, L. and Jetter, R. 2008. Sealing plant surfaces: Cuticular wax formation by epidermal cells. *Annu. Rev. Plant Biol.*, *59*: 683–707.

Shi, J.X., Adato, A., Alkan, N., He, Y., Lashbrooke, J., Matas, A.J., Meir, S., Malitsky, S., Isaacson, T. and Prusky, D. 2013. The tomato SlSHINE3 transcription factor regulates fruit cuticle formation and epidermal patterning. *New Phytol.*, *197*(2): 468–80.

Shi, J.X., Malitsky, S., De Oliveira, S., Branigan, C., Franke, R.B., Schreiber, L. and Aharoni, A. 2011. SHINE transcription factors act redundantly to pattern the archetypal surface of Arabidopsis flower organs. *PLoS Genet.*, *7*(5): e1001388.

Simon, G., Hrotkó, K. and Magyar, L. 2004. Fruit Quality of Sweet Cherry Cultivars Grafted on Four Different Rootstocks. *Int. J. Hortic. Sci.*, *10*: 365–70.

Singh, A., Shukla, A.K. and Meghwal, P.R. 2020. Fruit Cracking in Pomegranate: Extent, Cause, and Management: A Review. *Int. J. Fruit Sci.*, 1–20.

Teixeira, da Silva J.A., Rana, T.S., Narzary, D., Verma, N., Meshram, D.T. and Ranade, S.A. 2013. Pomegranate biology and biotechnology: A review. *Sci. Hortic.*, *160*: 85–107.

Thomidis, T. and Exadaktylou, E. 2013. Effect of a plastic rain shield on fruit cracking and cherry diseases in Greek orchards. *Crop. Prot.*, *52*: 125–29.

Vaidyanathan, S. and Harrigan, G. 2005. *Metabolome Analyses: Strategies for Systems Biology*. Springer: Berlin/Heidelberg, Germany.

Wang, J., Gao, X., Ma, Z., Chen, J. and Liu, Y. 2019. Analysis of the molecular basis of fruit cracking susceptibility in Litchi chinensis cv. Baitangying by transcriptome and quantitative proteome profiling. *J. Plant Physiol.*, 234–35.

Wang, S.Y. and Camp, M.J. 2000. Temperatures after bloom affect plant growth and fruit quality of strawberry. *Sci. Hortic.*, *85*: 183–99.

Wang, Y. and Long, L.E. 2015. Physiological and biochemical changes relating to postharvest splitting of sweet cherries affected by calcium application in hydrocooling water. *Food Chem.*, *181*: 241–47.

Wang, Y., Zhao, Y., Wu, Y., Zhao, X., Hao, Z., Luo, H. and Yuan, Z. 2022. Transcriptional profiling of long non-coding RNAs regulating fruit cracking in *Punica granatum* L. under bagging. *Front. Plant Sci.*, *13*: 943547.

Xue, L., Sun, M., Wu, Z., Yu, L., Yu, Q., Tang, Y. and Jiang, F. 2020. LncRNA regulates tomato fruit cracking by coordinating gene expression via a hormone-redox-cell wall network. *BMC Plant Biol.*, *20*(1): 162.

Yamaguchi, M., Sato, I. and Ishiguro, M. 2002. Influences of Epidermal Cell Sizes and Flesh Firmness on Cracking Susceptibility in Sweet Cherry (*Prunus avium* L.) Cultivars and Selections. *J. Jpn. Soc. Hortic. Sci.*, *71*: 738–46.

Yazici, K. and Ercisli, S. 2017. Characterization of hybrid pomegranate genotypes based on sunburn and cracking traits related to maturation time. *J. Appl. Bot. Food Qual.*, *90*: 132–39.

Yılmaz, C. and Özgüven, A.I. 2009. The effects of some plant nutrients, gibberellic acid, and pinolene treatments on the yield, fruit quality, and cracking in pomegranate. *Acta Hortic.*, *818*: 205–12.

Yuan, Z., Fang, Y., Zhang, T., Fei, Z., Han, F., Liu, C., Liu, M., Xiao, W., Zhang, W. and Wu, S. 2018. The pomegranate (*Punica granatum* L.) genome provides insights into fruit quality and ovule developmental biology. *Plant Biotechnol. J.*, *16*(7): 1363–74.

Zhang, C., Guan, L., Fan, X., Zheng, T., Dong, T., Liu, C. and Fang, J. 2020. Anatomical characteristics associated with different degrees of berry cracking in grapes. *Sci. Hortic.*, *261*: 108992 .

Zhang, T., Ding, Z., Liu, J., Qiu, B. and Gao, P. 2020. QTL mapping of pericarp and fruit-related traits in melon (*Cucumis melo* L.) using SNP-derived CAPS markers. *Sci. Hortic.*, *265*: 109243.

Zhu, M., Yu, J., Zhao, M., Wang, M. and Yang, G. 2020. Transcriptome analysis of metabolisms related to fruit cracking during ripening of a cracking-susceptible grape berry cv. Xiangfei (*Vitis vinifera* L.). *Genes and Genomics*, *42*(6): 639–50.

Zoffoli, J.P., Latorre, B.A. and Naranjo, P. 2008. Hairline, a postharvest cracking disorder in table grapes induced by sulfur dioxide. *Postharvest Biol. Tec.*, *47*(1): 90–97.

10

Elucidating the Molecular Basis of Pomegranate Salt Tolerance

Cuiyu Liu[1] and *Zhaohe Yuan*[2*]

Salinity affects the crucial processes involved in pomegranate growth and development, such as photosynthesis, protein synthesis, energy conversion, and ion balance. Pomegranate is considered to be moderately tolerant to salinity, and it can be grown in arid and semiarid regions. The salt-tolerance of plants is a trait that contributed to regulations of multiple genes. This chapter begins with an introduction to the transcriptome of pomegranate under salt stress, and then discovers the functional genes coding for abscisic acid (ABA) receptors, Ca^{2+}-sensors, heat shock proteins (HSPs), late embryogenesis abundant proteins (LEAs), aquaporins (AQPs) and peroxidases (PODs), high-affinity potassium transporters (HKTs), and different expression transcript factors (TFs) that responded to salinity. Finally, the typical signal transduction pathways associated with responses to salinity: ABA signaling pathway and Ca^{2+}-dependent signaling pathway that participate in sensing and transporting signals in cell are presented. This chapter covers current knowledge and advances in our understanding of the molecular mechanism of salt tolerance in pomegranate.

1. Introduction

Salt stress affects growth, development, and productivity of pomegranates. Researchers reported that seven-year-old 'Manfalouty', 'Nab-Elgamal', and

1 Research Institute of Subtropical Forestry, Chinese Academy of Forestry. Hangzhou, 311400, China.
2 College of Forestry, Nanjing Forestry University. Nanjing, 210037, China.
* Corresponding author: zhyuan88@hotmail.com

'Wonderful' pomegranate had higher reductions in growth, flowering, and yield after irrigating with saline water at an EC of 6.0 dS·m^{-1} than at an EC of 1.8 dS·m^{-1} (El-Khawaga et al., 2012). Also saline water (40, 80, or 120 mM NaCl) had reduced stem length, internode length and number, and leaf surface of "Malas Torsh" and "Alak Torsh" pomegranate when compared to control (Naeini et al., 2006). Reductions were also observed in leaf numbers, dry weight, flowering and fruit yield in pomegranate 'Manfalouty' and 'Nab-Elgamal' after irrigation with saline water (El-Agamy et al., 2010). The relative water content, electrical conductivity, stomatal conductance, chlorophyll content, and net photosynthetic rate of pomegranate leaves decreased significantly with increased levels of soil salinity (Sun et al., 2018). Considerable researchers have studied on pomegranate physiological and biochemical responses to salt stress, especially on its growth, ion balance, osmoregulation, and the scavenging of reactive oxygen species (ROS) (Karimi and Hasanpour 2014; Liu et al., 2018; Okhovatianardakani et al., 2010). However, there are only a few studies focused on the molecular mechanisms of pomegranate salt tolerance.

The salt-tolerance of plants is a trait that contributed to regulations of multiple genes. These multiple genes are identified and categorized into two types according to the functions of proteins (Munns and Tester, 2008). The first type is effector, which code for the functional proteins that directly involved in the physiological and biochemical responses to salt stress in plants. These effectors include the superoxide dismutase (SOD) (Attia et al., 2011), ascorbate peroxidase (APX) (Shafi et al., 2015), high-affinity potassium transporter (HKT) (Kumar et al., 2017), Na^+/H^+ antiporter (NHX) (Yokoi et al., 2010), aquaporin (AQP) (Hu et al., 2012), late embryogenesis abundant (LEA) (Duan and Cai, 2012), and H^+-ATPase (VHA) (Zhou et al., 2018), etc. The second type is a regulator that is involved in regulating the expressions of genes and the signal transduction pathways, such as transcription factors (TFs) and various kinases (Hasegawa et al., 2000). These well-characterized TFs include Apetala2/ethylene response factor (AP2/ERF), dehydration-responsive element-binding protein/C-repeat binding factor (DREB/CBF), WRKY, NAC, basic leucine zipper (bZIP), MYB, and basic helix-loop-helix (bHLH) family genes (Munns and Tester, 2008). These genes regulate the expressions of downstream genes via various ways, then may influence the plants salt-tolerance eventually (Deinlein et al., 2014).

Studies on pomegranate salt tolerance and the response mechanism to salinity, as well as mining for salt tolerance-related genes have become scientific problems urgently needed to be solved. Therefore, the key genes, cell components, metabolic pathways, and signal transduction pathways involved in pomegranate response to salinity must be addressed. The molecular basis salt tolerance and the key salt-tolerant genes will lay a foundation for molecular breeding and functional research in pomegranate.

2. Transcriptome of Pomegranate Tissues under Salt Stress

To elucidate the molecular responses to salt stress at mRNA level, 18 cDNA libraries of pomegranate roots and leaves from 0 day (controls, T0), 3 days (T1), and 6 days (T2) after 200 mM NaCl treatment were established. Using transcriptomic analysis, a global representation of potential candidate genes under salt stress was demonstrated. In total, 34,047 genes by mapping to genome were obtained, and then 2,255 DEGs (differentially expressed genes) were identified, including 1,080 upregulated and 1,175 downregulated genes, with little overlap of both up- and down-regulated genes between roots and leaves. Numbers of DEGs increased in pomegranate roots with the duration of salt-stress, which were opposite in leaves. The majority of DEGs were exclusively upregulated or downregulated in either tissue. Almost 85.9% of upregulated and 53.4% of downregulated DEGs were active in roots after 6-days salt stress, most of downregulated genes after 3 days and then recovered after 6 days in leaves, and only 22 (2.0%) genes were suppressed in leaves at two points of time.

Under salt stress, plants firstly experience osmotic stress due to a disorder of water uptake. Then a process of gradual recovery due to the partially or completely re-established uptake of water occurs within days after the salt treatment (Munns, 2005; Munns et al., 2006). More and more salt-related genes were upregulated in roots, while most suppressed genes recovered with salt-treating process in leaves. In contrast, the osmotic adjustment of leaves started only after roots had reached new water equilibrium (Monika et al., 2010). These tissue-specific and time-specific differences between roots and leaves have also been reported in *Arabidopsis thaliana* (Ma et al., 2006), *Populus euphratica* (Monika et al., 2010), and *Millettia pinnata* (Huang et al., 2012).

3. Gene Expressional Patterns under Salt Stress

All DEGs from pomegranate transcriptome were used to analyze the expressional patterns by STEM. 82.6% of DEGs in roots enriched into two upregulated patterns, and two downregulated patterns. 57.9% of DEGs in leaves were also clustered into two profiles, including one, first downregulated and then the upregulated pattern and one downregulated pattern. GO-term enrichment analysis of DEGs within upregulated patterns showed that the top subcategories of roots were involved in proteolysis and metabolic process, but that in cell wall organization, transmembrane transport, and oxidation-reduction process they were downregulated. Otherwise, most DEGs involved in oxidation-reduction process and ion transport in leaves were suppressed.

Metal ions/cations such as K^+, Ca^{2+}, Mg^{2+}, Fe^{2+}, Cu^{2+}, and Zn^{2+} act as cofactors participating in the catalytic activity of the bound enzymes. The upregulated DEGs involved in metal ion/cation binding might accelerate the catalytic activity. Genes were continuously upregulated with the extension of stress. The over-expressed genes in roots were enriched in proteolysis and metabolic process, carbohydrate metabolic process, and catabolic process, etc. The carbohydrates such as glycan, starch, sucrose, amino sugar have been demonstrated as osmolytes and energy resources in plant responses to salt stress, especially in halophyte species (Wang et al., 2013). The accelerated signal transduction, amino acid, and carbohydrate metabolisms in pomegranate can contribute to its adaption responses to salt stress. Similar expression patterns have been observed in *Oryza sativa* (Boriboonkaset et al., 2013), *Thellungiella halophila* (Wang et al., 2013) and *Helianthus tuberosus* (Zhang et al., 2018) under salt stress.

The top subcategories of the downregulated pattern were lipid biosynthetic process (GO:0008610), cell wall organization (GO:0071555), and external encapsulating structure organization (GO:0045229). The cell wall plays an important role in protecting the plant from salt toxicity (Le Gall et al., 2015). Many DEGs coding cell wall components were mostly suppressed in roots, including four extensions, two pectin acetylesterase, three polygalacturonates, and three xyloglucan endotransglucosylase/hydrolase proteins. These proteins are important components of the cell wall, which affect plant growth through mediating the cell enlargement and expansion (Le Gall et al., 2015).

Also, transmembrane transport (GO:0055085) and oxidation-reduction process (GO:0055114) were inhibited under salt stress. Many genes were identified in these biological processes, such as adenine/guanine permease, aquaporin, cationic amino acid transporter, sugar carrier, and zinc transporter. They work together to transport coenzymes, amino acids, carbohydrates, and ions under salt stress. The suppressed genes coding proteins, such as dehydrogenase, cytochrome P450s, and peroxidase involved in oxidation-reduction process, play essential roles in responding to salinity (Bushman et al., 2016; Fatehi et al., 2012). These results indicated that salinity affected pomegranate growth and development through inhibiting the cell division and transmembrane transport, as well as slowing down the redox reactions.

In total, the ion transport process was inhibited in roots and leaves. The restriction of ion transport indicate that the excess of Na^+ inhibited the uptake of mineral ions (Munns and Tester, 2008). The results correspond to the physiological responses in other pomegranate cultivars under salt stress (Sun et al., 2018). Pomegranate may cope with the toxic ions to some extent via decreasing uptake of ions from soil and increasing utilization of ions in cells. The toxic ions access into the leaf cells with the transpiration, but the detrimental effect is a slow process taking weeks or months, eventually causing salt toxicity in leaves (Aroca et al., 2011).

4. ABA Signaling Pathway

ABA is a critical hormone, which regulate plant growth, development, and responses to environmental stresses such as salinity, drought, heat, cold, wound, and pathogen (Finkelstein et al., 2002; Shinozaki and Yamaguchi-Shinozaki, 2000). Once ABA is produced, ABA-bound receptors bind ABA, inhibit *PP2Cs*, such as *ABI1* and *ABI2*, thereby activate *SnRK2s* (Fan et al., 2016). The *SnRK2s* are involved in plant responses to abiotic stresses, which can phosphorylate the ABA-responsive element binding factor (ABF) and trigger the expression of ABA-responsive genes (Zhang et al., 2016).

Three ABA receptors, *PYLs*, were significantly downregulated in pomegranate roots or leaves, which was consistent with results in tea (*Camellia sinensis*) (Wan et al., 2018) and grape (*Vitis vinifera*) (Boneh et al., 2012). The downregulation of *PYLs* can reduce plant sensitivity to ABA and help plants adapt to salinity. Thirteen *PP2Cs* had different expressional patterns in pomegranate roots and leaves under salt stress, which was like other plants. The result indicated that *PP2Cs* might participate in response to salinity via different strategies (Boneh et al., 2012; Cao et al., 2016). The over-expression of PP2Cs inhibited the activity of SnRK2 proteins, thus the activities of downstream stress response elements are affected, plants are more sensitive to stresses. Most downregulated PP2Cs in pomegranate leaves may accelerate the phosphorylation of ABA-response element ABF by SnRK2, the downstream genes will be inducted to adapt the stress conditions (Boneh et al., 2012; Cao et al., 2016). *PYR/PYLs* regulate *SnRK2s* directly or indirectly, the upregulated *SnRK2* may cause the accumulation of phosphorylated downstream ABFs and activation of ABA-response genes. However, whether *SnRK2s* responds to various abiotic stresses in an ABA-independent or ABA-dependent pathway needs further investigation.

5. Ca^{2+}-dependent Signaling Pathways

Ca^{2+}, referred as a second messenger, plays a very important role in many stress-related signaling transduction pathways. Under various abiotic stresses, over-accumulation of ions induce temporary fluctuations in plant cytosolic ($[Ca^{2+}]_{cyt}$) levels (McCormack et al., 2005). The genes involved in CIPKs and CDPKs pathways, such as Ca^{2+}-ATPases (ACAs), cation/H^+ antiporters (CAXs), glutamate receptor (GLRs), calcium-binding proteins (CaM/CMLs), CBL-interacting protein kinases (CIPKs), calcium-dependent protein kinases (CDPKs), etc., participate in transporting and binding Ca^{2+}, sensing and relaying signals in plant cell under salinity stress (Dang et al., 2013; Yang et al., 2019).

Thirty DEGs involved in Ca^{2+}-dependent signaling pathway have been identified from the transcriptome; these transcripts code function proteins include three *ACAs*, four *CAXs*, two cation/calcium exchangers (CCXs), four *GLRs*, 10 *CaM/CMLs*, two *CIPKs*, and five *CDPKs*. Expectedly, two *ACAs* and two *CAXs* located on vacuoles were upregulated in roots. Previous reports revealed a central role for *CaMs* in the regulation of Ca^{2+} channels and pumps, like CNGCs and ACAs (Cheval et al., 2013; Virdi et al., 2015). The negative effect of CaM on CNGC activity provides a direct feedback pathway to restrict the influx of Ca^{2+} into plant cells (Cheval et al., 2013). The CaM stimulates the activity of these *ACAs* by preventing their auto-inhibition (Cheval et al., 2013). Collectively, under salt stress, pomegranate plants restrain the excess influx of Ca^{2+} into root cells under salinity. Meanwhile, it copes with the excess Ca^{2+} via removing Ca^{2+} from the cytosol by *ACAs* and *CAXs* in plasma membrane, including efflux of excess Ca^{2+} into the outer rhizosphere and/or the influx into vacuoles (Apse et al., 2010; Huda et al., 2013). Most of Ca^{2+}-related genes were upregulated in roots and downregulated in leaves, indicating vacuolar compartmentalization of Ca^{2+} was accelerated to decrease the sensitivity of roots to salt stress. While the downregulated *CaMs*, *ACAs*, and *CAXs*, and the upregulated *GLRs* in leaves suggested that vacuolar compartmentalization of Ca^{2+} is inhibited, then the Ca^{2+} concentration in cells increase. The temporary fluctuations in plant cytosolic ($[Ca^{2+}]_{cyt}$) levels stimulate many signaling transduction pathways in pomegranate under salt stress, such as Ca^{2+}-related, abscisic acid (ABA), and MAPK signaling pathway, as well as the crosstalk networks among them, which play crucial roles in responses to salt stress (Guo et al., 2019; Munns and Tester, 2008).

6. Transcription Factors (TFs) in Pomegranate

A total of 1,346 TFs from transcriptome were classified into 55 putative families. Among these TFs, 27 TF families, including 151 genes expressed differently under salt stress compared to controls. The most abundant of differential expression TFs include NAC, ERF, MYB-related, C2H2, MYB, bHLH, GRAS, WRKY, LBD, B3, and *bZIP* family genes (Table 1).

Specifically, TFs such as NAC, MYB, AP2/ERF, bHLH, WRKY, GRAS family genes regulate the expression of downstream genes to cope with various stresses (Deinlein et al., 2014). Under NaCl stress, 12 of 19 *NAC* genes were upregulated, and 10 of 19 genes were downregulated in leaves or roots of pomegranate plants. Interestingly, 18 *PgNACs* significantly changed in roots when compared to controls, and four genes were upregulated or downregulated in roots, but reversed in leaves. These results suggested that many *NACs* were involved in salt stress, but there were different potential responding mechanisms of NAC domain genes to salinity. Thirteen *MYB* genes were induced by salt treatment in pomegranate. Among these genes, six in roots and four in leaves were upregulated, and four in roots and eight in leaves were downregulated. Moreover, 16 *MYB-related* genes were detected in our RNA-Seq analyses, 13 genes were downregulated, and seven genes were upregulated by salt stress (Table 1). But so far, little was reported that the MYB-related type proteins are related with the responses to salt stress.

The AP2/ERF superfamily is divided into three families: the AP2 family proteins containing two repeated AP2/ERF domains, the ERF family proteins containing a single AP2/ERF domain, and the RAV family proteins containing a B3 domain. There were 19 *PgERFs*, and three *PgAP2s* significantly expressed in treatments when compared to controls. Fifteen *ERFs* were downregulated, and all *AP2* were repressed by salinity. There were nine upregulated and eight downregulated *C2H2*, three upregulated and nine downregulated *bHLH*, five upregulated and three downregulated *WRKY* identified in pomegranate roots and leaves. Remarkably, nine *GRASs* were upregulated in the root, but there was no significant change in leaves (Table 1). The result is mainly due to fact that these proteins play roles in plant development, including root development, axillary shoot development, and maintenance of the shoot apical meristem (Bolle, 2004).

Table 1 The different expressional TFs in pomegranate under NaCl stress.

TFs	*Total Number*	*DEGs*	*T1R*		*T2R*		*T1L*		*T2L*	
			Up	*Down*	*Up*	*Down*	*Up*	*Down*	*Up*	*Down*
NAC	97	19	1	1	12	6	1	2	1	4
ERF	114	19	1	1	8	5	1	7		9
MYB_related	91	16	–	–	4	3	1	11	3	2
C2H2	91	13	–	–	8	3	–	1		2
MYB	79	13	–	1	6	4	–	8	4	2
bHLH	102	11	–	1	2	5	1	4	1	1
GRAS	50	9	1	–	9	–	–	–	–	–
LBD	38	7	2	2	2	2	–	2	–	1
WRKY	66	7	–	–	5		–	2	1	3
B3	49	3	–	–	1	1	–	1		1

7. The Salt-inducted Effectors in Pomegranate

There were 57 DEGs coding for salt-inducted effectors, including two *SODs*, two *APXs*, 19 *PODs*, five *LEAs*, five *AQPs*, 10 *HSFs*, three *AKTs*, and one *HKT*. Two *APXs* were upregulated and two *SODs* were downregulated in leaves. Fourteen and nine *PODs* were suppressed in roots and leaves, respectively. Most of *AQPs* (4 of 5) were downregulated, while all of *LEAs* (5 of 5) and majority of *HSFs* (9 of 10) were upregulated in pomegranate tissues under salt stress. *LEAs* and *HSPs* were reported to prevent protein denaturation and maintain cell membrane fluidity under stress (Hoekstra et al., 2001). Then the DEGs of *LEA* and *HSF* in pomegranate contributed to mitigate the salt stress. The expressions of three *AKTs* and one *HKT* were significantly downregulated, which suggested that the influxes of K^+ and Na^+ in the cell were restricted (Xu et al., 2018).

Briefly, there were significant differences between pomegranate roots and leaves after salt treatment, with little overlap of DEGs responding salinity when pomegranate plants were exposed to high salinity. Many salt-related genes played the differential and temporal expression patterns in pomegranate roots and leaves, which provided a clear picture of transcripts in response to salt stress. The expression patterns of most DEGs were tissue-specific and time-specific. Among DEGs of roots, genes associated with cell wall organization and transmembrane transport were suppressed, and most of metabolism-related genes were over-represented. In leaves, 41.29% of DEGs were first suppressed and then recovered, including genes related to ions/metal ions binding. Also, ion transport and oxidation-reduction process were restricted. In addition, many DEGs involved in ABA, Ca^{2+}-related and MAPK signaling transduction pathways, such as ABA-receptors, Ca^{2+}-sensors, MAPK cascades, TFs, and the downstream functional genes *HSPs*, *LEAs*, *AQPs*, *PODs*, *AKT1*, and *HKT* were found.

References

Apse, M.P., Sottosanto, J.B. and Blumwald, E. 2010. Vacuolar cation/H^+ exchange, ion homeostasis, and leaf development are altered in a T-DNA insertional mutant of AtNHX1, the Arabidopsis vacuolar Na^+/H^+ antiporter. *Plant J.*, *36*: 229–39.

Aroca, R., Porcel, R. and Ruiz-Lozano, J.M. 2011. Regulation of root water uptake under abiotic stress conditions. *J. Exp. Bot.*, *63*: 43–57.

Attia, H., Karray, N., Msilini, N. and Lachaâl, M. 2011. Effect of salt stress on gene expression of superoxide dismutases, and copper chaperone in *Arabidopsis thaliana*. *Biol. Plant.*, *55*: 159–63.

Bolle, C. 2004. The role of GRAS proteins in plant signal transduction and development. *Planta*, *218*: 683–92.

Boneh, U., Biton, I., Zheng, C., Schwartz, A. and Ben-Ari, G. 2012. Characterization of potential ABA receptors in *Vitis vinifera*. *Plant Cell Rep.*, *31*: 311–21.

Boriboonkaset, T., Theerawitaya, C., Yamada, N., Pichakum, A., Supaibulwatana, K., Cha-um, S., Takabe, T. and Kirdmanee, C. 2013. Regulation of some carbohydrate metabolism-related genes, starch and soluble sugar contents, photosynthetic activities, and yield attributes of two contrasting rice genotypes subjected to salt stress. *Protoplasma*, *250*: 1157–67.

Bushman, B.S., Amundsen, K.L., Warnke, S.E., Robins, J.G. and Johnson, P.G. 2016. Transcriptome profiling of Kentucky bluegrass (*Poa pratensis* L.) accessions in response to salt stress. *BMC Genomics*, *17*: 48.

Cao, J.M., Min, J., Peng, L. and Chu, Z.Q. 2016. Genome-wide identification and evolutionary analyses of the PP2C gene family with their expression profiling in response to multiple stresses in *Brachypodium distachyon*. *BMC Genomics*, *17*: 175.

Cheval, C., Aldon, D., Galaud, J.P. and Ranty, B. 2013. Calcium/calmodulin-mediated regulation of plant immunity. *Biochim. Biophys. Acta*, *1833*: 1766–71.

Dang, Z.H., Zheng, L.L., Jia, W., Zhe, G., Wu, S.B., Zhi, Q. and Wang, Y.C. 2013. Transcriptomic profiling of the salt-stress response in the wild recretohalophyte *Reaumuria trigyna*. *BMC Genomics*, *14*: 29.

Deinlein, U., Stephan, A.B., Horie, T., Luo, W., Xu, G. and Schroeder, J.I. 2014. Plant salt-tolerance mechanisms. *Trends Plant Sci.*, *19*: 371–79.

Duan, J. and Cai, W. 2012. OsLEA3-2: An abiotic stress induced gene of rice plays a key role in salt and drought tolerance. *PloS One*, *7*: e45117.

El-Agamy, S., Mostafa, R., Shaaban, M. and El-Mahdy, M. 2010. *In vitro* salt and drought tolerance of Manfalouty and Nab El-Gamal pomegranate cultivars. *Aust. J. Basic Appl. Sci.*, *4*: 1076–82.

El-Khawaga, A., Zaeneldeen, E. and Youssef, M. 2012. Responses of three pomegranate (*Punica granatum* L.) cultivars to salinity stress. *Middle East J. Agric. Res.*, *1*: 64–75.

Fan, W.Q., Zhao, M.Y., Li, S.X., Bai, X., Li, J., Meng, H.W. and Mu, Z.X. 2016. Contrasting transcriptional responses of *PYR1/PYL/RCAR* ABA receptors to ABA or dehydration stress between maize seedling leaves and roots. *BMC Plant Biol.*, *16*: 99.

Fatehi, F., Hosseinzadeh, A., Alizadeh, H., Brimavandi, T. and Struik, P.C. 2012. The proteome response of salt-resistant and salt-sensitive barley genotypes to long-term salinity stress. *Mol. Biol. Rep.*, *39*: 6387–97.

Finkelstein, R.R., Gampala, S.S. and Rock, C.D. 2002. Abscisic acid signaling in seeds and seedlings. *The Plant Cell*, *14*: S15–S45.

Guo, S.M., Tan, Y., Chu, H.J., Sun, M.X. and Xing, J.C. 2019. Transcriptome sequencing revealed molecular mechanisms underlying tolerance of *Suaeda salsa* to saline stress. *PloS One*, *14*: e0219979.

Hasegawa, P.M., Bressan, R.A., Zhu, J.K. and Bohnert, H.J. 2000. Plant cellular and molecular responses to high salinity. *Annu. Rev. Plant Physiol. Plant Mol. Biol.*, *51*: 463–99.

Hoekstra, F.A., Golovina, E.A. and Buitink, J. 2001. Mechanisms of plant desiccation tolerance. *Trends Plant Sci*,. *6*: 431–38.

Hu, W., Yuan, Q., Wang, Y., Cai, R., Deng, X., Wang, J., Zhou, S., Chen, M., Chen, L. and Huang, C. 2012. Overexpression of a Wheat Aquaporin Gene, *TaAQP8*, Enhances Salt Stress Tolerance in Transgenic Tobacco. *Plant Cell Physiol.*, *53*: 2127–41.

Huang, J., Xiang, L., Hao, Y., Chen, S., Zhang, W., Huang, R. and Zheng, Y. 2012. Transcriptome characterization and sequencing-based identification of salt-responsive genes in *Millettia pinnata*, a semi-mangrove plant. *DNA Res.*, *19*: 195–207.

Huda, K.M., Banu, M.S., Tuteja, R. and Tuteja, N. 2013. Global calcium transducer P-type Ca^{2+}-ATPases open new avenues for agriculture by regulating stress signaling. *J. Exp. Bot.*, *64*: 3099–3109.

Karimi, H.R. and Hasanpour, Z. 2014. Effects of salinity and water stress on growth and macro nutrients concentration of pomegranate (*Punica granatum* L.). *J. Plant Nutr.*, *37*: 1937–51.

Kumar, S., Beena, A.S., Awana, M. and Singh, A. 2017. Salt-induced tissue-specific cytosine methylation downregulates expression of *HKT* genes in contrasting wheat (*Triticum aestivum* L.) genotypes. *DNA Cell Biol.*, *36*: 283–94.

Le Gall, H., Philippe, F., Domon, J., Gillet, F., Pelloux, J. and Rayon, C. 2015. Cell wall metabolism in response to abiotic stress. *Plants*, *4*: 112–66.

Liu, C.Y., Yan, M., Huang, X.B. and Yuan, Z.H. 2018. Effects of salt stress on growth and physiological characteristics of pomegranate (*Punica granatum* L.) cuttings. *Pak. J. Bot.*, *50*: 457–64.

Ma, S., Gong, Q,. and Bohnert, H.J. 2006. Dissecting salt stress pathways. *J. Exp. Bot.*, *57*: 1097.

McCormack, E., Tsai, Y. and Braam, J. 2005. Handling calcium signaling: Arabidopsis *CaMs* and *CMLs*. *Trends Plant Sci.*, *10*: 383–89.

Monika, B., Mikael, B., Basia, V., Atef, A.O., Payam, F., Dennis, J., Ottow, E.A., Cullmann, A.D., Joachim, S. and Jaakko, K.R. 2010. Linking the salt transcriptome with physiological responses of a salt-resistant Populus species as a strategy to identify genes important for stress acclimation. *Plant Physiol.*, *154*: 1697–1709.

Munns, R. 2005. Genes and salt tolerance: Bringing them together. *New Phytol.*, *167*: 645–63.

Munns, R., James, R.A. and Läuchli, A. 2006. Approaches to increasing the salt tolerance of wheat and other cereals. *J. Exp. Bot.*, *57*: 1025–43.

Munns, R. and Tester, M. 2008. Mechanisms of salinity tolerance. *Annu. Rev. Plant Biol.*, *59*: 651–81.

Naeini, M., Khoshgoftarmanesh, A. and Fallahi, E. 2006. Partitioning of chlorine, sodium, and potassium and shoot growth of three pomegranate cultivars under different levels of salinity. *J. Plant Nutr.*, *29*: 1835–43.

Okhovatianardakani, A.R., Mehrabanian, M., Dehghani, F. and Akbarzadeh, A. 2010. Salt tolerance evaluation and relative comparison in cuttings of different pomegranate cultivars. *Plant Soil Environ.*, *56*: 176–85.

Shafi, A., Chauhan, R., Gill, T., Swarnkar, M.K., Sreenivasulu, Y., Kumar, S., Kumar, N., Shankar, R., Ahuja, P.S. and Singh, A.K. 2015. Expression of *SOD* and *APX* genes positively regulates secondary cell wall biosynthesis and promotes plant growth and yield in Arabidopsis under salt stress. *Plant Mol. Biol.*, *87*: 615–31.

Shinozaki, K. and Yamaguchi-Shinozaki, K. 2000. Molecular responses to dehydration and low temperature: Differences and cross-talk between two stress signaling pathways. *Curr. Opin. Plant Biol.*, *3*: 217–23.

Sun, Y., Niu, G., Masabni, J.G. and Ganjegunte, G. 2018. Relative Salt Tolerance of 22 Pomegranate (*Punica granatum*) Cultivars. *HortScience*, *53*: 1513–19.

Virdi, A.S., Singh, S. and Singh, P. 2015. Abiotic stress responses in plants: roles of calmodulin-regulated proteins. *Front. Plant Sci.*, *6*: 809.

Wan, S.Q., Wang, W.D., Zhou, T.D., Zhang, Y.H., Chen, J.F., Xiao, B., Yang, Y.J. and Yu, Y.B. 2018. Transcriptomic analysis reveals the molecular mechanisms of *Camellia sinensis* in response to salt stress. *Plant Growth Regul.*, *84*: 1–12.

Wang, X.C., Chang, L.L., Wang, B.C., Wang, D., Li, P.H., Wang, L.M., Yi, X.P., Huang, Q.X., Peng, M. and Guo, A.P. 2013. Comparative proteomics of *Thellungiella halophila* leaves from plants subjected to salinity reveals the importance of chloroplastic starch and soluble sugars in halophyte salt tolerance. *Mol. Cell. Proteomics*, *12*: 2174–95.

Xu, M., Chen, C., Cai, H. and Wu, L. 2018. Overexpression of *PeHKT1*; 1 Improves Salt Tolerance in Populus. *Genes*, *9*: 475.

Yang, Y., Zhang, C., Tang, R., Xu, H., Lan, W., Zhao, F. and Luan, S. 2019. Calcineurin B-Like proteins CBL4 and CBL10 mediate two independent salt tolerance pathways in Arabidopsis. *Int. J. Mol. Sci.*, *20*: 2421.

Yokoi, S., Quintero, F.J., Cubero, B., Ruiz, M.T., Bressan, R.A., Hasegawa, P.M. and Pardo, J.M. 2010. Differential expression and function of *Arabidopsis thaliana* NHX Na^+/H^+ antiporters in the salt stress response. *Plant J.*, *30*: 529–39.

Zhang, A., Han, D., Wang, Y., Mu, H., Zhang, T., Yan, X. and Pang, Q. 2018. Transcriptomic and proteomic feature of salt stress-regulated network in Jerusalem artichoke (*Helianthus tuberosus* L.) root based on *de novo* assembly sequencing analysis. *Planta*, *247*: 715–32.

Zhang, H., Li, W., Mao, X., Jing, R. and Jia, H. 2016. Differential Activation of the Wheat SnRK2 Family by Abiotic Stresses. *Front. Plant Sci.*, *7*: 420.

Zhou, A., Liu, E., Ma, H., Feng, S., Gong, S. and Wang, J. 2018. NaCl-induced expression of *AtVHA-c5* gene in the roots plays a role in response of Arabidopsis to salt stress. *Plant Cell Rep.*, *37*: 443–52.

11

Comprehensive Genomic Exploration and Expression Profiling of *MAPK* and *MAPKK* Gene Families in Pomegranate

Yuan Ren[1*] and *Zhaohe Yuan*[1]

Mitogen-activated protein kinase (MAPK) cascade is involved in the regulation of a series of biological processes in organisms, which are composed of *MAPKKKs*, *MAPKKs*, and *MAPKs*. Although genome-wide analyses of it have been well described in some species, little is known about *MAPK* and *MAPKK* genes in pomegranates. In this study, we identified 18 *PgMAPKs*, nine *PgMAPKKs* through a genome-wide search. Chromosome localization showed that 27 genes are distributed on seven chromosomes with different densities. Multiple sequence alignment and phylogenetic analysis revealed that *PgMAPKs* and *PgMAPKKs* could be divided into four subfamilies (groups A, B, C, and D), respectively. In addition, exon-introns structural analysis of each candidate gene has indicated high levels of conservation within and between phylogenetic groups. Cis-acting element analysis predicted that *PgMAPKs* and *PgMAPKKs* were widely involved in the growth, development, stress, and hormone response of pomegranate. Expression profile analyses of *PgMAPKs* and *PgMAPKKs* were performed in different tissues (root, leaf, flower, and fruit), and *PgMAPK13* was significantly expressed in all tissues. To our knowledge, this is the first genome-wide analysis of the *MAPK* and *MAPKK* gene family in pomegranate. This study provides valuable information for understanding the classification and functions of pomegranate MAPK signal.

1 College of Forestry, Nanjing Forestry University, Nanjing 210037, China.
* Corresponding author:

1. Introduction

In the long process of evolution, plants have evolved a complex signal network regulation mechanism to adapt to the biotic and abiotic stress during growth and development (Nakagami and Pitzschke et al., 2005; Raina and Wankhede et al., 2012; Shen and Liu et al., 2012; Zhan and Yue et al., 2017). The mitogen-activated protein kinase (MAPK) cascade is regarded as one of the representative signal transduction mechanism. As a broad and relatively conservative evolutionary pathway, it is involved in multiple metabolic pathways such as cell division, development process, and defense response in organism (Tena and Asai et al., 2001; Kong and Pan et al., 2013; Zhou and Ren et al., 2017). The typical MAPK cascade is composed of three specific kinases, MAPK kinase (MAPK), MAPK kinase (MAPKK), and MAPKKK kinase (MAPKKK) (Jonak and Ökrész et al., 2002). They are activated by phosphorylation and dephosphorylation at specific sites (Zhang and Liu et al., 2006). According to sequence alignment and phylogenetic analysis of model species *Arabidopsis thaliana* and rice, the *MAPK* genes can be divided into four groups (A, B, C, and D) and their MAPKs have a unique TxY phosphorylation motif, while the MAPKK genes are also divided into four groups (Liu and Zhang et al., 2017). Despite the highly conserved structure of plant MAPKs and MAPKKs, they have been shown to be extensively involved in cell division, growth, and hormonal response regulation processes, as well as diverse biological and abiotic stress, with the updating of bioinformatics methods (Kazuya Ichimura et al., and Ichimura et al., 2002; Moustafa and AbuQamar et al., 2014). To date, members of the MAPK cascade gene family have been identified in several species such as *Arabidopsis thaliana* (Andreasson and Ellis, 2010; Rasmussen and Roux et al., 2012), *Oryza sativa* (Pandhari and Misra et al.; Rao and Tambi et al., 2010), *Brachypdoium* (Feng and Liu et al., 2016), *Zea mays* (Liu and Zhang et al., 2013) and *Hordeum vulgare* (Cui and Yang et al., 2019). In addition, the study of MAPKs and MAPKKs function is gradually deepened. As for signal transmission under biological stress, studies have shown that the MAPK cascade pathway mekkl-mkk4/5-mpk3/6 involves the biosynthetic pathway of phytoalexins (Ren and Yang et al., 2006).

In response to the pathogen, *Arabidopsis thaliana* induced the expression of related defense genes by activating two MAPK cascade pathways, mekkl-mkk1/2-MPK4 and mekk1-mkk4/5-mpk3/6, in which the former played a negative regulatory role (Ichimura and Casais et al., 2006; Suarez-Rodriguez and Adams-Phillips et al., 2007), while the latter played a positive regulatory role (Asai and Tena et al., 2002). In terms of signaling under abiotic stress, studies have shown that a variety of MAPKs, including MAPK3, 4, and 6, can be rapidly activated in plants under abiotic stress (de Zelicourt and Colcombet et al., 2016). For example, the transcriptional level of ZmMPK3 in maize significantly increased and rapidly accumulated under abiotic stresses such as cold, drought, ultraviolet, salinity, heavy metal, and mechanical damage (Wang and Ding et al., 2010). Overexpression of OsMAPK5 in transgenic rice can increase its tolerance to drought, salt, and cold stress (Xiong and Yang, 2003) and RaMPK1 were significantly upregulated in response to low temperature stress along with RaMPK2 (Ghawana and Kumar et al., 2009). In the field of plant hormone signaling, MAPKs and MAPKKs are involved in the signaling of auxin, abscisic acid, and ethylene. MAPK activity has been found to be directly dependent on abscisic acid stimulation in different plants such as barley, tobacco, moss, peas, corn, and rice. OsMAPK5 (OsMAP1), OsMAPK2, OsMAPK44, and other MAPKs can be activated by ABA transcription (Danquah and de Zélicourt et al., 2015). MPK3 and MPK6 in *Arabidopsis* play their roles in the ethylene response pathway by promoting the stability of the key transcription regulator EIN3 (Hahn and Harter, 2009). It has been found that the mkk9-mpk3/MPK6 cascade controls EIN3 transcription to regulate ethylene signaling in plant cells, and links the interwoven MAPK cascade pathways to control the quantitative response and specificity in the signaling network (Yoo and Cho et al., 2008).

Punica granatum L. is an economic tree widely cultivated throughout the world (Harel-Beja and Sherman et al., 2015; Yan and Zhao et al., 2019; Liu and Zhao et al., 2020; Qin and Liu et al., 2020). Pomegranate, native to central Asia, is known for its bright red rinds, juicy kernels, and allagenic tannins (Zhao and Yuan et al., 2015; Yuan and Fang et al., 2018). With the further development of molecular biology, combining with bioinformatics, it is very important to study the functions and functions of signal transduction pathways in plants to resist stress from the molecular perspective, so as to regulate plant growth and development, comprehensive disease resistance, or environmental stress (Thomma and Penninckx et al., 2001). The MAPK and MAPKK gene families of pomegranates have not yet been identified, and their functions remain to be clarified. With the complete genome sequence of pomegranate, it is possible to study the gene family of pomegranate (Luo and Li et al., 2019).

This study used the latest release of pomegranate 'Tunisia' genomes; in the whole genome level systematically identified pomegranate MAPK and MAPKK gene families; the chromosome location, gene structure, cis-acting element, expression patterns were all analyzed, in order to determine the family information of pomegranate MAPK cascade genes; it will also be beneficial to further study the function of candidate genes.

In this study, the MAPK and MAPKK gene family members of pomegranate were systematically identified at the genome-wide level, using the pomegranate 'Tunisia' genomes. Physicochemical properties, chromosomal locations, gene structures, cis-acting elements, and expression patterns were analyzed to obtain the bioinformatics functions of pomegranate MAPKs and MAPKKs, which is conducive to further research on the application in pomegranate growth and development.

2. Materials and Methods

2.1. Genome Data Sources

Pomegranate genome-wide data (SAMN05193489) were downloaded from NCBI's official website (https://www.ncbi.nlm.nih.gov/biosample/SAMN05193489/). The

MAPK and MAPKK protein sequences of *Arabidopsis thaliana* were obtained from EnsemblPLants database (http://plants.ensembl.org/index.html) (Kersey and Allen et al., 2017).

2.2 Identification of PgMAPKs and PgMAPKKs in Pomegranate

The serine/threo-nine-protein kinetro-like domain sequence comparison file was downloaded fromPfam database(http://pfam.xfam.org/). Then the comparison file was converted into HMM model file by HMMER v3.2.1 software (Finn and Clements et al., 2011). Using the Selecthmm package screening was done to get pomegranate the protein sequence that contains the MAPK cascade genes sequence structure domain (https://github.com/Redpome/SelectHMM) (E-value 10-10 or less). Moreover, the protein domains were validated using SMART (http://smart.embl-heidelberg DE) (Schultz and Milpetz et al., 1998), NCBI CDD (https://www.ncbi.nlm. nih.gov/ CDD) (Marchler-Bauer and Bo et al., 2017). The isoelectric points and molecular weights of PgMAPKs and PgMAPKKs were obtained by the ExPASy Proteomics Server (http://expasy.org/) (Artimo and Jonnalagedda et al., 2012).

2.3 Phylogenetic Relationship and Conserved Motif Analysis

ClustalX v2.0 was used for multi-sequence alignment of identified proteins, with parameters set as default values. The results were then submitted to GeneDoc for visualization. Phylogenetic trees were constructed with the software MEGA to explore the phylogenetic relationships of PgMAPK and PgMAPKK family members. The input sequences were full-length protein sequences with a self-expansion value of 1,000 times. Protein structure domain and conservative motif were obtained by MEME online program (http://meme.nbcr.net/meme/intro.html) (Bailey and Boden et al., 2009) based on the expectation maximum (EM) algorithm. The introns and exons of pomegranate MAPKs and MAPKKs genes were analyzed and visualized by the TBtools software (Chen and Xia et al., 2018). Finally, the 2.0kb upstream DNA sequence of each gene was extracted, and submitted to Plantcare database (http://bioinformatics.psb.ugent.be/webtools/plantcare/html/) (Magali and Patrice et al., 2002) to predict the cis-acting elements. Cis-acting elements related to growth, stress, and hormones were retained, and then submitted to TBtools for drawing.

2.4 Chromosome Localization and Expression Profiles

The chromosome distribution information of *PgMAPKs* and *PgMAPKKs* was displayed by TBtools. RNA-seq data of six pomegranate cultivars were downloaded from NCBI database, including 'Dabenzi', 'Tunisia', 'Baiyushizi', 'Wonderful', 'Nana', and 'Black127' (Table 1). All RNA-Seq data were quality-controlled by fastp (Chen et al., 2018) to obtain clean reads. The index file was constructed

Table 1 RNA-Seq data of pomegranate.

Accession No.	*Cultivars*	*Sample Type*	*Library*	*Platform*
SRR5279396	'Dabenzi'	Root	Paired end	Illumina HiSeq 4000
SRR5279397	'Dabenzi'	Leaf	Paired end	Illumina HiSeq 4000
SRR5279395	'Dabenzi'	Flower	Paired end	Illumina HiSeq 4000
SRR5279391	'Dabenzi'	Inner seed coat (50 days after pollination)	Paired end	Illumina HiSeq 4000
SRR5279388	'Dabenzi'	Outer seed coat (50 days after pollination)	Paired end	Illumina HiSeq 4000
SRR5279394	'Dabenzi'	Pericarp (50 days after pollination)	Paired end	Illumina HiSeq 4000
SRR5678820	'Tunisia'	Inner seed coat (50 days after pollination)	Paired end	Illumina HiSeq 4000
SRR5678819	'Baiyushizi'	Inner seed coat (50 days after pollination)	Paired end	Illumina HiSeq 4000
SRR1055290	'nana'	Mixed samples of leaves, flowers, fruit, and roots	Single end	454 GS FLX Titanium
SRR1054190	'Black127'	Mixed samples of root, leaf, flower, and fruit	Single end	454 GS FLX Titanium
SRR5446598	'Tunisia'	Functional male flower (3.0– 5.0 mm)	Paired end	Illumina HiSeq 2500
SRR5446595	'Tunisia'	Functional male flower (5.1– 13.0 mm)	Paired end	Illumina HiSeq 2500
SRR5446592	'Tunisia'	Functional male flower (13.1– 25.0 mm)	Paired end	Illumina HiSeq 2500
SRR5446607	'Tunisia'	Female sterility flower (3.0– 5.0 mm)	Paired end	Illumina HiSeq 2500
SRR5446604	'Tunisia'	Female sterility flower (5.1– 13.0 mm)	Paired end	Illumina HiSeq 2500
SRR5446601	'Tunisia'	Female sterility flower (13.1– 25.0 mm)	Paired end	Illumina HiSeq 2500
SRR080723	'Wonderful'	Pericarp	Paired end	Illumina HiSeq 2000

by Kallisto V0.44.0 (Bray et al., 2016) and then the normalized expression units (Transcripts Per Kilobase of exon model per Million mapped reads (TPM) values) of each gene were extracted by kallisto quant command. The final expression levels were from the TPM values that were converted by Log_2 (TPM+1). Then the Heatmap package in RStudio V3.6.1 (Ophir et al., 2014) was used to draw the Heatmap of *PgMAPKs* and *PgMAPKKs*.

3. Results

3.1 *Identification and Sequence Analysis of PgMAPKs and PgMAPKKs*

According to the described method, 18 *PgMAPKs* and nine *PgMAPKKs* were obtained from pomegranate whole genome, respectively (Table 2). Furthermore, the number of amino acids, molecular weight, theoretical pI of *PgMAPKs* and *PgMAPKKs* were analyzed by ExPASy online tool. The results showed that the molecular weight was between 34910.05~70420.74 Da, the isoelectric point was between 4.94~9.38, the protein length was within the range of 314~619 amino acids, the number of amino acids varied significantly. These data were similar to economic trees such as grapes (Wang and Lovato et al., 2014), banana (Wang and Hu et al., 2017) and apples (Zhang and Xu et al., 2013). However, the variations in physicochemical properties among the *PgMAPKs* and *PgMAPKKs* indicated that these genes may have subfunctionalization and new functionalization. Chromosomal location analyses showed that 18 *MAPKs* and 9 *MAPKKs* presented on 7 chromosomes (chr1, chr2, chr3, chr4, chr6, chr7, chr8) (Table 2).

3.2 *Conserved Motif Analysis of PgMAPK and PgMAPKK Gene Family Members*

According to specific conserved motifs analysis, *PgMAPKs* and *PgMAPKKs* were found to have unique conserved motifs. The 'VGTxxYMSPER' conserved motifs were found in nine Pg*MAPKK* genes, and the 'T (E/D) YVxTRWYRAPE (L/V)' conserved motifs were contained in 18 *PgMAPK* genes. Through multiple sequence alignment of the protein sequences of *PgMAPK* family genes, 'T-loop' motifs were detected in the domain. There was an ATP phosphorylation site catalyzed by 'TxY', which plays an important role in the function of *PgMAPKs*. According to previous studies of model plants, the *PgMAPK* gene family can also be divided into four groups (A, B, C, and D) and can be divided into subtypes of 'TDY' and 'TEY' according to different 'TxY' phosphorylation sites (Table 1). In addition, phosphorylation domains with 'S/T-XXXXX-S/T' and an anchor site with '-D (L/I/V) K-' were found in *PgMAPKKs*. *PgMAPKKs* were also divided into four groups (A, B, C, and D) (Table 1) based on results reported in the *Arabidopsis thaliana* (Nicole and Hamel et al., 2006).

3.3 *Phylogenetic Relationship of PgMAPKs and PgMAPKKs*

To explore the evolutionary relationship toward PgMAPK and PgMAPKK proteins, the predicted protein sequences of 18 *PgMAPKs*, nine *PgMAPKKs*, *20 AtMAPKs*, and *10 AtMAPKKs* from *Arabidopsis* were subjected to a multiple sequence alignment using the MEGA7. The N-J (Neighbor-Joining) method was then used to construct an unrooted phylogenetic tree. The phylogenetic tree divided *PgMAPKs* and *PgMAPKKs* into four groups (A, B, C, D). The most abundant of MAPK gene

Table 2 The information of MAPK and MAPKK gene family in pomegranate.

Gene Name	*Protein ID*	*Genomic Position*	*Size (aa)*	*MW (Da)*	*PIs*	*Types*	*Group*
PgMAPK1	XP_031372466.1	chr8:25215947...25219536	615	68952.19	7.69	TDY	D
PgMAPK2	XP_031382378.1	chr2:26190661...26195979	391	44872.36	5.57	TEY	A
PgMAPK3	XP_031384487.1	chr3:27987357...27983062	484	55597.84	8.79	TDY	D
PgMAPK4	XP_031387033.1	chr3:6220495...6221628	377	42844.83	8.90	TGY	C
PgMAPK5	XP_031387034.1	chr3:6229201...6231072	435	48521.45	7.66	TGY	C
PgMAPK6	XP_031387035.1	chr3:6233547...6234680	377	42552.40	9.23	TGY	C
PgMAPK7	XP_031388383.1	chr3:6647361...6648512	383	43266.50	9.26	TGY	C
PgMAPK8	XP_031389841.1	chr4:494852...498050	374	42897.79	6.11	TEY	B
PgMAPK9	XP_031389944.1	chr4:8712267...8715865	380	43399.50	6.20	TEY	B
PgMAPK10	XP_031392146.1	chr4:18603577...18604728	383	43164.35	9.36	TGY	C
PgMAPK11	XP_031392360.1	chr4:34805345...34809858	617	70420.74	9.19	TDY	D
PgMAPK12	XP_031399464.1	chr6:27190634...27192917	377	43002.99	4.94	TEY	B
PgMAPK13	XP_031400536.1	chr6:27163862...27162051	375	42968.25	5.78	TEY	A
PgMAPK14	XP_031402771.1	chr1:54220111...54215519	566	64371.77	8.76	TDY	D
PgMAPK15	XP_031402779.1	chr1:54220111...54215513	564	64243.64	8.76	TDY	D

Contd.

Table 2 *Contd.*

Gene Name	*Protein ID*	*Genomic Position*	*Size (aa)*	*MW (Da)*	*PIs*	*Types*	*Group*
PgMAPK16	XP_031404075.1	chr7:28354650...28351083	597	68130.90	9.35	TDY	D
PgMAPK17	XP_031404901.1	chr7:21504171...21503431	372	42781.79	6.92	TEY	C
PgMAPK18	XP_031406973.1	chr7:3285226...3281535	619	69387.45	8.06	TDY	D
MAPKK1	XP_031374856.1	chr8:13616418...13612292	354	39870.90	5.96	DIK	A
MAPKK2	XP_031374857.1	chr8:13616418...13612292	354	39870.90	5.96	DIK	A
MAPKK3	XP_031381620.1	chr2:14357511...14354367	352	38956.51	5.49	DLK	A
MAPKK4	XP_031383136.1	chr2:2401332...2404194	518	58182.26	5.76	DIK	B
MAPKK5	XP_031383137.1	chr2:2401332...2404194	518	58182.26	5.76	DIK	B
MAPKK6	XP_031383138.1	chr2:2401664...2404194	436	49232.33	6.04	DIK	B
MAPKK7	XP_031388194.1	chr3:13155229...13156281	350	39123.69	9.38	DIK	C
MAPKK8	XP_031393241.1	chr4:1081579...1082574	331	36995.33	8.31	DIK	D
MAPKK9	XP_031406836.1	chr7:3593726...3594670	314	34910.05	8.35	DIK	D

family members were in group D (7), followed by group C (6), group B (3), and group A (2). This result was consistent with previous studies on *Arabidopsis*, rice, and tomato. The number of *MAPKK* gene family was less than that of *MAPKK*, so there was no significant difference in the number of each group. Three (*PgMAPKK1*, *PgMAPKK2*, *PgMAPKK3*), three (*PgMAPKK4*, *PgMAPKK5*, *PgMAPKK6*), one (*PgMAPKK7*), and two (*PgMAPKK8*, *PgMAPKK9*) *PgMAPKK* genes belonged to groups A, B, C, and D, respectively. The ratio of *PgMAPKKS* to *PgMAPKs* was about 1:2, which was in accordance with that in other plant (Table 3). In summary, compared with these model plants, the quantity and grouping status of *MAPKs* and *MAPKKs* were highly conserved in pomegranate (Table 3).

Table 3 The number of the MAPK and MAPKK gene family in Pomegranate and other plants.

Gene Family	*Species*	*GroupA*	*GroupB*	*GroupC*	*GroupD*	*Total*
MAPK	Pomegranate	2	3	6	7	18
	Chinese jujube	2	1	2	5	11
	Arabidopsis	3	5	4	8	20
	Rice	2	1	2	10	15
	Brachypodium distachyon	2	2	3	9	16
	Maize	4	2	2	11	10
	Apple	5	6	5	10	26
	Poplar	4	4	4	9	21
	Tomato	3	4	2	7	16
	Mulberry	2	3	2	3	10
	Bread wheat	7	3	8	36	54
MAPKK	Pomegranate	3	3	1	2	9
	Chinese jujube	2	1	0	2	5
	Arabidopsis	3	1	2	4	10
	Rice	2	1	2	3	8
	Brachypodium distachyon	2	3	2	5	12
	Apple	3	1	2	3	9
	Poplar	3	1	2	5	11
	Bread wheat	3	2	1	12	18

3.4 Gene Structure Analysis of PgMAPKs and PgMAPKKs

Gene structure analysis was supposed to provide valuable information about evolutionary and duplication events among gene families. We found that most members of the same group share similar exon/intron structures and conserved motifs. In the *PgMAPKs* gene family, each group showed different structural characteristics. In group A, *PgMPK2* and *PgMAPK13* were composed of six exons with the longer introns. Six exons were also found in the same group of *MAPKs* in *Arabidopsis*, poplar and tomato. *PgMPK8*, *PgMAPK9*, and *PgMAPK12* in group B had six exons, which was also consistent with the results of studies on tomato and poplar studies. In group C, *PgMAPK5* and *PgMAPK17* were only composed of two exons, and the size of each exon was strictly conservative. Compared with the highly conserved structural pattern in group C, the exon and intron distribution of *PgMPKs* in group D was more complex, exhibiting a diverse pattern. For example, *PgMPK1*, *PgMAPK18* had 11 exons, while *PgMAPK11*, *PgMAPK14*, *PgMAPK15*, *PgMAPK16* had 10 exons, and *PgMAPK3* only had nine. Although there were slight differences in exon lengths, it was clear that the structural patterns of exons were well conserved not only between close paralogs, but also between the *PgMAPKs* that apparently diverged following earlier duplication events.

The *PgMAPKK* gene showed two distinct structural patterns which were quite similar to the *MAPKK* gene in *Arabidopsis* and poplar. Pg*MAPKK7* in group C and *PgMAPKK8, PgMAPKK9* in group D both had less introns and extrons, whereas the *PgMAPKKs* in group A and group B possessed more exon and intron junctions. In group A, *PgMAPKK1* and *PgMAPKK2* had seven exons, while *PgMAPKK3* had eight exons. Consistent with other plants, *PgMAPKK1* and *PgMAPKK2* shared strong conservation of exonic length. There was a high consistency of gene structure among the three genes (*PgMAPKK4*, *PgMAPKK5*, *PgMAPKK6*) in group B.

3.5 Cis-acting Element Prediction of PgMAPKs and PgMAPKKs

We extracted upstream 2.0 KB of genome sequence for the cis-acting elements analysis. And hundreds of possible cis-acting elements downloaded from the Platcare database. After screening, 25 cis-acting elements with clear functional information were retained. Among them, we found that a total of six elements were related to stress, nine elements were related to growth and development, and the remaining 10 elements were related to hormones (**Table 4**). As shown in Table 3, elements named ARE, LTR, and MBS were present in 20, 18, and 17 genes, respectively. These results indicated that *PgMAPK* and *PgMAPKK* genes were widely involved in hypoxia, low temperature and drought stress of pomegranate. Furthermore, WUN-motif was found in six genes, indicating that these genes involved in the external damage mechanism of pomegranate. The elements related to plant growth and development mainly included circadian rhythm, meristem, palisade mesophytic differentiation, and endosperm. Notably, MBSI elements were found in five genes (*PgMAPK2*, *PgMAPK4*, *PgMAPK8*, *PgMAPK14*, and *PgMAPK15*), suggesting that these genes are involved in the mechanism of flavonoid synthesis in pomegranate. In addition, the cis-acting elements related to various hormone reactions (such as auxin, ethylene, salicylic acid, MeJA, GA3), indicating that *PgMAPKs* and *PgMAPKKs* were deeply involved in signal transduction of endogenous and exogenous hormone networks in pomegranate. To sum up, the various cis-acting elements in the gene promoter region suggested that the *PgMAPK* and *PgMAPKK* gene family plays a crucial role in its growth and development and stress adaptation (Table 4).

Table 4 Characteristics of cis-acting regulatory elements presented in the promoter regions of PgMAPK and PgMAPKK genes.

Function	*Promoter Name*	*Promoter Annotation*	*Total Number*
Motifs related to stress response	ARE	cis-acting regulatory element essential for the anaerobic induction	20
	LTR	cis-acting element involved in low-temperature responsiveness	18
	MBS	MYB binding site involved in drought-inducibility	17
	TC-rich repeats	cis-acting element involved in defense and stress responsiveness	6
	WUN-motif	wound-responsive element	12
	GC-motif	enhancer-like element involved in anoxic specific inducibility	9
Motifs related to growth and development	CAT-box	cis-acting regulatory element related to meristem expression	8
	GCN4_motif	cis-regulatory element involved in endosperm expression	5
	CCGTCC-box	cis-acting regulatory element related to meristem specific activation	10
	O2-site	cis-acting regulatory element involved in zein metabolism regulation	11
	HD-Zip 1	element involved in differentiation of the palisade mesophyll cells	4
	RY-element	cis-acting element involved in seed-specific regulation	1
	MBSI	MYB binding site involved in flavonoid biosynthetic genes regulation	5
	MSA-like	cis-acting element involved in cell-cycle regulation	1
	circadian	cis-acting regulatory element involved in circadian control	5
Motifs related to hormone response	ABRE	cis-acting element involved in the abscisic acid responsiveness	23
	AuxRR-core	cis-acting regulatory element involved in auxin responsiveness	1
	CGTCA-motif	cis-acting regulatory element involved in the MeJA-responsiveness	20
	TGACG-motif	cis-acting regulatory element involved in the MeJA-responsiveness	19
	ERE	ethylene-responsive element	20
	GARE-motif	gibberellin-responsive element	7
	P-box	gibberellin-responsive element	8
	TATC-box	cis-acting element involved in gibberellin-responsiveness	1
	TGA-element	auxin-responsive element	11
	TCA-element	cis-acting element involved in salicylic acid responsiveness	15

3.6 Expression Pattern of PgMAPKs and PgMAPKKs

We used the online RNA-seq data to explore the expression patterns of MAPK and MAPKK genes in pomegranates at different developmental stages. The results showed that these 27 genes were expressed in at least one tissue organ or stage of development. Compared with other genes, *PgMAPK13* was highly expressed in root, flower, and seed coat, while *PgMAPK4*, *PgMAPK5*, *PgMAPK6*, *PgMAPK7*, *PgMAPK8*, *PgMAPK10* had a less expression level. *PgMAPK9*, *PgMAPK12*, and *PgMAPKK7* was expressed in all except pericarp. A high *PgMAPKK9* and *PgMAPK13* expression level was found in inner seed coat after pollination for 50 days. *PgMAPKK3* and *PgMAPKK9* expressed significantly variously during the development of functional male flower and female sterility flower. The expression patterns demonstrated that the expression level of each gene was highly variable, suggesting that these genes play an indispensable role in regulating plant growth and development.

4. Discussion

Generally, plants responded to various biological and abiotic stresses through physiological, biochemical and even molecular changes to resist and adapt them (Bohnert and Nelson et al., 1995). As an important protein kinases group, *MAPKs* and *MAPKKs* can transfer extracellular signals to intracellular signals in plant growth and development, hormone regulation and stress response (Kovtun and Chiu et al., 1998; Zhang and Klessig, 2001). To date, *MAPK* and *MAPKK* gene family have been identified and functionally characterized in several model plants and numerous economic trees. However, the *MAPK* and *MAPKK* gene family in pomegranate has not been reported previously.

In this study, we first identified 18 *MAPKs* and 9 *MAPKKs* in pomegranate at genome-wide level. The numbers of these family members were similar to those found in other plant species, such as *Ziziphus jujuba* (11/5) (Liu et al., 2017), *Arabidopsis thaliana* (20/10) (Andreasson et al., 2009; Rasmussen et al., 2012), *Oryza sativa* (17/8) (Rao et al., 2010), *Brachypodium distachyon* (16/12) (Feng et al., 2016), *Malus Domestica* (26/9) (Zhang et al., 2013), *Populus* (21/11) (Nicolas et al., 2016), *Solanum lycopersicum* (16/5) (Kong et al., 2012), *Triticum aestivum* (54/18) (Zhan et al., 2017), *Capsicum annuum* (19/5) (Liu et al., 2015), *Cucumis sativus* (14/6) (Wang et al., 2015), *Citrullus lanatus* (15/6) (Song et al., 2015) and *Jatropha curcas* (12/5) (Wang et al., 2018).

The chromosome distribution pattern showed that the *PgMAPK* or *PgMAPKK* genes density was relatively high in certain chromosomes. Regions containing four or more genes in the range of 200kb or less can be regarded as gene clusters (Holub et al., 2001). Only one gene cluster with four members in *PgMAPK* gene family and no other gene cluster was found in *PgMAPKs*. However, given that our clustering criteria was relaxed to at least two genes, seven clusters would be found (63.6% of MAPK and MAPKK family members appeared in the cluster).

In addition, we also found that although the *PgMAPKK3* and the gene cluster consisting of *PgMAPKK4*, *PgMAPKK5*, and *PgMAPKK6* were all located on the same chromosome, they were physically far apart, which was consistent with the result of phylogenetic and multiple alignment analysis.

PgMAPKs were divided into three subtypes of 'TDY', 'TEY', and 'TGY' according to different 'TxY' phosphorylation sites by multiple sequence alignment (Table 2). Interestingly, all *PgMAPKs* in group C had a 'TGY' motif except *PgMAPK17*. The 'TGY' activation loop was similar to MAPKs in mammals, rather than the T(E/D)Y motif commonly in plants (Cui et al., 2019). As far as we know, the same results were also found in wheat, where *TaMPK25* also carries a TGY activating loop (Goyal et al., 2018). Furthermore, there were confirmed results *MAPKs* included in group B was mainly involved in cell division and environmental stress response. Therefore, *PgMAPK8*, *PgMAPK9*, and *PgMAPK12* may also display response to various stress. *MAPKKs* in pomegranate were phosphorylated by 'S/T-XXXXX-S/T' and anchored by '-D(L/I/V) K-'. Due to its wide range of functions, many *MAPKK* genes have been cloned recently, such as *AtMKK1* and *AtMKK2-5* of *Arabidopsis thaliana*, *SiMKK* of Alfalfa, *MEK1* of tomato, *NTMEK1-2* of tobacco, and *ZmMEK1* of maize (Zhang and Liu, 2006).

Since highly differentiated sequences may be responsible for regulatory diversity of MAPK and MAPKK protein, we constructed a phylogenetic tree based on the predicted MAPK and MAPKK proteins of pomegranate and *Arabidopsis thaliana*. Four groups were separated in phylogenetic tree, which were named following the previous studies (Kazuya Ichimura et al., and Ichimura et al., 2002). The members in each group shared similar intron-exon structures. Consistent with previous studies, these data indicate similar origins and evolutionary patterns of *MAPK* and *MAPKK* genes in plants. It is possible that the original family member had experienced replication events prior to the eudicot-monocot divergence (Zhang et al., 2013).

The MAPK cascade genes have been studied to interact with signaling pathways mediated by hormone responsive cis-acting elements such as SA, JA, and ethylene (Seyfferth et al., 2014). Hormone responsive elements such as MeJA-responsive element (CGTCA-motif and TGACG-motif) (Shariatipour et al., 2019), ethylene-responsive ERE element (Itzhaki et al., 1994), gibberellin-responsive element (GARE-motif, P-box element, TATC-box) (Gubler et al., 1992], abscisic acid-responsive ABRE (Guiltinan et al., 1990) and salicylic acid-responsive TCA element (Goldsbrough et al., 1993) were widely found in the promoters of *PgMAPK* and *PgMAPKK* genes in *Punica ganatum*. These results indicated that they were widely involved in pomegranate life cycles (Krysan and Colcombet et al., 2018).

Previous studies have confirmed that the response of plants to the development and biological process are greatly regulated at the transcription level. RNA-seq data showed that the expression levels of *PgMAPK*s and *PgMAPKK*s varied greatly in different tissues, which was consistent with the research of pepper and other species. Obviously, some *PgMAPK* and *PgMAPKK* genes were highly expressed while others expressed at a low level or not expressed, suggesting their diverse functions in pomegranate growth and development (Chardin and Schenk et al., 2017). For example, *MAPKs* in rice can regulate auxin signaling and cell cycle-related gene expression of root under cadmium stress (Zhao and Hu et al., 2013).

Our results will lay the foundation for the functional characterization of *MAPK* and *MAPKK* gene family, and further understanding of the structure-function relationship among gene members. In addition, our study also provides comprehensive information and new insights into the evolution and differentiation of *PgMAPK* and *PgMAPKK* genes. These studies may provide a molecular basis for key agronomic traits, developmental and other physiological processes in pomegranates.

Conclusions

In this study, a total of 18 *PgMAPK* and 9 *PgMAPKK* genes were identified in pomegranate and their phylogenetic relationships were explored. The gene structure of each group members was similar. *PgMAPKs* and *PgMAPKKs* may participate in the apical meristems, flower organ and fruit development, and the same group may have the similar expression pattern. These results lay the foundation for function studies on *PgMAPKs* and *PgMAPKKs* during their evolutionary process.

References

Andreasson, E. and Ellis, B. 2010. Convergence and specificity in the *Arabidopsis* MAPK nexus. *Trends Plant Sci.*,15, 106–113.

Artimo, P., Jonnalagedda, M., Arnold, K., Baratin, D., Csardi, G., de Castro, E., Duvaud, S., Flegel, V., Fortier, A., Gasteiger, E., et al. 2012. ExPASy: SIB bioinformatics resource portal. *Nucleic Acids Res.*, 40, 597–603.

Asai, T., Tena, G., Plotnikova, J., Willmann, M.R., Chiu, W., Gomez-Gomez, L., Boller, T., Ausubel, F.M. and Sheen, J. 2002. MAP kinase signalling cascade in *Arabidopsis* innate immunity. *Nature,* 415, 977–983.

Bailey, T.L., Boden, M., Buske, F.A., Frith, M., Grant, C.E., Clementi, L., Ren, J., Li, W.W. and Noble, W.S. 2009. MEME SUITE: Tools for motif discovery and searching. *Nucleic Acids Res.*, 37, 202–208.

Bray, N.L., Pimentel, H., Melsted, P. and Pachter, L. 2016. Near - optimal probabilistic RNA - seq quantification. *Nat.Biotechnol.*, 34, 525–527.

Bohnert, H., Nelson, D. and Jensen, R. 1995. Adaptations to environmental stresses. *Plant Cell,* 7, 1099–1111.

Chardin, C., Schenk, S., Hirt, H., Colcombet, J. and Krapp, A. 2017. Review: Mitogen-activated protein kinases in nutritional signaling in *Arabidopsis. Plant Sci.*, 260.

Chen, C., Xia, R., Chen, H. and He, Y. 2018. TBtools, a toolkit for biologists integrating various HTSdata handling tools with a user-friendly interface. *bioRxiv,* 289660.

Chen, S., Zhou, Y., Chen, Y. and Gu, J. 2018. fastp: An ultra-fast all-in-one FASTQ preprocessor. *Bioinformatics,* 34(17): 884–890.

Cui, L., Yang, G., Yan, J., Pan, Y. and Nie, X. 2019. Genome-wide identification, expression profiles and regulatory network of MAPK cascade gene family in barley. *BMC Gen.*, 20, 750.

Danquah, A., de Zélicourt, A., Boudsocq, M., Neubauer, J., Frei Dit Frey, N., Leonhardt, N., Pateyron, S., Gwinner, F., Tamby, J., Ortiz - Masia, D., et al. 2015. Identification and characterization of an ABA-activated MAP kinase cascade in *Arabidopsis thaliana. Plant J.*, 82, 232–244.

De Zelicourt, A., Colcombet, J. and Hirt, H. 2016. The role of MAPK modules and ABA during abiotic stress signaling. *Trends Plant Sci.*, 21, 677–685.

Feng, K., Liu, F., Zou, J., Xing, G., Deng, P., Song, W., Tong, W. and Nie, X. 2016. Genome - wide identification, evolution, and co - expression network analysis of mitogen-activated protein kinase kinase kinases in Brachypodium distachyon. *Front. Plant Sci.*, 7, 1400.

Finn, R.D., Clements, J. and Eddy, S.R. 2011. HMMER web server: Interactive sequence similarity searching. *Nucleic Acids Res.*, 39, 29–37.

Ghawana, S., Kumar, S. and Ahuja, P.S. 2009. Early low - temperature responsive mitogen activated protein kinases RaMPK1 and RaMPK2 from Rheum australe D. Don respond differentially to diverse stresses. *Mol. Biol Rep.*, 37, 933.

Gubler, F. and Jacobsen, J. 1992. Gibberellin-responsive elements in the promoter of a barley high-pl [alpha]-amylase gene. *Plant Cell*, 4, 1435–1441.

Goldsbrough, A.P., Albrecht, H. and Stratford, R. 1993. Salicylic acid - inducible binding of a tobacco nuclear protein to a 10 bp sequence which is highly conserved amongst stress-inducible genes. *Plant J.*, 3, 563–571.

Goyal, R., Tulpan, D., Chomistek, N., González - Peña Fundora, D., West, C., Ellis, B., Frick, M., Laroche, A. and Foroud, N. 2018. Analysis of MAPK and MAPKK gene families in wheat and related Triticeae species. *BMC Gen.*, 19, 178.

Guiltinan, M.J., Marcotte, W.R., Jr., Quatrano, R.S. 1990. A plant leucine zipper protein that recognizes an abscisic acid response element. *Science*, 250, 267–271.

Hahn, A. and Harter, K. 2009. Mitogen - activated protein kinase cascades and ethylene: Signaling, biosynthesis, or both? *Plant Physiol.*, 149, 1207–1210.

Harel - Beja, R., Sherman, A., Rubinstein, M., Eshed, R., Bar - Ya Akov, I., Trainin, T., Ophir, R. and Holland, D. 2015. A novel genetic map of pomegranate based on transcript markers enriched with QTLs for fruit quality traits. *Tree Genet. Genomes*, 11, 109.

Holub, E. 2001. The arms race is ancient history in Arabidopsis, the wildflower. *Nat. Rev.*, 2, 516–527.

Ichimura, K., Casais, C., Peck, S.C., Shinozaki, K. and Shirasu, K. 2006. MEKK1 is required for MPK4 activation and regulates tissue-specific and temperature-dependent cell death in Arabidopsis. *J. Biol. Chem.*, 281, 36969–36976.

Itzhaki, H., Maxson, J.M. and Woodson, W.R. 1994. An ethylene - responsive enhancer element is involved in the senescence-related expression of the carnation glutathione-S-transferase (GST1) gene. *Proc. Natl. Acad Sci. USA*, 91, 8925–8929.

Jonak, C., Ökrész, L., Bögre, L. and Hirt, H., 2002. Complexity, cross talk and integration of plant MAP kinase signalling. *Curr. Opin. Plant Biol.*, 5, 415–424.

Kazuya, I., Ichimura, K., Shinozaki, K., Tena, G., Sheen, J., Henry, Y., Champion, A., Kreis, M., Zhang, S., Hirt, H., et al. 2002. Mitogen - activated protein kinase cascades in plants: A new nomenclature. *Trends Plant Sci.*, 7, 301–308.

Kersey, P.J., Allen, J.E., Allot, A., Barba, M., Boddu, S., Bolt, B.J., Carvalho-Silva, D., Christensen, M.,Davis, P., Grabmueller, C., et al. 2017. Ensembl Genomes 2018: An integrated omics infrastructure for non-vertebrate species. *Nucleic Acids Res.*, 46, 802–808.

Kovtun, Y., Chiu, W., Zeng, W. and Sheen, J. 1998. Suppression of auxin signal transduction by a MAPK cascade in higher plants. *Nature*, 395, 716–720.

Kong, F., Wang, J., Cheng, L., Liu, S., Wu, J., Peng, Z. and Lu, G. 2012. Genome-wide analysis of the mitogen-activated protein kinase gene family in Solanum lycopersicum. *Gene*, 499, 108–120.

Kong, X., Pan, J., Zhang, D., Jiang, S., Cai, G., Wang, L. and Li, D. 2013. Identification of mitogenactivated protein kinase kinase gene family and MKK–MAPK interaction network in maize. Biochem. Biophys. *Res. Commun.*, 441, 964–969.

Krysan, P. and Colcombet, J. 2018. Cellular complexity in MAPK signaling in plants: Questions and emerging tools to answer them. Front. *Plant Sci.*, 9.

Liu, C., Zhao, Y., Zhao, X., Wang, J., Gu, M. and Yuan, Z. 2020. Transcriptomic profiling of pomegranate provides insights into salt tolerance. *Agronomy*, 10, 44.

Liu, Y., Zhang, D., Wang, L. and Li, D. 2013. Genome-wide analysis of mitogen-activated protein kinase gene family in maize. *Plant Mol. Biol. Rep.*, 31, 1446–1460.

Liu, Z., Zhang, L., Xue, C., Fang, H., Zhao, J. and Liu, M. 2017. Genome - wide identification and analysis of MAPK and MAPKK gene family in Chinese jujube (Ziziphus jujuba Mill.). *BMC Gen.*, 18, 855.

Liu, Z., Shi, L., Liu, Y., Tang, Q., Shen, L., Sheng, Y., Cai, J., Yu, H., Wang, R., Wen, J., et al. 2015. Genome - wide identification and transcriptional expression analysis of mitogen - activated protein kinase and mitogen-activated protein kinase kinase genes in Capsicum annuum. *Front. Plant Sci.*, 6.

Luo, X., Li, H., Wu, Z., Yao, W., Zhao, P., Cao, D., Yu, H., Li, K., Poudel, K., Zhao, D., et al. 2019. The pomegranate (Punica granatum L.) draft genome dissects genetic divergence between soft - and hard-seeded cultivars. *Plant Biotechnol. J.*, 18, 955–968.

Magali, L., Patrice, D., Gert, T., Kathleen, M., Yves, M., Yves, V.D.P., Pierre, R. and Stephane, R. 2002. PlantCARE, a database of plant cis-acting regulatory elements and a portal to tools for in silico analysis of promoter sequences. *Nucleic Acids Res.*, 30, 325–327.

Marchler-Bauer, A., Bo, Y., Han, L., He, J., Lanczycki, C.J., Lu, S., Chitsaz, F., Derbyshire, M.K., Geer, R.C.,Gonzales, N.R., et al. 2017. CDD/SPARCLE: Functional classification of proteins via subfamily domain architectures. *Nucleic Acids Res.*, 45, 200–203.

Moustafa, K., AbuQamar, S., Jarrar, M., Al - Rajab, A.J., Trémouillaux - Guiller, J. 2014. MAPK cascades and major abiotic stresses. *Plant Cell Rep.*, 33, 1217–1225.

Nakagami, H., Pitzschke, A. and Hirt, H. 2005. Emerging MAP kinase pathways in plant stress signalling. *Trends Plant Sci.*, 10, 339–346.

Nicolas, L.B., Harold, P., Páll, M. and Lior, P. 2016. Near-optimal probabilistic RNA-seq quantification. *Nature Biotech.*, 34: 525–27.

Nicole, M., Hamel, L., Morency, M., Beaudoin, N., Ellis, B.E. and Séguin, A. 2006. MAP-ping genomic organization and organ-specific expression profiles of poplar MAP kinases and MAP kinase kinases. *BMC Gen.*, 7, 223.

Pandhari, W.D., Misra, M., Singh, P., Sinha, A.K. and Provart, N.J. 2013. Rice mitogen activated protein kinase Kinase and mitogen activated protein kinase interaction network revealed by in-silico docking and yeast two-hybrid approaches. *PLoS ONE*, 8, e65011.

Qin, G., Liu, C., Li, J., Qi, Y., Gao, Z., Zhang, X., Yi, X., Pan, H., Ming, R. and Xu, Y. 2020. Diversity of metabolite accumulation patterns in inner and outer seed coats of pomegranate: Exploring their relationship with genetic mechanisms of seed coat development. *Hortic. Res.*, 7, 10.

Racine, J.S. 2012. RStudio: A platform-independent IDE for R and sweave. *J. Appl. Econ.*, 27, 167–172.

Raina, S.K., Wankhede, D.P., Jaggi, M., Singh, P., Jalmi, S.K., Raghuram, B., Sheikh, A.H. and Sinha, A.K. 2012. CrMPK3, a mitogen activated protein kinase from Catharanthus roseusand its possible role in stress induced biosynthesis of monoterpenoid indole alkaloids. *BMC Plant Biol.*, 12, 134.

Rasmussen, M., Roux, M., Petersen, M. and Mundy, J. 2012. MAP kinase cascades in Arabidopsis innate immunity. *Front. Plant Sci.*, 3, 169.

Rao, K.P., Tambi, R., Kundan, K., Badmi, R. and Krishna, S.A. 2010. In silico analysis reveals 75 members of mitogen-activated protein kinase kinase kinase gene family in rice. *DNA Res.*, 3, 139–153.

Ren, D., Yang, K.Y., Li, G.J., Liu, Y. and Zhang, S. 2006. Activation of Ntf4, a tobacco mitogen - activated protein kinase, during plant defense response and its involvement in hypersensitive response-like cell death. *Plant Physiol.*, 141, 1482–1493.

Schultz, J., Milpetz, F., Bork, P. and Ponting, C.P. 1998. SMART, a simple modular architecture research tool:Identification of signaling domains. *Proc. Natl. Acad. Sci. USA*, 95, 5857–5864.

Seyfferth, C. and Tsuda, K. 2014. Salicylic acid signal transduction: The initiation of biosynthesis, perception and transcriptional reprogramming. *Front. Plant Sci.*, 5, 697.

Shariatipour, N. and Heidari, B. 2020. Meta - analysis of expression of the stress tolerance associated genes and uncover their cis-regulatory elements in rice (Oryza sativa L.). *Open Biol. J.*, 13, 39–49.

Shen, H., Liu, C., Zhang, Y., Meng, X., Zhou, X., Chu, C. and Wang, X. 2012. OsWRKY30 is activated by MAP kinases to confer drought tolerance in rice. *Plant Mol. Biol.*, 80, 241–253.

Song, Q., Li, D., Dai, Y., Liu, S., Huang, L., Hong, Y., Zhang, H. and Song, F. 2015. Characterization, expression patterns and functional analysis of the MAPK and MAPKK genes in watermelon (Citrullus lanatus). *BMC Plant Biol.*, 15, 298.

Suarez-Rodriguez, M.C., Adams-Phillips, L., Liu, Y., Wang, H., Su, S., Jester, P.J., Zhang, S., Bent, A.F. and Krysan, P.J. 2007. MEKK1 is required for flg22 - induced MPK4 activation in Arabidopsis plants. *Plant Physiol.*, 143, 661–669.

Tena, G., Asai, T., Chiu, W. and Sheen, J. 2001. Plant mitogen - activated protein kinase signaling cascades. *Curr. Opin. Plant Biol.*, 4, 392–400.

Thomma, B.P., Penninckx, I.A., Cammue, B.P. and Broekaert, W.F. 2001. The complexity of disease signaling in Arabidopsis. *Curr. Opin. Immunol.*, 13, 63–68.

Wang, G., Lovato, A., Polverari, A., Wang, M., Liang, Y., Ma, Y. and Cheng, Z. 2014. Genome-wide identification and analysis of mitogen activated protein kinase kinase kinase gene family in grapevine (Vitis vinifera). *BMC Plant Biol.*, 14, 219.

Wang, H., Gong, M., Guo, J., Xin, H., Gao, Y., Liu, C., Dai, D. and Tang, L. 2018. Genome - wide identification of Jatropha curcas MAPK, MAPKK, and MAPKKK gene families and their expression profile under cold stress. *Sci. Rep.*, 8, 16163.

Wang, J., Pan, C., Wang, Y., Ye, L., Wu, J., Chen, L., Zou, T. and Lu, G. 2015. Genome - wide identification of MAPK, MAPKK, and MAPKKK gene families and transcriptional profiling analysis during development and stress response in cucumber. *BMC Gen.*, 16, 386.

Wang, L., Hu, W., Tie, W., Ding, Z., Ding, X., Liu, Y., Yan, Y., Wu, C., Peng, M., Xu, B., et al. 2017. The MAPKKK and MAPKK gene families in banana: Identification, phylogeny and expression during development, ripening and abiotic stress. *Sci. Rep.*, 7, 1159.

Wang, J., Ding, H., Zhang, A., Ma, F., Cao, J. and Jiang, M. 2010. A novel mitogen-activated protein kinase gene in maize (Zea mays), ZmMPK3, is involved in response to diverse environmental cues. *J. Integr. Plant Biol.*, 52, 442–452.

Xiong, L. and Yang, Y. 2003. Disease resistance and abiotic stress tolerance in rice are inversely modulated by an abscisic acid-inducible mitogen-activated protein kinase. *Plant Cell*, 15, 745–759.

Yan, M., Zhao, X., Zhou, J., Huo, Y., Ding, Y. and Yuan, Z. 2019. The complete chloroplast genomes of Punica granatum and a comparison with other species in lythraceae. *Int. J. Mol. Sci.*, 20, 2886.

Yuan, Z., Fang, Y., Zhang, T., Fei, Z., Han, F., Liu, C., Liu, M., Xiao, W., Zhang, W., Wu, S., et al. 2018. The pomegranate (Punica granatum L.) genome provides insights into fruit quality and ovule developmental biology. *Plant Biotechnol. J.*, 16, 1363–1374.

Yoo, S., Cho, Y., Tena, G., Xiong, Y. and Sheen, J. 2008. Dual control of nuclear EIN3 by bifurcate MAPK cascades in C2H4 signaling. *Nature*, 451, 789–795.

Zhan, H., Hong, Y., Zhao, X., Wang, M., Weining, S. and Nie, X. 2017. Genome-wide identification and analysis of MAPK and MAPKK gene families in bread wheat (Triticum aestivum L.). *Genes*, 8, 10.

Zhang, S. and Klessig, D.F. 2001. MAPK cascades in plant defense signaling. *Trends Plant Sci.*, 6, 520–527.

Zhang, S., Xu, R., Luo, X., Jiang, Z. and Shu, H. 2013. Genome - wide identification and expression analysis of MAPK and MAPKK gene family in Malus domestica. *Genes*, 531, 377–387.

Zhang, T., Liu, Y., Yang, T., Zhang, L., Xu, S., Xue, L. and An, L. 2006. Diverse signals converge at MAPK cascades in plant. *Plant Physiol. Biochem.*, 44, 274–283.

Zhao, F.Y., Hu, F., Zhang, S.Y., Wang, K., Zhang, C.R. and Liu, T. 2013. MAPKs regulate root growth by influencing auxin signaling and cell cycle-related gene expression in cadmium-stressed rice. Environ. *Sci.Pollut. Res.*, 20, 5449–5460.

Zhao, X., Yuan, Z., Feng, L. and Fang, Y. 2015. Cloning and expression of anthocyanin biosynthetic genes in red and white pomegranate. *J. Plant Res.*, 128, 687–696.

Zhou, H., Ren, S., Han, Y., Zhang, Q., Qin, L. and Xing, Y. 2017. Identification and analysis of mitogenactivated protein kinase (MAPK) cascades in *Fragaria vesca. Int. J. Mol. Sci.*, 18, 1766.

12

Unveiling the Genomic Landscape of MIKC-Type MADS-Box Genes in *Punica granatum* L.

A Genome-Wide Exploration

Yujie Zhao[1] and *Zhaohe Yuan*[2*]

MADS-box is a critical transcription factor regulating the development of floral organs and plays essential roles in the growth and development of floral transformation, flower meristem determination, the development of male and female gametophytes, and fruit development. In this study, 36 MIKC-type MADS-box genes were identified in the 'Taishanhong' pomegranate genome. By utilizing phylogenetic analysis, 36 genes were divided into 14 subfamilies. Bioinformatics methods were used to analyze the gene structure, conserved motifs, *cis*-acting elements, and the protein interaction networks of the MIKC-type MADS-box family members in pomegranate, and their expressions pattern in different tissues of pomegranate were analyzed. Tissue-specific expression analysis revealed that the E-class genes (*PgMADS03*, *PgMADS21*, and *PgMADS27*) were highly expressed in floral tissues, while *PgMADS29* was not expressed in all tissues, indicating that the functions of the E-class genes were differentiated. *PgMADS15* of the C/D-class was the key gene in the development network of pomegranate flower organs, suggesting that *PgMADS15* might play an essential role in the peel and inner seed coat development of pomegranate. The results in this study will provide a reference for the classification, cloning, and functional research of pomegranate MADS-box genes.

1 College of Horticulture, Henan Agricultural University, Zhengzhou 450002, China.
2 College of Forestry, Nanjing Forestry University, Nanjing 210037, China.
* Corresponding author: zhyuan88@hotmail.com

1. Introduction

MADS-box, its name from Minichromosome maintenance 1 (MCM1), AGAMOUS (AG), DEFICIENS (DEF), and serum response factor (SRF)) genes encode transcription factors, have been widely found in plants, fungi and animals, and contain the highly conservative MADS-domain composed of approximately 55 amino acids (Becker and Theissen, 2003; Messenguy and Dubois, 2003). On the basis of genetic structures and phylogenetic analysis, the MADS-box family can be divided into two phylogenetically distinct groups: Type I and Type II (Alvarezbuylla et al., 2000). Most of the well-studied plant genes are type II genes that have three more domains than those of type I genes: a more conservative intervening domain (I-domain), a moderately conservative keratin-like coiled-coil domain (K-domain), and a variable C-terminal domain (C-domain) (Smaczniak et al., 2012). These genes are considered as the MIKC-type and are specific to plants.

The plant-specific MIKC-type MADS-box genes were first identified as floral organ determinant genes in *Arabidopsis thaliana* and *Antirrhinum majus* (Sommer et al., 1990; Yanofsky et al., 1990). MADS-box family genes are involved in plant growth and development, flower transformation, flower meristem decision, male/female gametophyte development, fruit development and maturation, and somatic embryogenesis (Ma, 2005; Chi et al., 2011; Galimba and Stilio, 2015; Gao et al., 2020). Type I only contain MADS-domains without K-domains and are called M-type (Parenicova et al., 2003). Type I genes are divided into four subfamilies: Mα, Mβ, Mγ, and Mδ. Based on some research, the Mδ subfamily is referred to as Type II, so it is also called MIKC*-type (Arora et al., 2007; Xu et al., 2014). Type II genes can be divided into two subfamilies: MIKC* and MIKC^{C}. MIKC^{C} subfamily can be divided into 13 branches: AG, AGL6 (AGAMOUS-LIKE, AGL), AGL12, AGL15, AGL17, AP1(APETALA, AP)/FUL (FRUITFULL), AP3/PI (PISTILLATA), FLC (FLOWERING LOCUS C), SOC1 (SUPPRESSOR OF OVEREXPRESSION OF CONSTANS1), SEP (SEPALLATA), SVP (SHORT VEGETATIVE PHASE), BS (B-SISTER), and TM8 (Henschel, et al., 2002; Díaz-Riquelme et al., 2009). In *Arabidopsis*, MIKC^{C}-type embraces 12 subfamilies, without Tomato MADS 8 (TM8). So, there is little research on TM8's function (Heijmans et al., 2012).

According to the ABCDE model of flower development, the five class genes of A, B, C, D, and E coordinately regulate the development of sepals, petals, stamens, carpels, and ovules. Arabidopsis has two A-class genes (*AP1* and *AP2*), two B-class genes (*PI* and *AP3*), C/D-class genes (*AG*, *AGL11* and *STK* (*SEEDSTICK*)), and a single E-class gene (*SEP*), of which only *AP2* is not a MADS-box gene (Theisen et al., 2016). The model suggests that A + E control meristem and sepals' formation, A + B + E regulate petals formation, B + C + E control stamens' formation, C + E control carpels' formation, and C + D + E regulate ovules development (Theisen et al., 2016). Martín-Pizarro et al. (2019) identified two B-class genes *FaAP3* and *FaTM6* in strawberries (*Fragaria × ananassa*) and used CRISPR/Cas9 technology for the first time to study the function of *FaTM6* in flower development. They found that *FaTM6* played a key role in the development of strawberry petals and anthers;

the expression pattern of *FaTM6* was similar to *AP3* in *Arabidopsis* (Martín-Pizarro et al., 2019). *AG* is the first reported MADS-box gene to regulate stamens and carpels development (Winter et al., 1999). C-class genes regulate the establishment of floral meristems by inhibiting the activity of the A-class gene (Chen, 2004). In the control carpels, *STK*, *SHP1* and *SHP2* exist functional redundancy to some extent. *SHP1*, *SHP2* and *STK* jointly regulate the development of ovules, with *SHP1/SHP2* sharing some function of C- and D-class (Xu and Zhong, 2015). Ehlers et al. (2016) found that *shp1* and *shp2* could not coordinate the development of ovules, indicating that *SHP1* and *SHP2* might be the key genes regulating ovules development (Ehlers et al., 2016). There are two AG subfamily genes *TAG1* and *TAGL1* in tomato. Pan et al. (2010) found that the stamens and carpel of *TAGL1* RNAi strain were normal, and the fruit ripening was inhibited, speculating that *TAGL1* shows new functions (Pan et al., 2010). The development of stamens and carpel in *TAG1* RNAi were significantly inhibited, indicating that *TAG1* plays a major role in regulating the development of stamens and carpel (Pan et al., 2010). Xu et al. (2008) cloned three *SEP*-like genes, *PrpMADS2*, *PrpMADS5*, and *PrpMADS7*, in peach (Xu et al., 2008). They found that the transgenic plant of *PrpMADS2* had a similar phenotype with the wild. A strain with an overexpressed *PrpMADS5* flowered earlier than the control plant, and it was involved in the differentiation of flower primordia. *PrpMADS7* transgenic plants showed weak growth potential and earlier flowering than the wild, indicating new functions of the *SEP-like* gene (Xu et al., 2008). SEP participates in the formation of tetrameric protein complexes, and forms the complexes with A, B, and C-type proteins (Melzer and Theissen, 2009; Pan et al., 2014). These complexes help to transform leaf primordia into floral organ primordium, of which ectopic expression of A, B, and E leads to the transformation of leaves into petals. The expression of B, C and E gene causes the transformation of leaves into stamen organs (Honma and Goto, 2001).

Pomegranate (*Punica granatum* L.), belonging to the Lythraceae family, is an excellent fruit tree with economic, nutritional, medicinal, ornamental, and ecological values (Yuan, 2015). Pomegranate is mostly cultivated in tropics and subtropics regions, such as the Mediterranean basin, Asia, Australia, and North America (Holland et al., 2009). In China, the total scale of pomegranate cultivation is about 128,000 hm^2, with an annual output of about 1.7 million tons (Yuan, 2015). Pomegranate contains abundant polyphenols, tannins, anthocyanins, vitamins, and minerals that are responsible to reduce blood pressure, and act against serious diseases such as cancer (Huang et al., 2005; Lansky and Newman, 2007; Basu and Penugonda, 2009). These have led to an increasing demand for the consumption of fresh fruit, juice, tea, and other pomegranate products. The completion of genome sequencing of pomegranate and the publication of data (Qin et al., 2017; Yuan et al., 2018; Luo et al., 2020) provided necessary data support for the promotion of research on gene function of pomegranate. In this study, we aim to identify the MIKC-type MADS-box members of pomegranate by bioinformatics methods, trying to analyze their gene structure, *cis*-acting elements, and tissue-specific expression to explore their functions in the development of pomegranate flowers.

The results of this study would lay a foundation for the cloning and functional research of *PgMADS* genes.

2. Materials and Methods

2.1 Data Collection for Expression Analysis

The transcription data of pomegranate were obtained from the NCBI database (https://www.ncbi.nlm.nih.gov/), including leaves, roots, bisexual flowers, functional male flowers, exocarp, pericarp, and other tissues. The respective accession sequence numbers are 'Dabenzi' SRR5678820, SRR5279388–SRR5279397; 'Baiyushizi' SRR5678819; 'Tunisian' SRR5446590–SRR5446607; 'Black127' SRR1054190; 'Nana' SRR1055290; and 'Wonderful' SRR080723 (Table S1 (Qin et al., 2017; Chen et al., 2017; Ono et al., 2011; Ophir et al., 2014).

MIKC-type MADS-box family protein sequences of *Arabidopsis thaliana* were downloaded from the Arabidopsis Information Source (https://www.arabidopsis.org/), and that of poplar (*Populus trichocarpa*) and apple (*Malus domestica*) were downloaded from Plant Transcription Factor Database (http://planttfdb.cbi.pku.edu.cn/index.php).

2.2 Identification and Sequence Analysis of Pomegranate MIKC-type MADS-box Transcription Factors

The HMM (Hidden Markov Models) profile of the Pfam MADS-domain (PF00319) was downloaded from the pfam database (http://pfam.xfam.org/). Then, the HMM profile was performed against pomegranate protein databases (GCA_002864125) using the HMMER 3.0 software package (The European Bionfirmatics Institute, Hinxton, Cambridgeshire, UK), (E-value < $1e^{-5}$), and redundant sequences were removed manually. Moreover, all obtained MADS protein sequences were further analyzed on the CDD website (https://www.ncbi.nlm.nih.gov/cdd) (Marchler-Bauer et al., 2017) to verify the presence of the MADS-domain. Identified MADS-box protein sequences were categorized using the PlantTFDB to follow-up analysis for M-type and MIKC-type MADS-box members (Jin et al., 2014; Gao et al., 2017).

The published MADS-box protein sequences of Arabidopsis, poplar and apple were used as queries to perform BLAST against the pomegranate protein database (E-value < $1e^{-10}$, identity > 50%), duplication was removed. Having predicted and divided members into MIKC-type and M-type using the PlantTFDB website (Jin et al., 2014; Gao et al., 2017), MIKC-type MADS-box protein sequences were screened and selected.

ExPASy Proteomics Server (https://web.expasy.org/protparam/) was used to predict the physicochemical properties of MIKC-type MADS-box protein, including amino acid sequence length, molecular weight, and isoelectric point.

2.3 Phylogenetic Analysis of Pomegranate MIKC-type MADS-box Family

Multiple sequences alignment of pomegranate, Arabidopsis (45 members) and Poplar (51 members) MIKC-type MADS-box proteins were carried out using ClustalX 2.1 software (http://www.clustal.org/omega/), and artificially corrected them. An unrooted neighbor-joining (NJ) phylogenetic tree was constructed with all of the MIKC-type MADS-box protein sequences from pomegranate, Arabidopsis, poplar using MEGA 7.0 (Penn State-A, State College, PA, USA) (Kumar et al., 2016). Bootstrap analysis was performed using 1,000 repetitions with the following parameters: complete deletion and Poisson model. The phylogenetic tree was beautified with EvolView (https://www.evolgenius.info/evolview/#login).

The subfamily was classified according to the support rate of subfamily branches (> 70) and the structural integrity of phylogenetic tree. Subfamily classification was obtained by Díaz-Riquelme et al. (2009).

2.4 Gene Structure Analysis and Motif Identification of Pomegranate MIKC-type MADS-box Family

The protein sequences of the MIKC-type MADS-box were extracted from the pomegranate genome. Motifs of MIKC-type MADS-box proteins were identified using MEME online tools (http://meme-suite.org/tools/meme) with a default parameter, and the motif characteristics of the MIKC-type MADS-box proteins were obtained. The SMART program (http://smart.embl.de/) was used to further analysis for the conservative motifs.

The annotation information of 36 MIKC-type MADS-box genes was extracted from the pomegranate genome annotation file, then the results were submitted to Gene Structure Display Server (GSDS 2.0: http://gsds.cbi.pku.edu.cn) for online genetic structure analysis and drawing Schematic diagram of gene structures.

2.5 Prediction Promoter Elements and Protein Interaction Network of Pomegranate MIKC-type MADS-box Family

To identify putative *cis*-acting elements in promoters, 1,500 bp upstream sequences of the initiation codon were obtained from the pomegranate genome sequence by TBtools (GitHub Inc., San Francisco, CA, USA). The *cis*-acting elements were analyzed by PlantCARE (http://bioinformatics.psb.ugent.be/webtools/plantcare/html/) (Jin et al., 2014). The result was presented as images using Tbtools.

The protein–protein interaction network of the MIKC-type MADS-box family was analyzed by String (https://string-db.org/).

2.6 Expression Analysis for Pomegranate MIKC-type MADS-box Family

To study the expression patterns of MIKC-type MADS-box genes in pomegranate different tissues, the published transcriptome data were downloaded as the primary data source for expression analysis from the NCBI database (http://www.ncbi.nlm.nih.gov/). Firstly, all RNA-Seq data were qualitatively controlled by fastp (Chen et al., 2018) to obtain cleaned reads. Then the sequences were indexed with pomegranate transcriptome data. Through the use of Kallisto 0.44.0 software (https://pachterlab.github.io/kallisto/download.html) (Nicolas et al., 2016), and the gene expression levels was further calculated and analyzed. The corresponding expression level (TPM (Transcripts Per Million) value) of MIKC-type MADS-box genes was thus obtained. TPM value was transformed into Log_2 (TPM + 1). Finally, the thermal map was drawn with Tbtools.

3. Results

3.1 Identification and Sequence Analysis of Pomegranate MIKC-type MADS-box Family

A total of 81 MADS-box family members were identified from the 'Taishanhong' pomegranate genome using the HMMER search program. Through CDD online verification, all members contained the MADS-domain. After classifying of M-type and MIKC-Type family on PlantTFDB, 36 MIKC-type MADS-box candidate members were obtained.

Subsequently, 71 MADS-box family members were identified using the local blast (ftp://ftp.ncbi.nlm.nih.gov/blast/executables/blast+/LATEST/). It was found that eight members contained the type II specific K-domain, without MADS-domain, so they were discarded. Through classification analysis on PlantTFDB, 36 MIKC-type MADS-box members were obtained.

Based on the identification and classification of above two methods, a total of 36 MIKC-type MADS-box genes were identified in pomegranate, which were renamed as *PgMADS01~PgMADS36* according to the gene sequence number (Xu et al., 2014), for convenience in subsequent analysis (Table 1).

The physicochemical properties of 36 MIKC-type MADS-box amino acid sequences were analyzed. The results were shown in Table 1. The length of the pomegranate MIKC-type genes varied greatly. The shortest length of the encoded protein was 167 aa, the longest was 657 aa. The corresponding protein molecular mass was between 19,349.0~74,414.5 Da, the average molecular weight was 28553.23 Da. The isoelectric point (PI) ranged from 4.9 (PgMADS24) to 10.89 (PgMADS32), which contained nine acidic proteins and 27 basic proteins. These results would provide a theoretical basis for PgMADS protein purification and functional research.

3.2 Phylogenetic Analysis of Pomegranate MIKC-type MADS-box Family

To study the phylogenetic relationships in MIKC-type MADS-box proteins, these protein sequences of pomegranate, Arabidopsis and poplar, were used to construct the phylogenetic tree. In this study, the phylogenetic tree was divided into 14

Table 1 The basic information of the pomegranate MIKC-type MADS-box gene family.

Gene Name	*Gene ID*	*Location*	*CDS (bp)*	*Exon No.*	*Strand*	*Amino Acid Residues*	*MW(Da)*	*PI*	*Subfamily*
PgMADS01	Pg000170.1	scaffold1:4730801:4734403	720	7	–	239	27591.1	9.76	A(AP1-FUL)
PgMADS02	Pg001270.1	scaffold10:781136:787463	576	5	–	191	22251.1	8.76	SOC1
PgMADS03	Pg002695.1	scaffold11:837228:841996	846	7	–	281	31958.1	8.19	E(SEP)
PgMADS04	Pg002696.1	scaffold11:827349:833260	600	7	–	199	22764.8	9.58	FLC
PgMADS05	Pg002697.1	scaffold11:821950:824737	603	6	–	200	22793.1	10.17	FLC
PgMADS06	Pg003192.1	scaffold111:252873:258483	522	6	–	173	20047.1	9.59	C/D(AG)
PgMADS07	Pg003228.1	scaffold111:423542:426294	645	7	+	214	24079.5	8.95	AGL12
PgMADS08	Pg003757.1	scaffold119:139771:142418	639	7	–	212	24900.2	9.28	B(AP3-PI)
PgMADS09	Pg005184.1	scaffold13:3171597:3176055	741	8	–	246	28261	8.58	E(SEP)
PgMADS10	Pg005185.1	scaffold13:3153310:3158145	807	8	–	268	30475.7	8.8	A(AP1-FUL)
PgMADS11	Pg006006.1	scaffold136:390163:393471	906	9	–	301	34973.4	9.9	BS
PgMADS12	Pg008155.1	scaffold16:570614:573248	885	7	–	294	33598.8	6.39	AGL15
PgMADS13	Pg008465.1	scaffold16:2554235:2556059	642	7	+	213	23721.7	8.48	SVP
PgMADS14	Pg008776.1	scaffold17:2387895:2392008	687	8	–	228	25779.9	5.52	SVP
PgMADS15	Pg009904.1	scaffold184:267398:273943	765	5	–	254	28775.2	8.88	C/D(AG)
PgMADS16	Pg014226.1	scaffold24:1271362:1277266	504	5	+	167	19349.0	5.97	MIKC*
PgMADS17	Pg014283.1	scaffold24:2253081:2257895	687	8	+	228	25738.3	6.53	SVP
PgMADS18	Pg014451.1	scaffold25:814656:818749	717	8	–	238	27124.6	9.52	AGL6
PgMADS19	Pg014630.1	scaffold25:796521:803849	654	7	+	217	24695	7.93	SOC1
PgMADS20	Pg016044.1	scaffold3:3612017:3628588	612	7	–	203	23056.1	4.99	FLC

Contd.

Table 1 *Contd.*

Gene Name	*Gene ID*	*Location*	*CDS (bp)*	*Exon No.*	*Strand*	*Amino Acid Residues*	*MW(Da)*	*PI*	*Subfamily*
PgMADS21	Pg016673.1	scaffold3:3637896:3641838	768	8	–	255	29008.8	8.85	E(SEP)
PgMADS22	Pg017747.1	scaffold33:1901905:1904548	759	8	+	252	28454	7.62	AGL15
PgMADS23	Pg018533.1	scaffold37:1603815:1615806	837	7	–	278	31322	6.89	FLC
PgMADS24	Pg020204.1	scaffold40:966778:969371	699	7	+	232	26761	4.9	B(AP3-PI)
PgMADS25	Pg020205.1	scaffold40:971279:973422	708	7	+	235	27089.6	6.31	B(AP3-PI)
PgMADS26	Pg022433.1	scaffold5:1599117:1601542	675	7	–	224	25754	9.07	B(AP3-PI)
PgMADS27	Pg023007.1	scaffold50:465601:468168	750	7	+	249	28634.4	8.57	E(SEP)
PgMADS28	Pg023008.1	scaffold50:480602:487666	699	7	+	232	26732.3	9.43	A(AP1-FUL)
PgMADS29	Pg024984.1	scaffold60:1113804:1120065	753	8	–	250	28461.2	9.5	E(SEP)
PgMADS30	Pg026995.1	scaffold7:90410:99792	1974	11	+	657	74414.5	9.56	C/D(AG)
PgMADS31	Pg027190.1	scaffold71:566889:578438	654	7	+	217	24935.5	9.15	SOC1
PgMADS32	Pg027284.1	scaffold72:255144:257303	528	6	–	175	20534.3	10.89	TM8
PgMADS33	Pg028066.1	scaffold79:321847:329529	672	8	–	223	25020.5	10.39	SVP
PgMADS34	Pg028180.1	scaffold8:2827697:2833438	663	7	–	220	25172.5	10.04	SOC1
PgMADS35	Pg029246.1	scaffold86:465224:468460	762	7	–	253	28671.7	8.89	AGL17
PgMADS36	Pg029338.1	scaffold86:426938:437880	1443	13	–	480	55016.3	6.61	AGL17

Note: CDS-coding sequence; MW-molecular weight.

subfamilies, according to Díaz-Riquelme et al. (2009) and Leseberg et al. (2012). Among the 36 type II members, only PgMADS16 was classified as MIKC*. The other 35 members were classified as MIKC^C, which were divided into 13 subclasses. Among the 13 subfamilies, E (SEP) contained five members; SOC1, B (AP3-PI), and SVP each contained four members; C/D (AG), A (FUL), and FLC contained three member genes; AGL17 had two members. The group of AGL12, AGL6, BS, AGL15, and TM8 subfamilies had only one gene.

3.3 Gene Structure Analysis and Motif Identification of Pomegranate MIKC-type MADS-box Family

In the prediction of conservative motifs of MIKC-type MADS-box members, 10 conservative motifs were identified in pomegranate. According to the distribution and number of motifs, the number of conservative motifs of MIKC-type MADS-box protein varied greatly in pomegranate. Only PgMADS16 contained two conservative motifs, while PgMADS36 contained 12 motifs. The SMART analysis showed that motif 1 was the typical MADS-domain, consisting of about 60 amino acids, and motif 2 was located in the K-box domain. All PgMADS proteins contain motif 1. Except for PgMADS05, PgMADS20, PgMADS16, and PgMADS6, other members contained motif 3 with unknown functions. Besides PgMADS28, PgMADS01, PgMADS04, PgMADS16, and PgMADS02, other members contained conservative motif 4. Motif 5 only existed in A, E, AGL6, and AGL12 subfamilies. Motif 8 was shared by BS, SOC1, and AGL17. It was worth noting that the K-box domain was not found in PgMADS16, but there was a specific K-box motif 2.

The gene structures were investigated on the gene structure display server (GSDS). The length of genes ranged from 2 kb–17 kb. Among 36 *PgMADSs*, there were 18 genes composed of six introns and seven exons, accounting for 50%. *PgMADS36* with 12 introns contains the most exons. Comparative analysis showed that the introns of TM8, BS, AGL15, and B subgroup genes were shorter, resulting in the shorter length of their genes than other subfamilies. Most FLC members had longer introns than other subfamily genes. The number of exons was seven in the B subfamily. The number of introns of the C/D class ranged from four (*PgMADS15*) to 10 (*PgMADS30*).

3.4 Prediction Promoter Elements of the Pomegranate MIKC-type MADS-box Family

In this study, the *cis*-element analysis showed that promoters of *PgMADSs* contained a wide variety of *cis*-acting elements, ranging from 18 *cis*-acting elements in *PgMADS29* to 30 in *PgMADS30*. *PgMADSs* responded to different plant hormones and abiotic stress signals, including abscisic acid response elements (ABRE) (Hobo et al., 1999), gibberellin response elements (GARE-motif, TATC-box and P-box) (Gubler and Jacobsen, 1992), auxin response elements (TGA-element, AuxRR-core, TGA-box and AuXRE) (Khan et al., 2012), methyl jasmonate response elements

(TGACG-motif and CGTCA-motif) (Rouster et al., 1997), salicylic acid response elements (TCA-element) (Shah and Klessig, 1996), and stress-related response elements (ARE, TC-rich repeats, WUN-motif, LTR, and MBS) (Yoshida et al., 1998). The ABRE element was abundant in pomegranate, with 19 genes containing two or more ABRE. *PgMADS13* promoter sequence contained up to 10 ABRE, indicating a strong response to the abscisic acid signal. Many members contained gibberellin response elements, such as the GARE-motif of *PgMADS01* in subfamily A, the TATC-box of *PgMADS09* in subfamily E, and the P-box of *PgMADS07* in subfamily AGL12. Except for AGL12 and TM8 subfamilies without LTR, other genes contained LTR elements. *PgMADS21* and *PgMADS35* contained two drought-induced MYB binding sites MBS. There were 15 genes containing auxin response elements, among which *PgMADS28* and *PgMADS12* contained two TGA-elements, while *PgMADS05* contained both TGA-element and AuxRE. *PgMADSs* were rich in methyl jasmonate and salicylic acid response elements. In addition to the AGL12 subfamily, the remaining genes of the AGL12 family contained one or more methyl jasmonate and salicylic acid response elements.

3.5 Protein Interaction Networks of the Pomegranate MIKC-type MADS-box Family

The protein interaction between MIKC-type PgMADSs was analyzed to know potential functions and signal transduction or metabolic pathways. The result showed a close interaction among A, B, C/D, and E types, among which the C/D genes (PgMADS30, PgMADS15, and PgMADS06) also had an interaction relationship (Figure 1). PgMADS15 was located at the core of the interaction network and interacted with A, B, and E types genes. PgMADS15 had stronger interaction with PgMADS06, PgMADS09, PgMADS27, and PgMADS29. Class B genes can form homologous dimers, but they steadily play function when they have formed heterodimers with AP3 and PI (Winter et al., 2002). There was an interaction between AP3 (PgMADS24, PgMADS25 and PgMADS26) and PI (PgMADS08) in pomegranate. E class genes could interact with multiple genes. When MIKC-type MADS-box protein formed a polymer, E genes acted as a 'binder' binding the dimer complex formed to form a polymer (Melzer and Theissen, 2009; Pan et al., 2014). PgMADS03 and PgMADS21 belonging to E interacted with A, B, and C/D classes genes, indicating that E genes might play an essential role in the formation of MIKC complexes related to the organ development of pomegranate flower.

According to the PgMADSs family protein interaction network, PgMADS15 was at the core of the interaction network. PgMADS15 had high similarity with Arabidopsis AtAG (bit score 247.7, e-value $7.9E^{-66}$). AG interacted with LFY (LEAFY) and AP2 proteins to regulate the flowering transition. AG, BEL1 and SUP coordinated to regulate ovule development. AG interacted with uncertain function proteins such as HUA1, ICU2, and LUG. It is predicted that PgMADS15 had the function of regulating flowering, ovule, and other flower organ development.

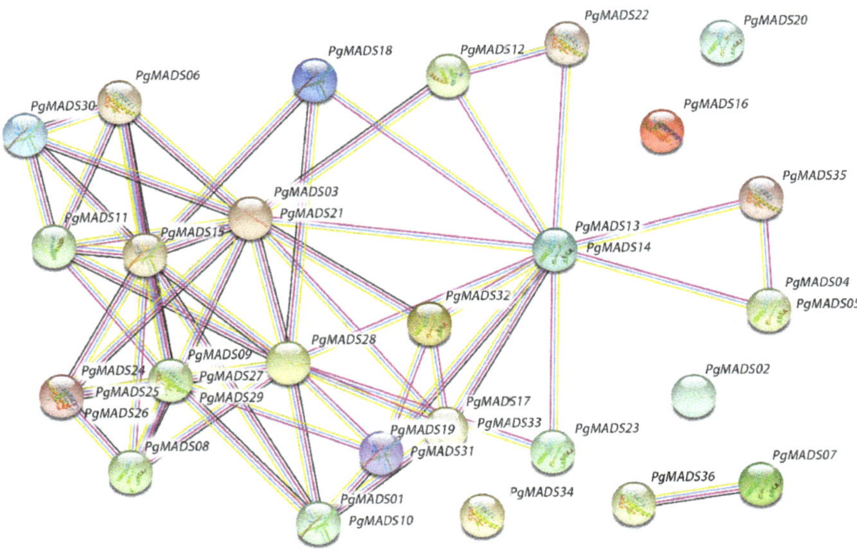

Fig. 1 Protein-protein interaction network of MIKC-type MADS-box family in pomegranate.

3.6 Expression Analysis for Pomegranate MIKC-type MADS-box Family

According to the published transcriptome data, MIKC-type MADS-box genes are expressed in leaf, root, flower, fruit, endocarp, and pericarp. There were remarkable differences in tissue-specific expression and expression quantity. Significant differences were also found in the expression quantities of different subfamily genes of MIKC-type MADS-box. The expression levels of E genes were the highest in flower organs, followed by C/D subfamily genes. Most genes of other subfamilies were low or not expressed in leaves and roots. The analysis found that the expression levels of the same subfamily genes in different tissues exhibited obvious differences, such as E-type genes. *PgMADS03* and *PgMADS21* were not expressed in root and leaf, while they were highly expressed in all tissues of pomegranate flower organs. *PgMADS09* was not expressed in leaves, roots and inner/outer seed skins, while *PgMADS29* was almost not expressed in all tissues. These results indicated that the function of E genes may occur with differentiation. *PgMADS06* and *PgMADS30* of C/D subfamily genes showed similar tissue-specific expression patterns, but *PgMADS15* was not expressed in functional male flowers and bisexual flowers. *PgMADS17*, *PgMADS19*, *PgMADS31*, and *PgMADS34* were only expressed in leaves and roots, indicating that the C/D subfamily genes of MADS-box expression were highly specific.

The obtained 36 genes can be divided into three groups. The genes in group I were highly expressed in flowers, mixtures, seed coats and peels, indicating that these genes play essential roles in the development of flowers and fruits.

The expression levels of group II genes were lower than that in group I and III genes, and the genes of the second subgroup in group II were highly expressed in roots and leaves. The expression levels of group III genes were higher than in other groups in floral organ development. Cluster expression analysis showed that different subfamilies genes might have similar expression patterns, such as the E gene (*PgMADS03*, *PgMADS21,* and *PgMADS27*) and C/D gene (*PgMADS06* and *PgMADS30*). They were not expressed or behaved at a low expression level in roots and leaves, but were highly expressed in other tissues.

4. Discussion

In the plants, MIKC-type MADS-box transcription factors contain MADS-domain and K-box domain, and play important roles in flowering and floral organ development (Folter et al., 2006; Mondragonpalomino and Theisen, 2011). In recent years, the identification and evolutionary analysis of the MIKC-type MADS-box family have been reported in many species, such as 32 members in *Prunus mume* (Xu et al., 2014), 33 members in *Fragaria vesca* (Zhou et al., 2014), 38 members in *Vitis vinifera* (Díaz-Riquelme et al., 2009), 39 members in *Prunus persica* (Well et al., 2015), 72 members in *Glycine max* (Shu et al., 2013), and 82 genes in *Malus domestica* (Tian et al., 2014). Genomic data of three pomegranate varieties have been published on the NCBI database, but the identification and analysis of the MIKC-type MADS-box family have not been reported. In this study, a total of 36 candidate MIKC-type MADS-box genes were identified in pomegranate. Members were encoded with 167–657 amino acids. The 36 MIKC-type MADS-box members in pomegranate were far fewer than 201 in *Triticum aestivum* and 94 in *Nicotiana tabacum* (Bai et al., 2019; Schilling et al., 2020). The differences in the number of members may be due to the fact that tobacco and wheat are polyploid, while pomegranate being polyploid (Holland et al., 2009; Xu et al., 2008). According to the phylogenetic relationships of MIKC-type MADS-box genes between Arabidopsis and poplar, *PgMADS* genes were divided into 13 distinct clades, including TM8 subfamily which was not found in Arabidopsis.

According to the analysis of gene structure, *PgMADS* genes contained 5–12 introns, and seven exons were the main structure of PgMADSs. The genetic structure of PgMADSs was consistent with that of Arabidopsis, apples and soybeans (Parenicova et al., 2003; Shu et al., 2013; Tian et al., 2014), indicating that MIKC-type MADS-box gene structure is relatively conserved. Conservative sequence analysis of 36 MIKC-type MADS-box genes showed that members of the same subfamily had the same number of motifs and similar sequences, while there were some differences in the number of motifs between subfamilies. Also, members of the same subfamily have certain similarities in gene structure, and these results support the findings of the subfamily classification of pomegranate MIKC-type MADS-box family. The conservative motif 1 encodes the MEF2-type MADS-domain, which is the most conserved domain. Motif 2 encodes the K-box domain, which is the second most conservative in the PgMADS family. In general, the K-box motif exists only in the MIKC^C subfamily (Gao et al., 2020), but there is motif 2 in MIKC* type PgMADS16. The gene structure of MIKC* and MIKC^C family is similar. Mδ of type I genes is also treated as a type II gene in Arabidopsis and rice, so it is called the MIKC* type gene (Arora et al., 2007; De et al., 2003). MIKC*-type genes share the characteristics of both M-type and MIKC^C family. Furthermore, the evolutionary tree of Mδ is different from Mα, Mβ, and Mγ, but it is similar to that of MIKC-type (Xu et al., 2014; Tang et al., 2019). It was speculated that the MIKC* subfamily might be the transitional form in the evolution of MADS-box genes. Xu et al. (2014) put forward a similar view that MIKC^C subfamily lost K-domain, while retained more introns, making it become MIKC* subfamily (Xu et al., 2014). With the further loss of introns, MIKC* subfamily became an M-type subfamily with shorter sequences and fewer introns (Xu et al., 2014).

Cis-acting elements exist in the gene promoters and form specific binding with transcription factors and play important roles in regulating the expression of target genes (Riechmann et al., 2000). The promoter sequence of the MIKC-type MADS-box gene contains multiple *cis*-acting elements related to hormone response and abiotic stress, which are rich in methyl jasmonate response elements and abscisic acid response elements. This result is consistent with that by research reports on grape and *Phyllostachys edulis* (Zhang et al., 2018; Cui et al., 2017), indicating that the promoter regulatory element of the MADS-box gene has a certain conservatism among different species. Previous studies have found that ABRE is associated with drought, ABA induction, and high salt stress in plants (Hundertmark and Hincha, 2008; Li et al., 2014). There were many ABRE elements in *PgMADS13*, *PgMADS03*, *PgMADS16*, *PgMADS30,* and *PgMADS25* which were speculated to be related to pomegranate response to drought, ABA induction or high salt stress. MIKC-type MADS-box gene also contains other elements such as TGA-element, TATC-box, TGACC-motif, and TCA-element, indicating that the MIKC-type MADS-box gene can respond to the regulation of hormones such as auxin, gibberellin, jasmonic acid, and salicylic acid, which are involved in plant growth and development and plant stress response. These results suggest that MIKC-type MADS-box family genes may play essential roles in growth and development of pomegranate and in response to abiotic stress.

Studies have shown that the MADS-box gene play roles in the development of floral organs by forming homologous or heterologous complexes (Kaufmann et al., 2005). An analysis of the protein network of MIKC-type MADS-box family showed that there was the interaction among the MIKC-type MADS-box members, indicating that they might jointly regulate flower development by forming heterologous complexes. We also found that PgMADS15 occupied a vital position in the protein interaction network, indicating PgMADS15 played a crucial role in the regulation network of flower development.

Analysis of the expression pattern of MIKC-type MADS-Box gene found that *PgMADS17*, *PgMADS19*, *PgMADS31,* and *PgMADS34* were only expressed in leaves and roots, while other members were expressed in flower tissues. The results indicated that the MIKC-type MADS-box gene might have participated in pomegranate development regulation of different tissues. E subfamily genes (*PgMADS03*, *PgMADS21 PgMADS27*) and C/D subfamily genes (*PgMADS06* and *PgMADS30*) were highly expressed in various tissues of pomegranate flower organs, but not in leaves and roots, indicating that C/D and E subfamily genes were highly expressed specifically in pomegranate. Studies have shown that AP1-FUL, AP3-PI, AG, and AGL6 subfamilies genes regulate floral organ development in Arabidopsis, and the different subfamilies genes of MIKC-type MADS-box present similar expression patterns in different species (Henschel et al., 2002; Ma and Depamphilis, 2000; Yoo et al., 2011). AP1-FUL, AP3-PI, AG, and AGL6 subfamilies genes have high expression levels in the pomegranate bisexual flowers, proving that these genes might be involved in the regulation of floral organ development in pomegranate. In *Arabidopsis thaliana*, the fruit differentiation and development are regulated by the C/D (*AGL11*, *SHP1*/*SHP2*) and A(AP1-ful) class genes (Ferrandiz et al., 2000; Pinyopich et al., 2003). Reports found that *FaMADS9*, a member of the strawberry E(*SEP*) class, was also involved in fruit development and maturation (Seymour et al., 2011). E class genes (*PgMADS03*, *PgMADS21*, *PgMADS27*, *PgMADS09*), C/D class (*PgMADS06*, *PgMADS15*, *PgMADS30*), and A class (*PgMADS01*) were highly expressed in pomegranate peel, indicating that these genes might play important roles in pomegranate fruit development. The study showed that the same subfamily genes had different expression patterns, and the C/D genes regulated the development of carpel and ovule (Xu and Zhong, 2015; Theißen, 2001). In pomegranate, the C gene *PgMADS06* and D gene *PgMADS30* were highly expressed at three development stages of functional male and bisexual flower, while the D gene *PgMADS15* was not expressed, suggesting that *PgMADS15* may lose some functions in the evolution and development of species.

Thirty-six MIKC-type MADS-box genes were identified in pomegranate, and the necessary information concerning these genes was analyzed by bioinformatics to explore their phylogenetic relationship and their functions in the development of flower organs. The gene structure and conserved domain of 36 PgMADSs were analyzed, showing that the same subfamilies' gene structures were relatively similar. *Cis*-acting element analysis showed that promoter sequences of *PgMADS* genes contained multiple hormone response and abiotic stress related elements, suggesting that *PgMADSs* might be closely related to plant hormone signal transduction and adverse situations. It was speculated that *PgMADSs* might be related to the growth and development of pomegranate. Protein interaction network analysis found that PgMADS15 was located at the core of the interaction network. Combining the analysis of tissue specific expressions, *PgMADS15* was highly expressed in the peel and inner seed coat. It was concluded that *PgMADS15* might play an essential role in the peel and inner seed coat development of pomegranate. Our conclusions will lay the foundation for further research on the function of the MIKC-type MADS-box gene in pomegranate, primarily providing a reference for the analysis of the gene function regulating the development of flower organs.

References

Alvarezbuylla, E.R., S. Pelaz, S.J. Liljegren, S.E. Gold, C. Burgeff, G.S. Ditta, L.R. De Pouplana, L.P. Martinezcastilla and M.F. Yanofsky. 2000. An ancestral MADS-box gene duplication occurred before the divergence of plants and animals. *Proc. Natl. Acad. Sci. USA*, *97*: 5328–33.

Arora, R., P. Agarwal, S. Ray, A.K. Singh, V.P. Singh, A.K. Tyagi and S. Kapoor. 2007. MADS-box gene family in rice: Genome-wide identification, organization, and expression profiling during reproductive development and stress. *BMC Genomics*, *8*: 242.

Bai, G., D. Yang, P. Cao, H. Yao, Y. Zhang, X. Chen, B. Xiao, F. Li, Z. Wang and J. Yang. 2019. Genome-wide identification, gene structure, and expression analysis of the MADS-Box gene family indicate their function in the development of tobacco (*Nicotiana tabacum* L.). *Int. J. Mol. Sci.*, *20*: 5043.

Basu, A. and K. Penugonda. 2009. Pomegranate juice: A heart healthy fruit juice. *Nutr. Rev.*, *67*: 49–56.

Becker, A. and G. Theissen. 2003. The major clades of MADS-box genes and their role in the development and evolution of flowering plants. *Mol. Phylogenet. Evol.*, *29*: 464–89.

Chen, X. 2004. A MicroRNA as a Translational Repressor of *APETALA2* in Arabidopsis Flower Development. *Science*, *303*: 2022–25.

Chen, L.N., J. Zhang, H.X. Li, J. Niu, H. Xue, B.B. Liu, Q. Wang, X. Luo, F.H. Zhang, D.G. Zhao, et al., 2017. Transcriptomic analysis reveals candidate genes for female sterility in pomegranate flowers. *Front. Plant Sci.*, *8*: 1430.

Chen, C.J., Xia, R., Chen, H., He, Y.H., 2018. TBtools: A Toolkit for biologists integrating various HTS-data handling tools with a user-friendly interface. *BioRxiv*. doi: 10.1101/289660.

Chen, S., Y. Zhou, Y. Chen and J. Gu. 2018. fastp: An ultra-fast all-in-one FASTQ preprocessor. *Bioinformatics*, *34*(17): i884–i890.

Chi, Y., F. Huang, H. Liu, S. Yang and D. Yu. 2011. An *APETALA1-like* gene of soybean regulates flowering time and specifies floral organs. *J. Plant. Physiol.*, *168*: 2251–59.

Cui, M., C. Wang, W. Wu, J. Wang, W. Tang, W. Zhang, X. Zhu, H. Jia, W. Shen and J. Fang. 2017. Expression characteristic analysis of MADS-box family members of grape transcription factors in berry development and ripening process. *J. of Fruit Sci.*, *34*: 1497–1508.

De Bodt, S., J. Raes, Y.V. De Peer and G. Theisen. 2003. And then there were many: MADS goes genomic. *Trends Plant Sci.*, *8*: 475–83.

Díaz-Riquelme, J., D. Lijavetzky, J.M. Martinezzapater and M.J. Carmona. 2009. Genome-wide analysis of MIKC^{C}-Type MADS-box genes in grapevine. *Plant Physiol.*, *149*: 354–69.

Ehlers, K., A.S. Bhide, D.G. Tekleyohans, B. Wittkop,R.J. Snowdon and A. Becker. 2016. The MADS-box genes *ABS*, *SHP1*, and *SHP2* are essential for the coordination of cell divisions in ovule and seed coat development and for endosperm formation in *Arabidopsis thaliana*. *PLoS One*, *11*: e0165075.

Ferrandiz, C., S.J. Liljegren and M.F. Yanofsky. 2000. Negative Regulation of the SHATTERPROOF Genes by FRUITFULL During Arabidopsis fruit development. *Science*, *289*: 436–38.

Folter, S.D., A.V. Shchennikova, J. Franken, M. Busscher, R. Baskar, U. Grossniklaus, G.C. Angenent and R.G.H. Immink. 2006. A Bsister MADS-box gene involved in ovule and seed development in petunia and Arabidopsis. *Plant J.*, *47*: 934–46.

Galimba, K.D. and V.S.D. Stilio. 2015. Sub-functionalization to ovule development following duplication of a floral organ identity gene. *Dev. Biol.*, *405*: 158–72.

Gao, Y., J. Sun, Z. Sun, Y. Xing, Q. Zhang, K. Fang, Q. Cao and L. Qin. 2020. The MADS-box transcription factor CmAGL11 modulates somatic embryogenesis in Chinese chestnut (*Castanea mollissima* Blume). *J. of Integr. Agr.*, *19*: 1033–43.

Gao, H.H., Y.X. Zhang, S.W. Hu and Y. Guo. 2017. Genome-wide survey and phylogenetic analysis of MADS-box gene family in *Brassica napus*. *Chin. Bull. Botany*, *52*: 699–712.

Gubler, F. and J.V. Jacobsen. 1992. Gibberellin-responsive elements in the promoter of a barley high-pI alpha-amylase gene. *Plant Cell*, *4*: 1435–41.

Heijmans, K., P. Morel and M. Vandenbussche. 2012. MADS-box genes and floral development: The Dark Side. *J. Exp. Bot.*, *63*: 5397–5404.

Henschel, K., R. Kofuji, M. Hasebe, H. Saedler, T. Munster and G. Theissen. 2002. Two ancient classes of MIKC-type MADS-box genes are present in the Moss Physcomitrella Patens. *Mol. Biol. Evol.*, *19*: 801–14.

Hobo, T., M. Asada, Y. Kowyama and T. Hattori. 1999. ACGT-containing abscisic acid response element (ABRE) and coupling element 3 (CE3) are functionally equivalent. *Plant J.*, *19*: 679–89.

Honma, T. and Goto, K. 2001. Complexes of MADS-box proteins are sufficient to convert leaves into floral organs. *Nature*, *409*: 525–29.

Holland, D., K. Hatib and I. Bar-Yaakov. 2009. Pomegranate: Botany, Horticulture, Breeding. *In*: *Horticultural Rev.*, *35*: 127–91. John Wiley & Sons Inc., USA.

Huang, T.H., G. Peng, B.P. Kota, G.Q. Li, J. Yamahara, B.D. Roufogalis and Y. Li. 2005. Anti-diabetic action of *Punica granatum* flower extract: Activation of PPAR-gamma and identification of an active component. *Toxicol. Appl. Pharm.*, *207*: 160–69.

Hundertmark, M. and D.K. Hincha. 2008. LEA (Late Embryogenesis Abundant) proteins and their encoding genes in *Arabidopsis thaliana. BMC Genomics*, *9*: 118.

Jin, J.P., H. Zhang, L. Kong, G. Gao and J.C. Luo. 2014. PlantTFDB 3.0: A portal for the functional and evolutionary study of plant transcription factors. *Nucleic Acids Res.*, *42*: 1182–87.

Khan, M.R., J.Y. Hu and G.M. Ali. 2012. Reciprocal loss of CArG-boxes and auxin response elements drives expression divergence of MPF2-Like MADS-box genes controlling calyx inflation. *PloS One*, *7*(8): e42781, doi: 10.1371/journal.pone.0042781.

Kaufmann, K., R. Melzer and G. Theissen. 2005. MIKC-type MADS-domain proteins: Structural modularity, protein interactions, and network evolution in land plants. *Gene*, *347*: 183–98.

Kumar, S., G. Stecher and K. Tamura. 2016. MEGA7: Molecular evolutionary genetics analysis version 7.0 for bigger datasets. *Mol. Biol. and Evol.*, *33*: 1870–74.

Lansky, E.P. and R.A. Newman. 2007. *Punica granatum* (pomegranate) and its potential for prevention and treatment of inflammation and cancer. *J. Ethnopharmacol.*, *109*: 177–206.

Leseberg, C.H., A.L. Li, H. Kang, M. Duvall and L. Mao. 2006. Genome-wide analysis of the MADS-box gene family in *Populus trichocarpa. Gene*, *378*: 84–94.

Li, P.C., S.W. Yu, J. Shen, Q.Q. Li, D.P. Li, D.Q. Li, C.C. Zheng and H.R. Shu. 2014. The transcriptional response of apple alcohol acyltransferase (MdAAT2) to salicylic acid and ethylene is mediated through two apple MYB TFs in transgenic tobacco. *Plant Mol. Biol.*, *85*: 627–38.

Luo, X., H. Li, Z. Wu, W. Yao, P. Zhao, D. Cao, H. Yu, K. Li, K.L. Poudel and D. Zhao. 2020. The pomegranate (*Punica granatum* L.) draft genome dissects genetic divergence between soft- and hard-seeded cultivars. *Plant Biotechnol. J.*, *18*: 955–68.

Ma, H. and C.W. Depamphilis. 2000. The ABCs of floral evolution. *Cell*, *101*: 5–8.

Ma, H. 2005. Molecular genetic analyses of microsporogenesis and microgametogenesis in flowering plants. *Annu. Rev. Plant. Biol.*, *56*: 393–434.

Martín-Pizarro, C., J.C. Triviño and D. Posé. 2019. Functional analysis of TM6 MADS-box gene in the octoploid strawberry by CRISPR/Cas9 directed mutagenesis. *J. Exp. Bot.*, *70*: 885–95.

Marchler-Bauer, A., Y. Bo, L. Han, J. He, C.J. Lanczycki, S. Lu, F. Chitsaz, M.K. Derbyshire, R.C. Geer, N.R. Gonzales, et al., 2017. CDD/SPARCLE: Functional classification of proteins via subfamily domain architectures. *Nucleic Acids Res.*, *45*(D1): D200–D203.

Melzer, R. and G. Theissen. 2009. Reconstitution of "floral quartets" *in vitro* involving class B and class E floral homeotic proteins. *Nucleic Acids Res.*, *37*: 2723–36.

Messenguy, F. and E. Dubois. 2003. Role of MADS-box proteins and their cofactors in combinatorial control of gene expression and cell development. *Gene*, *316*: 1–21.

Mondragonpalomino, M. and G. Theisen. 2011. Conserved differential expression of paralogous DEFICIENS- and GLOBOSA-like MADS-box genes in the flowers of Orchidaceae: Refining the "orchid code". *Plant J.*, *66*: 1008–19.

Nicolas, L.B., P. Harold, M. Páll and P. Lior. 2016. Near-optimal probabilistic RNA-seq quantification. *Nature Biotech.*, *34*: 525–27.

Ono, N.N., M.T. Britton, J.N. Fass, C.M. Nicolet, D. Lin and L. Tian. 2011. Exploring the transcriptome landscape of pomegranate fruit peel for natural product biosynthetic gene and SSR marker discovery. *J. Integr. Plant Biol.*, *53*: 800–813.

Ophir, R., A. Sherman, M. Rubinstein, R. Eshed, M.S. Schwager, R. Harel-Beja, I. Bar-Ya'akov and D. Holland. 2014. Single-nucleotide polymorphism markers from *de novo* assembly of the pomegranate transcriptome reveal germplasm genetic diversity. *PLoS One*, *9*: e88998.

Pan, I.L., R. McQuinn, J.J. Giovannoni and V.F. Irish. 2010. Functional diversification of AGAMOUS lineage genes in regulating tomato flower and fruit development. *J. Exp. Bot.*, *61*: 1795–1806.

Pan, Z.J., Y.Y. Chen, J.S. Du, Y.Y. Chen, M.C. Chung, W.C. Tsai, C. Wang and H. Chen. 2014. Flower development of phalaenopsis orchid involves functionally divergent SEPALLATA-like genes. *New Phytol.*, *202*: 1024–42.

Parenicova, L., S. De Folter, M. Kieffer, D.S. Horner, C. Favalli, J. Busscher, H. Cook, R. Ingram, M.M. Kater and B. Davies. 2003. Molecular and phylogenetic analyses of the complete MADS-box transcription factor family in Arabidopsis: New Openings to the MADS World. *Plant Cell*, *15*: 1538–51.

Pinyopich, A., G.S. Ditta, B. Savidge, S.J. Liljegren, E. Baumann, E. Wisman and M.F. Yanofsky. 2003. Assessing the redundancy of MADS-box genes during carpel and ovule development. *Nature*, *424*: 85–88.

Qin, G., C. Xu, R. Ming, H. Tang, R. Guyot, E.M. Kramer, Y. Hu, X. Yi, Y. Qi and X. Xu. 2017. The pomegranate (*Punica granatum* L.) genome and the genomics of punicalagin biosynthesis. *Plant J.*, *91*: 1108–28.

Riechmann, J.L., J.E. Heard, G.M. Martin, L. Reuber, C. Jiang, J. Keddie, L. Adam, O. Pineda, O. Ratcliffe and R. Samaha. 2000. Arabidopsis transcription factors: Genome-wide comparative analysis among eukaryotes. *Science*, *290*: 2105–10.

Rouster, J., R. Leah, J. Mundy and V. Cameron-Mills. 1997. Identification of a methyl jasmonate-responsive region in the promoter of a lipoxygenase 1 gene expressed in barley grain. *Plant J.*, *11*: 513–23.

Schilling, S., A. Kennedy, S. Pan, L.S. Jermiin and R. Melzer. 2020. Genome-wide analysis of MIKC-type MADS-box genes in wheat: Pervasive duplications, functional conservation and putative neofunctionalization. *New Phytol.*, *225*: 511–29.

Seymour, G.B., C.D. Ryder, V. Cevik, J.P. Hammond, A. Popovich, G.J.W. King, J. Vrebalov, J.J. Giovannoni and K. Manning. 2011. A SEPALLATA gene is involved in the development and ripening of strawberry (Fragaria x ananassa Duch.) fruit, a non-climacteric tissue. *J. Ex. Bot.*, *62*: 1179–88.

Shah, J. and D.F. Klessig. 1996. Identification of a salicylic acid-responsive element in the promoter of the tobacco pathogenesis-related beta-1,3-glucanase gene, PR-2d. *Plant* J., *10*: 1089–1101.

Shu, Y., D. Yu, D. Wang, D. Guo and C. Guo. 2013. Genome-wide survey and expression analysis of the MADS-box gene family in soybean. *Mol. Biol. Rep.*, *40*: 3901–11.

Smaczniak, C., R.G.H. Immink, G.C. Angenent and K. Kaufmann. 2012. Developmental and evolutionary diversity of plant MADS-domain factors: Insights from recent studies. *Development*, *139*: 3081–98.

Sommer, H., J.P. Beltran, P. Huijser, H. Pape, W. Lonnig, H. Saedler and Z.D. Schwarzsommer. 1990. A homeotic gene involved in the control of flower morphogenesis in Antirrhinum majus: The protein shows homology to transcription factors. *The EMBO Journal*, *9*: 605–13.

Tang, W., Y. Tu, X. Cheng, L. Zhang, H. Meng, X. Zhao, W. Zhang and B. He. 2019. Genome-wide identification and expression profile of the MADS-box gene family in Erigeron breviscapus. *PLoS One*, *14*: e0226599.

Theißen, G. 2001. Development of floral organ identity–Stories from the MADS house. *Curr. Opin. Plant Biol.*, *4*: 75–85.

Theisen, G., R. Melzer and F. Rumpler. 2016. MADS-domain transcription factors and the floral quartet model of flower development: Linking plant development and evolution. *Development*, *143*: 3259–71.

Tian, Y., Q. Dong, Z. Ji, F. Chi and Z. Zhou. 2014. Genome-wide identification and analysis of the MADS-box gene family in apple. *Gene*, *555*: 277–90.

Wells, C.E., E. Vendramin, S.J. Tarodo, I. Verde and D.G. Bielenberg. 2015. A genome-wide analysis of MADS-box genes in peach [*Prunus persica* (L.) Batsch]. *BMC Plant Biol., 15*: 41.

Winter, K., C. Weiser, K. Kaufmann, A. Bohne, C. Kirchner, A. Kanno, H. Saedler and G. Theissen. 2002. Evolution of class B floral homeotic proteins: Obligate heterodimerization originated from homodimerization. *Mol. Biol. Evol., 19*: 587–96.

Winter, K., A. Becker, T. Munster, J.T. Kim, H. Saedler and G. Theissen. 1999. MADS-box genes reveal that gnetophytes are more closely related to conifers than to flowering plants. *Proc. Natl. Acad. Sci. USA, 96*: 7342–47.

Xu, Z., Q. Zhang, L. Sun, D. Du, T. Cheng, H. Pan, W. Yang and J. Wang. 2014. Genome-wide identification, characterization, and expression analysis of the MADS-box gene family in Prunus mume. *Mol. Genet. Genomics, 289*: 903–20.

Xu, Z. and K. Zhong. 2015. *Plant Cell Differentiation and Organogenesis*, 246–62. Science Press: Biejing, China. Xu, Y., Z. Ding, Y. Yao, R. Gong and Y. Wen. 2008. Karyotype analysis of four pomegranate cultivars. *Nonwood Forest Res., 26*: 47–52.

Xu, Y., L. Zhang, H. Xie, Y. Zhang, M.M. Oliveira and R. Ma. 2008. Expression analysis and genetic mapping of three SEPALLATA-like genes from peach (*Prunus persica* (L.) Batsch). *Tree Genet. Genomes, 4*: 693–703.

Yanofsky, M.F., H. Ma, J.L. Bowman, G.N. Drews, K.A. Feldmann and E.M. Meyerowitz. 1990. The protein encoded by the Arabidopsis homeotic gene agamous resembles transcription factors. *Nature, 346*: 35–39.

Yoshida, H., K. Haze, H. Yanagi, T. Yura and K. Mori. 1998. Identification of the cis-acting endoplasmic reticulum stress response element responsible for transcriptional induction of mammalian glucose-regulated proteins. Involvement of basic leucine zipper transcription factors. *J. Biol. Chem., 273*: 33741–49.

Yoo, S.K., X. Wu, J.S. Lee and H.A. Ji. 2011. *AGAMOUS-LIKE 6* is a floral promoter that negatively regulates the FLC/MAF clade genes and positively regulates FT in Arabidopsis. *Plant J., 65*: 62–76.

Yuan, Z.H. 2015. *Chinese Fruit Tree Science and Practice: Pomegranate*, 1–10.– Shaanxi Science and Technology Press: Xi'an, China.

Yuan, Z., Y. Fang, T. Zhang, Z. Fei, F. Han, C. Liu, M. Liu, W. Xiao, W. Zhang and S. Wu. 2018. The pomegranate (*Punica granatum* L.) genome provides insights into fruit quality and ovule developmental biology. *Plant Biotechnol. J., 16*: 1363–74.

Zhang, Y., D. Tang, X. Lin, M. Ding and Z. Tong. 2018. Genome-wide identification of MADS-box family genes in moso bamboo (*Phyllostachys edulis*), and a functional analysis of *PeMADS5* in flowering. *BMC Plant Biol., 18*: 176.

Zhou, H., G. Li, X. Zhao, Z. Kang and A. Guo. 2014. Genome-wide sequence identification of MADS-box transcription factor gene family from Fragaria vesca and cloning of FaMADS1 gene from *F. ananassa* Fruit. *Plant Phys. J., 50*: 1630–38.

13

Decoding the Genome
Comprehensive Identification and Expression Analysis of the TALE Gene Family in Pomegranate

Yuying Wang[1] and *Zhaohe Yuan*[1*]

The three-amino-acid-loop-extension (TALE) gene family is a pivotal transcription factor that regulates the development of flower organs, flower meristem formation, organ morphogenesis, and fruit development. A total of 17 genes of pomegranate TALE family were identified and analyzed in pomegranate via bioinformatics methods, which provided a theoretical basis for the functional research and utilization of pomegranate TALE family genes. The results showed that the *PgTALE* family genes were divided into eight subfamilies (KNOX-I, KNOX-II, KNOX-III, BELL-I, BELL-II, BELL-III, BELL-IV, and BELL-V). All *PgTALEs* had a KNOX domain or a BELL domain, and their structures were conservative. The 1500 bp promoter sequence had multiple cis-elements in response to hormones (auxin, gibberellin) and abiotic stress, indicating that most of PgTALE were involved in the growth and development of pomegranates and stress. Function prediction and protein-protein network analysis showed that PgTALE may participate in regulating the development of apical meristems, flowers, carpels, and ovules. Analysis of gene expression patterns showed that the pomegranate TALE gene family had a particular tissue expression specificity. In conclusion, the knowledge of the TALE gene gained in pomegranate may be applied to other fruit as well.

1. Introduction

A homeobox (HB) encodes the transcriptional regulatory factors with homeodomain, which play considerable roles in the development of plants and animals (Di Giacomo

[1] College of Forestry, Nanjing Forestry University, Nanjing 210037, China.
* Corresponding author: zhyuan88@hotmail.com

et al., 2013). A typical homeobox domain consists of 60 amino acids, to form three-helix regions, the first and second helixes form a loop structure, and the second and third helices form a helix-corner-helical structure (Billeter et al., 1993). Plant homeobox genes have been divided into distinct subgroups. Bharathan et al. (1997) divided them into seven classes, including KNOTTED-like homeobox (KNOX/KNAT), BEL1-like homeobox (BELL/BLH), *Zea mays* homeobox (ZM-HOX), homeobox from *Arabidopsis thaliana* 1 (HAT1), homeobox from *A. thaliana* 2 (HAT2), *A. thaliana* homeobox 8 (ATHB8), and GL2. Mukherjee et al. (2009) classified them into 14 classes, containing homeodomain-leucine zipper I–IV (HD-ZIP I–IV), BELL, KNOX, plant zinc finger (PLINC), wuschel homeobox (WOX), plant homeodomain (PHD), DDT, nodulin homeobox genes (NDX), luminidependens (LD), SAWADEE and PINTOX. While Bürglin and Affolter (2016) classified them into 11 classes, including HD-ZIP I–IV, WOX, NDX, PHD, PLINC, LD, DDT, SAWADEE, PINTOX, KNOX, and BELL. In Plant TFDB, the homeobox genes consist of five families: HD-ZIP, TALE, WOX, HB-PHD, and HB-other (Jin et al., 2017).

Based on protein sequence and evolution, BELL and KNOX belong to the TALE gene family (Arnaud et al., 2014; Ma et al., 2019). Except for some homeobox genes, TALE encodes an atypical structure forming two helices and three additional amino acid residues (P-Y-P) (Chen et al., 2003; Hay and Tsiantis, 2010; Bertolino et al., 1995). The TALE family plays a vital role in regulating plant growth and development (Mahajan et al., 2012; Sakakibara, 2013; Lin et al., 2013; Furumizu et al., 2015), regulating the sporophyte program (Ruiz-Estévez et al., 2017), the formation of plant meristems (Arnaud et al., 2014), and the maintenance of organ morphology (Belles-Boix et al., 2006), organ position (Aida et al., 1999), hormone regulation (Shani et al., 2006), signal transduction (Cnops et al., 2006), and tuber formation (Kondhare et al., 2019). Studies have shown that BELL and KNOX proteins specifically recognize and bind to form the BELL-KNOX heterodimer protein (Bhatt et al., 2004), which is essential for the nuclear localization of two transcription factor proteins and the activity of binding target gene (Kim et al., 2013; Smith et al., 2002). TALE can form complexes to regulate ovule development (Brambilla et al., 2007). After binding to the OVATE family protein (OFP), the BELL-KNOX dimer protein is reversely transferred from the nucleus to the cytoplasm to negatively regulate ovule development (Hackbusch et al., 2005). BELL proteins comprise two highly conserved domains: a POX domain (POX is composed of SKY and BEL), and a homeodomain. The BELL plays essential roles in ovule development, frond development, and fruit development (Byrne et al., 2003; Meng et al., 2018). *BEL1* is expressed in the ovule and controls the ovule integument identity. The dimer formed by the *A. thaliana* homeobox 1 (ATH1) protein and the shoot meristemless (STM) protein participates in the development of plant meristems (Rutjens et al., 2009), while the dimer formed by the ATH1 protein and the KNOTTED-like from *A. thaliana* 2 (KNAT2) protein regulates the development of plant inflorescence tissue (Li et al., 2012). The interaction of BLH6 and KNAT7 affects the development of secondary cell walls (Li et al., 2014). The

KNOX gene family contains KNOX1, KNOX2, ELK, and homeodomain, except for a novel gene *KNATM* without the homeodomain (Di Giacomo et al., 2013; Magnani et al., 2008; Hamant and Pautot, 2010). In addition, KNOX1 and KNOX2 domains merge to form a MEINOX domain. KNOX1 is expressed in the meristem, which is necessary for meristem development and maintenance. Studies have shown that the KNOX2 gene is involved in regulating the secondary growth of plant cell walls and plays a crucial regulatory role in the development of roots, stems, seed coats, and heartwood (Zhong et al., 2008; Bhargava et al., 2010; Li et al., 2011).

Myrtales, the myrtle order of flowering plants, is placed in the Angiosperm Phylogeny Group IV (APG IV) botanical classification system (Byng et al., 2016). Pomegranate (*Punica granatum* L.) is a considerable economic fruit tree of the *Lythraceae* family and widely cultivated worldwide. It was that pomegranate and the related species *Eucalyptus grandis* H., belonging to the order Myrtales, shared the paleotetraploidy event (Yuan et al., 2018). Studying the function and regulatory mechanism of TALE genes in pomegranate helps regulate pomegranate growth patterns, flower and fruit development. The completion of pomegranate genome data provided momentous data support for the study of pomegranate gene function (Yuan et al., 2018; Qin et al., 2017; Ophir et al., 2014). In this study, the members of the TALE gene family were identified based on the genome sequence of 'Taishanhong', and their physical and chemical properties, protein structure, cis-elements, phylogenetic relationship, and gene tissue expression were analyzed. Through the systematic identification of PgTALE, the result lays a foundation for further study of the function of TALE genes in pomegranate.

2. Materials and Methods

2.1 Genome and Transcriptome Data Sources

Pomegranate genome sequences (ASM286412v1), protein sequences, and transcriptome data were downloaded from NCBI (http://www.ncbi.nlm.nih.gov/), and the *A. thaliana* TALE protein sequence was downloaded from the *A. thaliana* database (http: //www.arabidopsis.org), *E. grandis*, *Populus trichocarpa*, *Malus domestica*, *Vitis vinifera*, and *Solanum lycopersicum* TALE protein sequence were downloaded from the PlantTFDB database (http: //planttfdb.cbi.pku.edu.cn) (Jin et al., 2017).

2.2 Identification and Sequence Analysis of PgTALE Gene Family Members

The hidden Markov model file of the TALE family (E-value < $1e^{-5}$) was constructed by using TALE (PF00046) in the Pfam database (El-Gebali et al., 2019) (https://pfam.xfam.org/) and hmmsearch program in HMMER 3.0 software package (Virginia, USA) (Finn et al., 2011). The candidate PgTALE protein conserved domains were searched, and the BELL domains (POX, homeodomain) or KNOX domains (KNOX1, KNOX2, ELK, homeodomain) were TALE gene family conserved. At the same time, using the published 'Taishanhong' protein sequences (Yuan et al., 2018) and the TALE family protein sequences of six species (*A. thaliana*, *E. grandis*, *P. trichocarpa*, *M. domestica*, *V. vinifera*, and *S. lycopersicum*) as baits to make a local BLASTP alignment (E-value < e^{-5}, identity > 50%), the repetition was removed, the candidate TALE protein sequences were screened. In addition, *A. thaliana* contains a member of the TALE gene family *KNATM* without homeodomain; we added a pomegranate gene homologous to *KNATM* (Magnani and Hake, 2008), and then the target protein domains were detected by SMART and CDD (Schultz et al., 1998; Marchler-Bauer et al., 2017). The sequences without the TALE domain were removed. The online tool ExPASy Proteomics Server (https://web.expasy.org/protparam/) was used to predict the physical and chemical properties of PgTALE protein, such as amino acid sequence length, molecular weight, isoelectric point, grand average of hydropathicity (Artimo et al., 2012). Signal peptide of the PgTALE proteins was performed by SignalP 5.0 Sever (http://www.cbs.dtu.dk/services / SignalP). Subcellular localization of the PgTALE proteins was performed using CELLO (http://cello.life.nctu.edu.tw/) (Yu et al., 2004).

2.3 Construction of Phylogenetic Tree of PgTALE Gene Family

Multiple sequence alignments of candidate proteins with *A. thaliana*, *E. grandis*, *P. trichocarpa*, and *V. vinifera* TALE gene family proteins were performed using MAFFT (Katoh and Standley, 2016). The phylogenetic tree was constructed by using RAxML-NG (Kozlov et al., 2019) with Bootstrap 1,000 repeats and the best model of JTT + F + I + G4 selected by ModelFinder (Kalyaanamoorthy et al., 2017). Then, the phylogenetic tree was beautified by using the online software tool EvolView (http://www.evolgenius.info/) (Subramanian et al., 2019).

2.4 Analysis of PgTALE Conserved Motifs and Gene Structure

The motif type and sequence of the PgTALE family were analyzed by MEME (http://meme-suite.org/tools/meme) (Bailey et al., 2009), and the motif characteristics of PgTALE were obtained. According to the protein sequence and gene sequence of the PgTALE gene, the gene structure information of pomegranate TALE was obtained by Perl script, including intron, exon, and upstream and downstream sequence. In addition, a combined figure of the phylogenetic tree, conserved motifs, and gene structure was drawn by TBtools (Chen et al., 2020).

2.5 Analysis of PgTALE Protein Structure

Protein sequence similarity of more than 35% as a template, the tertiary structure and homologous modelling of PgTALE proteins were analyzed using the SWISS-MODEL (https://swissmodel.expasy.org/) (Waterhouse et al., 2018), and Ramachandran Plots were used to display protein properties.

2.6 Analysis of Cis-elements and Protein-protein Interaction Network of PgTALE Gene Family

To analyze the cis-elements of the promoter region, the 1500 bp sequence upstream of the start codon was obtained from the pomegranate genome sequence by Perl script, and the sequence was searched by PlantCARE (http://bioinformatics.psb.ugent.be/webtools/plantcare/html/) (Lescot et al., 2002). The protein–protein interaction network of the TALE family was analyzed by String (https://string-db.org/) (Szklarczyk et al., 2019).

2.7 Expression Analysis of PgTALE Gene Family

RNA-Seq data of tissues and organs closely related to pomegranate were downloaded from the NCBI database (Table 1). Subsequently, Kallisto version 0.44.0 software (California, USA) (Bray et al., 2016) was used to index the sequence with the 'Taishanhong' transcriptome file to calculate further and analyze gene expression. The corresponding expression levels (TPM values) of the TALE family members were obtained, and the obtained TPM values were converted by Log2 (TPM + 1). Finally, a heat map of the TALE gene was drawn by using the R package heatmap.

3. Results

3.1 Identification and Sequence Analysis of PgTALE Gene Family Members

In this study, 74 homebox gene family members were identified by using the hmmsearch method. The homebox family consists of five families (HD-ZIP, TALE, WOX, HB-PHD, and HB-other), and they share a PF number (PF00046). As TALE encodes an atypical structure forming two helices and three additional amino acid residues, 16 candidate members of the TALE gene family were identified, and all candidate proteins were identified to belong to the TALE protein family. Twenty-three candidate members of the TALE gene family were identified by BLASTP. *Pg001623.1*, *Pg009439.1*, *Pg011532.1*, *Pg017964.1*, *Pg017965.1*, *Pg022249.1*, and *Pg027515.1* were removed because they did not contain TALE conserved domain. Our result showed that there were 17 members of the TALE gene family in pomegranate. *PgTALE* gene family was renamed, the results as shown in Table 2.

The physical and chemical properties of the PgTALEs was analyzed using the ExPASy online tool. The results showed that the length of the 17 PgTALE gene coding regions ranged from 465 bp (PgTALE17) to 2400 bp (PgTALE16). The amino acid length of the TALE protein ranged from 154 aa (PgTALE17) to 799 aa (PgTALE16), and the protein molecular weight ranged from 17605.66 Da (PgTALE17) to 87381.25 Da (PgTALE16). The pI ranged from 5.13 (PgTALE7) to 8.78 (PgTALE16). Among them, the pI of three PgTALE proteins were higher than seven, suggesting that proteins were slightly alkaline; the other 14 PgTALEs were acidic proteins. The grand average of hydropathicity (GRAVY) was between

Table 1 RNA-Seq data of pomegranate.

Accession No.	*Cultivars*	*Sample Type*	*Library*	*Platform*	*Reference*	*Note*
SRR5279396	'Dabenzi'	root	Paired end	Illumina HiSeq 4000	(Qin et al., 2017)	
SRR5279397	'Dabenzi'	leaf	Paired end	Illumina HiSeq 4000	(Qin et al., 2017)	
SRR5279395	'Dabenzi'	flower	Paired end	Illumina HiSeq 4000	(Qin et al., 2017)	
SRR5279391	'Dabenzi'	Inner seed coat (50 days after pollination)	Paired end	Illumina HiSeq 4000	(Qin et al., 2017)	
SRR5279388	'Dabenzi'	Outer seed coat (50 days after pollination)	Paired end	Illumina HiSeq 4000	(Qin et al., 2017)	
SRR5279394	'Dabenzi'	Pericarp (50 days after pollination)	Paired end	Illumina HiSeq 4000	(Qin et al., 2017)	
SRR5446598	'Tunisia'	flower (3.0–5.0 mm)	Paired end	Illumina HiSeq 2500	(Chen et al., 2017)	Functional male flower
SRR5446595	'Tunisia'	flower (5.1–13.0 mm)	Paired end	Illumina HiSeq 2500	(Chen et al., 2017)	Functional male flower
SRR5446592	'Tunisia'	flower (13.1–25.0 mm)	Paired end	Illumina HiSeq 2500	(Chen et al., 2017)	Functional male flower
SRR5446607	'Tunisia'	flower (3.0–5.0 mm)	Paired end	Illumina HiSeq 2500	(Chen et al., 2017)	Female sterility
SRR5446604	'Tunisia'	flower (5.1–13.0 mm)	Paired end	Illumina HiSeq 2500	(Chen et al., 2017)	Female sterility
SRR5446601	'Tunisia'	flower (13.1–25.0 mm)	Paired end	Illumina HiSeq 2500	(Chen et al., 2017)	Female sterility
SRR5678820	'Tunisia'	Inner seed coat (50 days after pollination)	Paired end	Illumina HiSeq 4000	(Qin et al., 2017)	
SRR5678819	'Baiyushizi'	Inner seed coat (50 days after pollination)	Paired end	Illumina HiSeq 4000	(Qin et al., 2017)	
SRR080723	'Wonderful'	pericarp	Paired end	Illumina HiSeq 2000	(Ono et al., 2011)	
SRR1055290	'nana'	Mixed samples of leaves, flowers, fruit, and roots	Single end	454 GS FLX Titanium	(Ophir et al., 2014)	
SRR1054190	'Black127'	Mixed samples of root, leaf, flower,and fruit	Single end	454 GS FLX Titanium	(Ophir et al., 2014)	

Table 2 The basic information of the TALE gene family in pomegranate.

Gene Name	*Gene ID*	*Location*	*Exon No.*	*CDS*	*AA*	*MW(Da)*	*PI*	*GRAVY*	*Subcellular Localization*
PgTALE1	Pg002952.1	scaffold11:2366550:2369265	4	2097	698	75,172.93	7.80	−0.538	Nuclear
PgTALE2	Pg005682.1	scaffold13:4182457:4188943	5	1035	344	38,810.15	5.25	−0.795	Nuclear
PgTALE3	Pg009001.1	scaffold17:1307284:1312800	4	1959	652	70,713.74	6.51	−0.557	Nuclear
PgTALE4	Pg011533.1	scaffold2:5274088:5277780	4	1932	643	70,086.66	6.33	−0.566	Nuclear
PgTALE5	Pg014946.1	scaffold26:1035323:1039874	5	1065	354	40,403.90	6.15	−0.969	Nuclear
PgTALE6	Pg015766.1	scaffold29:588196:590712	4	1905	634	70,031.46	5.87	−0.717	Nuclear
PgTALE7	Pg022248.1	scaffold49:1377908:1386181	5	1065	354	39,800.27	5.13	−0.751	Nuclear
PgTALE8	Pg024529.1	scaffold6:3685906:3688228	4	1563	520	58,385.62	6.10	−0.449	Nuclear
PgTALE9	Pg024817.1	scaffold6:2019598:2022376	5	1275	424	46,087.05	5.94	−0.635	Nuclear
PgTALE10	Pg026506.1	scaffold7:1735075:1737746	4	2103	700	76,325.21	7.82	−0.647	Nuclear
PgTALE11	Pg027513.1	scaffold73:612910:617019	4	1023	340	37,775.80	6.53	−0.534	Nuclear
PgTALE12	Pg028434.1	scaffold8:2877271:2880879	5	909	302	33,829.16	6.27	−0.635	Nuclear
PgTALE13	Pg028770.1	scaffold81:128287:132801	5	2136	711	78,483.78	6.54	−0.513	Nuclear
PgTALE14	Pg029909.1	scaffold9:754612:757823	4	1851	616	68,481.09	6.27	−0.707	Nuclear
PgTALE15	Pg030082.1	scaffold9:3730507:3733807	6	1203	400	45,378.59	5.30	−0.730	Nuclear
PgTALE16	Pg030621.1	scaffold96:200629:218877	6	2400	799	87,381.25	8.78	−0.549	Nuclear
PgTALE17	Pg005241.1	scaffold13:2266264:2268869	3	465	154	17,605.66	5.18	−0.730	Nuclear

−0.969 to −0.449, suggesting that PgTALEs are all hydrophilic proteins. The number of exons of the PgTALEs was 3–6. Besides, the signal peptide prediction showed that there were no signal peptides in all PgTALE proteins, which belonged to non-secreted proteins. Subcellular localization prediction suggested that all PgTALE proteins were distributed on the nucleus.

3.2 Phylogenetic Tree Analysis of PgTALE Gene Family

To clarify the evolutionary relationship and possible biological functions of members of the PgTALE gene family, the phylogenetic tree of the TALE gene was constructed based on the amino acid sequences of the pomegranate, *A. thaliana, E. grandis, P. trichocarpa*, and *V. vinifera*. Based on the classification of *A. thaliana* TALE gene family (BELL and KNOX family), the pomegranate BELL proteins were classified into five subfamilies: BELL-I (one member), BELL-II (two), BELL-III (one), BELL-IV (two), and BELL-V (three), and KNOX proteins were classified into three subfamilies: KNOX-I (five), KNOX-II (two), KNOX-III (five). In each clade, there are branches from the same species, which may be caused by gene duplications (Zhang et al., 2019).

3.3 Analysis of Conserved Motifs and Gene Structures of PgTALE Gene Family

The conserved motifs of PgTALE were identified. Ten conserved motifs, in which Motif 1 represents the homeodomain (homeobox domain, HOX), Motif 4 represents the ELK domain, and Motif 7 represents SKY domain. The location information of the PgTALE protein domain was analyzed. The results showed that the six members of KNOX subfamily contained KNOX1, KNOX2, ELK, and HOX domains, and only *PgTALE17* did not contain HOX domains. In the BELL subfamily all contained POX and HOX domains.

Structural analysis showed that the gene structure of *PgTALE* was a similarity, and there were little differences in the number of exons and introns among the *PgTALE* genes. The number of exons and introns of the *PgTALE* genes were 3–6 and 2–5, respectively. Members of the same subfamily of *PgTALE* showed similar gene structure and protein conserved motif distribution. For example, five members of KNOX group (*PgTALE2*, *PgTALE5*, *PgTALE7*, *PgTALE9*, and *PgTALE12*) contained five exons and four introns, and only one member (*PgTALE17*) contained three exons and two introns. Seven members of BELL group (*PgTALE1, PgTALE3, PgTALE4, PgTALE6, PgTALE8, PgTALE10*, and *PgTALE14*) contained four exons and three introns. The above results indicated that the *PgTALE* gene family had a certain degree of conservation regardless of its genetic structure or protein conserved motifs.

3.4 Protein Structure Analysis and Protein Interaction Networks of Pomegranate TALE Gene Family

The spatial structure of proteins plays a role in the biological function of proteins. The tertiary structure of the protein was analyzed, in which it was found that the structure of the PgTALE family members was similar, except for PgTALE17 without a template (protein sequence similarity of less than 35%) that we cannot predict protein tertiary structure. The protein is a multi-chain folded protein, mainly α-helix. The calculation test showed that the Ramachandran Favoured value of the PgTALE family was above 90%, and PgTALE2 and PgTALE15 reached 100%, except PgTALE9 which was only 87.27%. The results showed that the PgTALE protein had a stable spatial structure.

Protein function prediction suggested that PgTALE2, PgTALE7, and PgTALE15 played roles in meristem function, contributing to the shoot apical meristem (SAM) maintenance and organ separation. They may also be involved in maintaining cells in a meristematic state. In addition, PgTALE14 might be involved in the regular pattern of organ initiation. PgTALE11 may be required for SAM formation in embryogenesis. PgTALE12 may be involved in secondary cell wall biosynthesis. PgTALE13 might be required for the SAM to respond appropriately to floral inductive signals.

The protein-protein interaction of PgTALE was analyzed for predicting its potential function, signal transduction, and metabolic pathways. It was predicted that there were interactions between PgTALE14 and AG, SEP3, KNAT1, INO, and other proteins to regulate ovule development. In addition, BEL1 can form heterodimers with KNAT1; it predicted that PgTALE14 (BELL family) may interact with PgTALE5 (KNOX family) to form heterodimers. PgTALE8 might interact with STM and KNAT6 and enhance the apical meristem of these genes.

From the gene co-expression, we can see the level of co-expression of *KNAT1/KNAT3/KNAT6/STM/BEL1/BLH6/ATH1* and other genes. Among them, *KNAT1* and *KNAT6*, *KNAT1*, and *STM* were higher than that of other genes. They may participate in or respond to a biological or abiotic stress process, and it may be inferred that *PgTALE2/PgTALE5/PgTALE11* may also have similar functions.

3.5 Analysis of Cis-elements of PgTALE Gene Family

In this study, the upstream 1500 bp sequence of *PgTALE* gene was extracted, the possible cis-elements in the promoter region were found. Thirteen cis-elements related to abiotic stress were found, which were ABRE, ARE, AuxRR-core, CAAT-box, CGTCA-motif, GARE-motif, LTR, MBS, P-box, TATC-box, TCA-element, TGA-element, and TGACG-motif. AuxRR-core and TGA-element are auxin-responsive elements. CGTCA-motif and TGACG-motif are MeJA-responsiveness elements, while GARE-motif, P-box, and TATC-box are gibberellin response elements. The *PgTALE* genes contain the enhancer response element CAAT-box.

64.7% of the *PgTALE* genes contain ABA response element ABRE and the cold stress response element LTR. 70.6% of the *PgTALE* genes contain the antioxidant response element ARE. 41.2% of the *PgTALE* gene contains MeJA-responsiveness response elements CGTCA-motif and TGACG-motif, the salicylic acid response element TCA-element; 29.4%, 23.5%, 35.3% of the *PgTALE* genes contain gibberellin response elements GARE-motif, P-box, TATC-box; 29.4% of the *PgTALE* genes contain the drought stress response element MBS. Besides that, only the *PgTALE8* gene contains the auxin response element AuxRR-core, and *PgTALE12* and *PgTALE13* contain the auxin response element TGA-element.

3.6 Expression Analysis of PgTALE Gene Family

To further analyze the characteristics and function of the *PgTALE* genes, the tissue-specific expression of the TALE gene was analyzed. The results showed that the vast majority of *PgTALE* genes were expressed in different tissues, but PgBLH8 was expressed in trace or no expression in all tissues.

PgTALE5, *PgTALE12*, and *PgTALE15* are expressed during functional male flower development, indicating that these genes may be involved in the female and male organ differentiation; *PgTALE1* and *PgTALE9* are higher expressed in leaves, bisexual and functional male flower, indicating that they may be related to the differentiation of male and female organs of pomegranate flowers and regulating leaf development. There are also differences in the expression of different *PgTALE* genes in different tissue, such as *PgTALE2* is not expressed in the inner seed coat, outer seed coat, and pericarp. The expression of *PgTALE9* is the highest in the functional male flower (5.1 m–13.0 mm), and the expression of *PgTALE10* is the highest in the pericarp. However, there are some differences in the expression of different *PgTALE* genes in different pomegranate varieties, such as *PgTALE7* and *PgTALE14* in the varieties of 'Dabenzi', 'Tunisia', and 'Baiyushizi'. In the same pomegranate variety 'Dabenzi', there are also significant differences in tissue expression between leaves and outer seed coat. For example, the expression of *PgTALE16* is higher in leaves, but the lowest in the outer seed coat.

4. Discussion

The TALE gene family is found in plant meristems and is related to the differentiation and signal transduction of meristems, for example, it can inhibit the expression of the critical enzyme gene *ga20ox1* in the GA pathway (Chen et al., 2004). In other important fruits belonging to the Rosaceae family, TALE are involved in the rootstock responding to apple cold stress (Wang et al., 2017), the cherry anthesis (Wen et al., 2019). In addition, it regulated tomato fruit development (Meng et al., 2018). Currently, the TALE gene family has been found in many plants: 33 *AtTALE* genes in *A. thaliana*; 40 *LjTALE* genes in *Lotus japonicas* K. (Qiu et al., 2019); 46 *GaTALE* genes in *Gossypium arboretum* L.; 47 *GrTALE* genes in *G. raimondii* L;,

88 *GbTALE* genes in *G. barbadense* L.; 94 *GhTALE* genes *G. hirsutum* L. (Ma et al., 2019); 35 *PtTALE* genes in poplar (Zhao et al., 2019); and seven *VsTALE* genes in *Vandenboschia speciose* G. (Ruiz-Estévez et al., 2017). Therefore, the copy number of the TALE gene family in different species is different. At present, the genomic data of three pomegranate varieties have been released in China, but there are no reports on the identification and analysis of pomegranate TALE family genes. In this study, for the first time, 17 TALE genes were identified in the pomegranate. Through the analysis of the physicochemical properties of the protein (Table 1), it was found that pomegranate TALE proteins are all hydrophilic proteins that are consistent with studies in Popular and *L. japonicas* (Qiu et al., 2019; Zhao et al., 2019). Domain differences may represent the regulatory effects of promoting or inhibiting. In addition, the *PgTALE* genes are divided into eight subfamilies, which is consistent with the *A. thaliana* and cotton TALE gene subfamily classification (Ma et al., 2019).

The cis-elements exist at the gene promoter site and specifically binds transcription factors to regulate gene transcription. This study found that the PgTALE promoter sequence contained multiple cis-elements related to hormonal response and abiotic stress, which are rich in methyl jasmonate response element, abscisic acid response element, and gibberellin response element, which is similar to antecedent studies (Ma et al., 2019). It indicated that the promoter of the TALE gene has a certain conservative. Previous studies have found that ABRE is associated with plant drought, ABA induction, and high salt stress in plants (Li et al., 2014; Hundertmark and Hincha, 2008). In addition, there are a series of elements related to stress, such as ARE, MBS, and LTR. The results indicate that PgTALE plays a role in pomegranate abiotic stress. Gene function prediction and protein-protein network analysis also show that the PgTALE family plays a significant role in regulating ovule and inflorescence development. Gene functional prediction and protein-protein network analysis also showed that there are some interactions between PgTALE14, AG, and KNAT1 in floral organs; the results were consistent with the previous study (Arnaud and Pautot, 2014; Brambilla et al., 2007).

The tissue expression analysis of the *PgTALE* found that most of them were expressed in diverse tissues and varieties, but diverse *PgTALE* genes were expressed in different tissue varieties and showed specific differences. It was similar to the results of TALE genes in *A. thaliana* (Liberman et al., 2015). It can be speculated that the TALE family of pomegranate has similar functions to this family in other plants. According to the function of *BEL1* in *A. thaliana*, we speculated that its homologous gene *PgTALE14* has an important regulatory significance in the development of pomegranate ovules (Ray et al., 1994; Bencivenga et al., 2012); *PgTALE8*, as the homologous gene of *ATH1*, controls inflorescence development (Gómez-Mena and Sablowski, 2008; Proveniers et al., 2007). The specificity of tissue and variety expression is speculated to be closely related to its gene function. For example, *PgTALE1, PgTALE6, PgTALE9, PgTALE10, PgTALE12*, and *PgTALE14* had high expression levels in the functional male flowers, bisexual flowers, and fruit tissues.

It is predicted that *PgTALE* may have roles in maintaining flower organ and fruit development. However, due to the inconsistency of some sequencing platforms (Illumina and 454) in RNA-seq data, to a certain extent, it may lead to the uneven sequencing depth among tissue samples and the gap in reading length, which has a certain impact on the analysis results, while the difference in pomegranate varieties also has a certain error on the expression analysis results. After the normalization of RNA-seq data, the error may be reduced.

Conclusions

In this study, 17 PgTALE members were identified in pomegranate and explored their phylogenetic relationships. The *PgTALE* gene structure of all members of the subfamily is very similar. *PgTALE* may participate in the apical meristems, flower organ, and fruit development, and the subfamily genes may have the same expression pattern. These conclusions are the foundation for the function research of the *PgTALE* gene and provide a reference for exploring its evolutionary process.

References

Aida, M., Ishida, T. and Tasaka, M. 1999. Shoot apical meristem and cotyledon formation during Arabidopsis embryogenesis: Interaction among the *CUP-SHAPED COTYLEDON* and *SHOOT MERISTEMLESS* genes. *Development, 126*(8): 1563–70.

Arnaud, N. and Pautot, V.R. 2014. Ring the BELL and tie the KNOX: Roles for TALEs in gynoecium development. *Front. Plant Sci., 5*: 93.

Artimo, P., Jonnalagedda, M., Arnold, K., Baratin, D., Csardi, G., De Castro, E., Duvaud, S., Flegel, V., Fortier, A. and Gasteiger, E. 2012. ExPASy: SIB bioinformatics resource portal. *Nucleic Acids Res., 40*: 597–603.

Bailey, T.L., Bodén, M., Buske, F.A., Frith, M., Grant, C.E., Clementi, L., Ren, J., Li, W.W. and Noble, W.S. 2009. MEME SUITE: Tools for motif discovery and searching. *Nucleic Acids Res., 37*: W202–W208.

Belles-Boix, E., Hamant, O., Witiak, S.M., Morin, H., Traas, J. and Pautot, V. 2006. *KNAT6*: An Arabidopsis homeobox gene involved in meristem activity and organ separation. *Plant Cell, 18*(8): 1900–1907.

Bencivenga, S., Simonini, S., Benkova, E. and Colombo, L. 2012. The transcription factors BEL1 and SPL are required for cytokinin and auxin signaling during ovule development in Arabidopsis. *Plant Cell*, 24(7): 2886–97.

Bertolino, E., Reimund, B., Wildtperinic, D. and Clerc, R.G. 1995. A novel homeobox protein which recognizes a TGT core and functionally interferes with a retinoid-responsive motif. *J. Biol. Chem., 270*(52): 31178–88.

Bharathan, G., Janssen, B.J. and Kellogg, E.A. 1997. Did homeodomain proteins duplicate before the origin of angiosperms, fungi, and metazoa? *Proc. Natl. Acad. Sci. U.S.A., 94*(25): 13749–53.

Bhargava, A., Mansfield, S.D., Hall, H., Douglas, C.J. and Ellis, B.E. 2010. MYB75 functions in regulation of secondary cell wall formation in the Arabidopsis inflorescence stem. *Plant Physiol. 154*(3): 1428–38.

Bhatt, A.M., Etchells, J.P., Canales, C., Lagodienko, A., Dickinson, H.G. 2004. VAAMANA, a BEL1-like homeodomain protein, interacts with KNOX proteins BP and STM, and regulates inflorescence stem growth in Arabidopsis. *Gene, 328*: 103–11.

Billeter, M., Qian, Y.Q., Otting, G., Müller, M., Gehring, W. and Wüthrich, K. 1993. Determination of the nuclear magnetic resonance solution structure of an Antennapedia homeodomain-DNA complex. *J. Mol. Biol.*, *234*(4): 1084–97.

Brambilla, V., Battaglia, R., Colombo, M., Masiero, S., Bencivenga, S., Kater, M.M. and Colombo, L. 2007. Genetic and molecular interactions between BELL1 and MADS Box factors support ovule development in Arabidopsis. *Plant Cell*, *19*(8): 2544–56.

Bray, N., Pimentel, H., Melsted, P. and Pachter, L. 2016. Near-optimal probabilistic RNA-seq quantification. *Nat. Biotechnol.*, *34*(5): 525–27.

Bürglin, T.R. and Affolter, M. 2016. Homeodomain proteins: An update. *Chromosoma*, *125*(3): 497–521.

Byng, J.W., Chase, M.W., Christenhusz, M.J.M., Fay, M.F., Judd, W.S., Mabberley, D.J., Sennikov, A.N., Soltis, D.E., Soltis, P.S. and Stevens, P.F. 2016. An update of the angiosperm phylogeny group classification for the orders and families of flowering plants: APG IV. *Bot. J. Linn. Soc.*, *181*(1): 1–20.

Byrne, M.E., Groover, A., Fontana, J.R. and Martienssen, R.A. 2003. Phyllotactic pattern and stem cell fate are determined by the Arabidopsis homeobox gene *BELLRINGER*. *Development*, *130*(17): 3941–50.

Chen, C., Chen, H., Zhang, Y., Thomas, HR., Frank, M.H., He, Y. and Xia, R. 2020. TBtools: An integrative toolkit developed for interactive analyses of big biological data. *Mol. Plant*, *13*(8): 1194–1202.

Chen, H., Banerjee, A.K. and Hannapel, D.J. 2004. The tandem complex of BEL and KNOX partners is required for transcriptional repression of ga20ox1. *Plant J.*, *38*(2): 276–84.

Chen, H., Rosin, F.M., Prat, S. and Hannapel, D.J. 2003. Interacting transcription factors from the three-amino acid loop extension superclass regulate tuber formation. *Plant Physiol.*, *132*(3): 1391–1404.

Chen, L., Zhang, J., Li, H., Niu, J., Xue, H., Liu, B., Wang, Q., Luo, X., Zhang, F. and Zhao, D. 2017. Transcriptomic analysis reveals candidate genes for female sterility in pomegranate flowers. *Front Plant Sci.*, *8*: 1430.

Cnops, G., Neyt, P., Raes, J., Petrarulo, M., Nelissen, H., Malenica, N., Luschnig, C., Tietz, O., Ditengou, F.A. and Palme, K. 2006. The *TORNADO1* and *TORNADO2* genes function in several patterning processes during early leaf development in *Arabidopsis thaliana*. *Plant Cell*, *18*(4): 852–66.

Di Giacomo, E., Iannelli, M.A. and Frugis, G. 2013. TALE and shape: How to make a leaf different. *Plants*, *2*(2): 317–42.

El-Gebali, S., Mistry, J., Bateman, A., Eddy, S.R., Luciani, A., Potter, S.C., Qureshi, M., Richardson, L.J., Salazar, G.A. and Smart, A. 2019. The Pfam protein families' database in 2019. *Nucleic Acids Res.*, *47*(D1): D427–D432.

Finn, R.D., Clements, J. and Eddy, S.R. 2011. HMMER web server: Interactive sequence similarity searching. *Nucleic Acids Res.*, *39*: 29–37.

Furumizu, C., Alvarez, J.P., Sakakibara, K. and Bowman, J.L. 2015. Antagonistic roles for KNOX1 and KNOX2 genes in patterning the land plant body plan following an ancient gene duplication. *PLoS Genet.*, *11*(2): e1004980.

Gómez-Mena, C. and Sablowski, R. 2008. *ARABIDOPSIS THALIANA HOMEOBOX GENE1* establishes the basal boundaries of shoot organs and controls stem growth. *Plant Cell*, *20*(8): 2059–72.

Hackbusch, J., Richter, K., Muller, J., Salamini, F. and Uhrig, J.F. 2005. A central role of *Arabidopsis thaliana* ovate family proteins in networking and subcellular localization of 3-aa loop extension homeodomain proteins. *Proc. Natl. Acad. Sci. U.S.A.*, *102*(13): 4908–12.

Hamant, O. and Pautot, V. 2010. Plant development: A TALE story. *C.R. Biol.*, *333*(4): 371–81.

Hay, A. and Tsiantis, M. 2010. KNOX genes: Versatile regulators of plant development and diversity. *Development*, *137*(19): 3153–65.

Hundertmark, M. and Hincha, D.K. 2008. LEA (Late Embryogenesis Abundant) proteins and their encoding genes in *Arabidopsis thaliana*. *BMC Genom.*, *9*: 118.

Jin, J., Tian, F., Yang, D., Meng, Y., Kong, L., Luo, J. and Gao, G. 2017. PlantTFDB 4.0: Toward a central hub for transcription factors and regulatory interactions in plants. *Nucleic Acids Res.*, *45*(D1): D1040–D1045.

Kalyaanamoorthy, S., Minh, B.Q., Wong, T.K.F., von Haeseler, A. and Jermiin, L.S. 2017. ModelFinder: Fast model selection for accurate phylogenetic estimates. *Nat. Methods*, *14*(6): 587–89.

Katoh, K. and Standley, D.M. 2016. A simple method to control overalignment in the MAFFT multiple sequence alignment program. *Bioinformatics*, *32*(13): 1933–42.

Kim, D., Cho, Y., Ryu, H., Kim, Y., Kim, T. and Hwang, I. 2013. BLH1 and KNAT3 modulate ABA responses during germination and early seedling development in Arabidopsis. *Plant J.*, *75*(5): 755–66.

Kondhare, K.R., Vetal, P.V., Kalsi, H.S. and Banerjee, A.K. 2019. BEL1-like protein (StBEL5) regulates CYCLING DOF FACTOR1 (StCDF1) through tandem TGAC core motifs in potato. *J. Plant Physiol.*, *241*: 153014.

Kozlov, A.M., Darriba, D., Flouri, T., Morel, B. and Stamatakis, A. 2019. RAxML -NG: A fast, scalable and user-friendly tool for maximum likelihood phylogenetic inference. *Bioinformatics*, *35*(21): 4453–55.

Lescot, M., Dhais, P., Thijs, G., Marchal, K., Moreau, Y., Peer, Y.V., Rouz, P. and Rombauts, S. 2002. PlantCARE: A database of plant cis-acting regulatory elements and a portal to tools for *in silico* analysis of promoter sequences. *Nucleic Acids Res.*, *30*(1): 325–27.

Li, E., Bhargava, A., Qiang, W., Friedmann, M., Forneris, N., Savidge, R., Johnson, L., Mansfield, S., Ellis, B. and Douglas, C. 2012. The Class II KNOX gene *KNAT7* negatively regulates secondary wall formation in Arabidopsis and is functionally conserved in Populus. *New Phytol.*, *194*(1): 102–15.

Li, E., Wang, S., Liu, Y., Chen, J. and Douglas, C.J. 2011. *OVATE FAMILY PROTEIN4* (*OFP4*) interaction with *KNAT7* regulates secondary cell wall formation in *Arabidopsis thaliana*. *Plant J.*, *67*(2): 328–41.

Li, P., Yu, S., Shen, J., Li, Q., Li, D., Li, D., Zheng, C. and Shu, H. 2014. The transcriptional response of apple alcohol acyltransferase (MdAAT2) to salicylic acid and ethylene is mediated through two apple MYB TFs in transgenic tobacco. *Plant Mol. Biol.*, *85*(6): 627–38.

Liberman, L.M., Sparks, E.E., Morenorisueno, M.A., Petricka, J.J. and Benfey, P.N. 2015. MYB36 regulates the transition from proliferation to differentiation in the Arabidopsis root. *Proc. Natl. Acad. Sci. U.S.A.*, *112*(39): 12099–12104.

Lin, T., Sharma, P., Gonzalez, D.H., Viola, I.L. and Hannapel, D.J. 2013. The impact of the long-distance transport of a BEL1-Like Messenger RNA on development. *Plant Physiol.*, *161*(2): 760–72.

Ma, Q., Wang, N., Hao, P., Sun, H. and Yu, S. 2019. Genome-wide identification and characterization of TALE superfamily genes in cotton reveals their functions in regulating secondary cell wall biosynthesis. *BMC Plant Biol.*, *19*(1): 432.

Magnani, E. and Hake, S. 2008. KNOX lost the OX: The Arabidopsis *KNATM* gene defines a novel class of KNOX transcriptional regulators missing the homeodomain. *Plant Cell*, *20*(4): 875–87.

Mahajan, A., Bhogale, S., Kang, I.H., Hannapel, D.J. and Banerjee, A.K. 2012. The mRNA of a Knotted1-like transcription factor of potato is phloem mobile. *Plant Mol. Biol.*, *79*(6): 595–608.

Marchler-Bauer, A., Bo, Y., Han, L., He, J., Lanczycki, C.J., Lu, S., Chitsaz, F., Derbyshire, M.K., Geer, R.C. and Gonzales, N.R. 2017. CDD/SPARCLE: Functional classification of proteins via subfamily domain architectures. *Nucleic Acids Res.*, *45*(D1): D200–D203.

Meng, L., Fan, Z., Zhang, Q., Wang, C., Gao, Y., Deng, Y., Zhu, B., Zhu, H., Chen, J. and Shan, W. 2018. BEL1-LIKE HOMEODOMAIN 11 regulates chloroplast development and chlorophyll synthesis in tomato fruit. *Plant J.*, *94*(6): 1126–40.

Mukherjee, K., Brocchieri, L. and Bürglin, T.R. 2009. A comprehensive classification and evolutionary analysis of plant homeobox genes. *Mol. Biol. Evol.*, *26*(12): 2775–94.

Ono, N.N., Britton, M.T., Fass, J.N., Nicolet, C.M., Lin, D. and Tian, L. 2011. Exploring the transcriptome landscape of pomegranate fruit peel for natural product biosynthetic gene and SSR marker discovery. *J. Integr. Plant. Biol.*, *53*(10): 800–813.

Ophir, R., Sherman, A., Rubinstein, M., Eshed, R., Sharabi Schwager, M., Harel-Beja, R., Bar-Ya'Akov, I. and Holland, D. 2014. Single-nucleotide polymorphism markers from *de novo* assembly of the pomegranate transcriptome reveal germplasm genetic diversity. *PLoS One*, *9*(2): e88998.

Proveniers, M., Rutjens, B., Brand, M. and Smeekens, S. 2007. The Arabidopsis TALE homeobox gene *ATH1* controls floral competency through positive regulation of *FLC*. *Plant J.*, *52*(5): 899–913.

Qin, G., Xu, C., Ming, R., Tang, H., Guyot, R., Kramer, E.M., Hu, Y., Yi, X., Qi, Y. and Xu, X. 2017. The pomegranate (*Punica granatum* L.) genome and the genomics of punicalagin biosynthesis. *Plant J.*, *91*(6): 1108–28.

Qiu, R., Zhang, T., Yang, S., Song, L. and Zhao, D. 2019. Genome-wide identification and bioinformatics analysis of TALE transcription factor family in *Lotus japonicas*. *J. Plant Genet. Resour.*, *20*: 466–75.

Ray, A., Robinsonbeers, K., Ray, S., Baker, S.C., Lang, J.D., Preuss, D., Milligan, S.B. and Gasser, C.S. 1994. Arabidopsis floral homeotic gene *BELL* (*BEL1*) controls ovule development through negative regulation of *AGAMOUS* gene (*AG*). *Proc. Natl. Acad. Sci. U.S.A.*, *91*(13): 5761–65.

Ruiz-Estévez, M., Bakkali, M., Martinblazquez, R. and Garridoramos, M.A. 2017. Identification and characterization of TALE homeobox genes in the endangered fern Vandenboschia speciosa. *Genes*, *8*(10): 275.

Rutjens, B., Bao, D., Van Eckstouten, E., Brand, M., Smeekens, S. and Proveniers, M. 2009. Shoot apical meristem function in Arabidopsis requires the combined activities of three BEL1-like homeodomain proteins. *Plant J.*, *58*(4): 641–54.

Sakakibara, K., Ando, S., Yip, H.K., Tamada, Y., Hiwatashi, Y., Murata, T., Deguchi, H., Hasebe, M. and Bowman, J.L. 2013. KNOX2 genes regulate the haploid-to-diploid morphological transition in land plants. *Science*, *339*(6123): 1067–70.

Schultz, J., Milpetz, F., Bork, P. and Ponting, C.P. 1998. SMART, a simple modular architecture research tool: Identification of signaling domains. *Proc. Natl. Acad. Sci. U.S.A.*, *95*(11): 5857–64.

Shani, E., Yanai, O. and Ori, N. 2006. The role of hormones in shoot apical meristem function. *Curr. Opin. Plant Biol.*, *9*(5): 484–89.

Smith, H.M., Boschke, I. and Hake, S. 2002. Selective interaction of plant homeodomain proteins mediates high DNA-binding affinity. *Proc. Natl. Acad. Sci. U.S.A.*, *99*(14): 9579–84.

Subramanian, B., Gao, S., Lercher, M.J., Hu, S. and Chen, W.H. 2019. Evolview v3: A webserver for visualization, annotation and management of phylogenetic trees. *Nucleic Acids Res.*, *47*(W1): W270–W275.

Szklarczyk, D., Gable, A.L., Lyon, D., Junge, A., Wyder, S., Huerta-Cepas, J., Simonovic, M., Doncheva, N.T., Morris, J.H. and Bork, P. 2019. STRING v11: Protein-protein association networks with increased coverage, supporting functional discovery in genome-wide experimental datasets. *Nucleic Acids Res.*, *47*(D1): D607–D613.

Wang, H., Cheng, L., He, P., Chang, Y. and Li, L. 2017. Identification of genes encoding transcription factors of apple cold-resistant dwarfing rootstock in response to cold stress. *Plant Physiol. J.*, *53*(8): 1468–78.

Waterhouse, A., Bertoni, M., Bienert, S., Studer, G., Tauriello, G., Gumienny, R., Heer, F.T., de Beer, T.A.P., Rempfer, C. and Bordoli, L. 2018. SWISS-MODEL: Homology modeling of protein structures and complexes. *Nucleic Acids Res.*, *46*(W1): W296–W303.

Wen, B., Song, W., Sun, M., Chen, M., Mu, Q., Zhang, X., Wu, Q., Chen, X., Gao, D. and Wu, H. 2019. Identification and characterization of cherry (*Cerasus pseudocerasus* G. Don) genes responding to parthenocarpy induced by GA3 through transcriptome analysis. *BMC Genet.*, *20*(1): 65.

Yu, C., Lin, C. and Hwang, J. 2004. Predicting subcellular localization of proteins for Gram-negative bacteria by support vector machines based on n-peptide compositions. *Protein Sci.*, *13*(5): 1402–06.

Yuan, Z., Fang, Y., Zhang, T., Fei, Z., Han, F., Liu, C., Liu, M., Xiao, W., Zhang, W. and Wu, S. 2018. The pomegranate (*Punica granatum* L.) genome provides insights into fruit quality and ovule developmental biology. *Plant Biotechnol. J.*, *16*(7): 1363–74.

Zhang, T., Liu, C., Huang, X., Zhang, H. and Yuan, Z. 2019. Land-plant phylogenomic and pomegranate transcriptomic analyses reveal an evolutionary scenario of CYP75 genes subsequent to whole genome duplications. *J. Plant Biol.*, *62*: 48–60.

Zhao, K., Zhang, X., Cheng, Z., Yao, W., Li, R., Jiang, T. and Zhou, B. 2019. Comprehensive analysis of the three-amino-acid-loop-extension gene family and its tissue-differential expression in response to salt stress in poplar. *Plant Physiol. Bioch.*, *136*: 1–12.

Zhong, R., Lee, C., Zhou, J., Mccarthy, R.L. and Ye, Z.H. 2008. A battery of transcription factors involved in the regulation of secondary cell wall biosynthesis in Arabidopsis. *Plant Cell*, *20*(10): 2763–82.

14

Genomic Insights into Flower Development
A Comprehensive Study of YABBY Gene Family Identification and Expression in *Punica granatum* L.

Yujie Zhao[1] and *Zhaohe Yuan*[2*]

The plant-specific YABBY transcription factors have important biological roles in plant morphogenesis, growth and development. In this study, we identified six YABBY genes in pomegranate and characterized their expression pattern during flower development. Six PgYABBY genes were divided into five subfamilies (YAB1/3, YAB2, INO, CRC, and YAB5), based on protein sequence, motifs, and similarity of exon-intron structure. Next, analysis of putative *cis*-acting element showed that *PgYABBY*s contained lots of hormone response and stress response elements. Subsequently, gene function prediction and protein-protein network analysis showed that PgYABBYs were associated with the development of apical meristem, flower, carpel, and ovule. Analysis of PgYABBY genes expression in various structures and organs suggested that *PgYABBYs* were highly activated in flower, leaf, and seed coat. Analysis of expression during flower development in pomegranate showed that *PgINO* might play a critical role in regulating the differentiation of flowers. This study provided a theoretical basis for function research and utilization of YABBY genes in pomegranate.

1. Introduction

The YABBY is unique transcription factor in plants, belonging to the zinc finger protein superfamily. YABBY proteins have two highly conserved domains:

1 College of Horticulture, Henan Agricultural University, Zhengzhou 450002, China.
2 College of Forestry, Nanjing Forestry University, Nanjing 210037, China.
* Corresponding author: zhyuan88@hotmail.com

N-terminal C2C2 type zinc finger domain, and C-terminal YABBY domain (Golz et al., 2004; Sieber et al., 2004). The YABBY transcription factors play significant roles in regulation of diverse developmental processes, such as formation of adaxial-adaxial polarity, lamina expansion, and floral organ development (Eckardtn et al., 2010; Ha et al., 2010; Tanaka et al., 2012). It has been reported that the YABBY plays different roles in lateral organ development, such as leaves (Eckardtn et al., 2010; Ha et al., 2010), floral organs (Yamada et al., 2011; Tanaka et al., 2012; Fourquin et al., 2014), and fruit development (Han et al., 2015).

The YABBY transcription factors have been extensively studied in the dicotyledon plant. There are six YABBY members in *Arabidopsis*, *YABBY1* (*FIL, FILAMENTOUS FLOWER*), *YABBY2* (*YAB2*), *YABBY3* (*YAB3*), *YABBY4* (*INO, INNER NO OUTER*), *YABBY5* (*YAB5*), and *CRC* (*CRABS CLAW*). *FIL*, *YAB2*, and *YAB3* are always expressed in the primordia of lateral organs, which determine the abaxial cell fates (Siegfried et al., 1999) and participate in floral organ formation and leaf development (Sawa et al., 1999; Eshed et al., 2004; Stahle et al., 2009), and they act redundantly to promote vegetative organ development (Bowman 2000; Eckardtn et al., 2010; Sarojam et al., 2010). *INO* and *CRC* are expressed in specific tissues: *INO* participates in the development of ovule outer integument, while *CRC* have been reported to be associated with polarity of nectary and carpel (Bowman and Smyth, 1999; Zhang et al., 2013). Twenty-three YABBY genes have been identified in cotton (*Gossypium hirsutum*). All *GhYABBYs* are expressed in bud, flower and stem meristem tissues. The YABBY genes in cotton may be involved in the development of the above-mentioned tissues (Xu et al., 2015). In tomato (*Solanum lycopersicum*), nine YABBY genes have been located on seven chromosomes. The expression pattern of *SlYABBY* is similar to that of Arabidopsis (Huang et al., 2013). As a homolog of *INO*, *LeYABBYB* have been cloned from tomato, which is critical for the development of abaxial cells of leaf primordium (Kim et al., 2003). Cong et al. (2008) found that FAS protein was encoded by a *YABBY-like* gene in tomato, which mainly regulated the number of carpels during flowering and fruit development. Bartley and Ishida (2002, 2003, 2007) found that *LeYAB2*, independent of ethylene signaling pathway, was induced by low temperature to express in sepals of cv.VFNT cherry tomato. *LeYAB2* is highly expressed in the epidermis of tomato fruit during ripening period, suggesting that *LeYAB2* may affect the abaxial cell of pericarp and act a pivotal part in fruit ripening. Previous studies proved that homologous gene *YABBY5* played a key role in regulating the fruit morphogenesis of Fingered Citron (*Citrus medica* L. var. *sarcodactylis* Swingle) (Han, 2014). *CmsYABBY5* is highly expressed in pistils and stamens of Fingered Citron, but it is significantly low expressed in leaves. It suggests that *CmsYABBY5* may be involved in pistil and stamen development of Fingered Citron (Liao et al., 2016).

Many YABBY genes have been studied in monocotyledon plants, such as *Oryza sativa*, *Zea mays* and *Triticum aestivum* (Zhao et al., 2006). Thirteen YABBY genes have been described in the maize genome (Ge et al., 2014), eight YABBY genes in rice (Toriba et al., 2007). *OsYABBY4* play a critical role in the development of vascular structure (Liu et al., 2007), *DL* (*DROOPING LEAF*) is orthologous with *CRC*, which is mainly related to the formation of leaf mid-vein and flower development in rice (Yamaguchi et al., 2004). In *dl* mutant, carpel is transformed to stamen (Toriba et al., 2007). *OsYAB1* has high homology with *AtYAB2* and *AtYAB5*, it participates in the development of stamen, carpel and meristem (Toriba et al., 2007). Tanaka et al. (2017) found that *OsTOB*1, *OsTOB*2 and *OsTOB*3 were orthologous with *FIL*, which participated in the development of reproductive meristem. *OsTOB*1 acts a vital role to promote development of lateral organs and maintenance of meristem in rice spikelets (Tanaka et al., 2012). *OsYAB3* participates in the development of leaves, while *OsYAB4* is mainly expressed in rice vascular system (Dai et al., 2007; Liu et al., 2007). Juarez et al. (2004) cloned *zyb9* and *zyb14* of maize, based on conservative domain sequences of *FIL*, *YAB2* and *YAB3*. *zyb9* and *zyb14* are expressed only in the paraxial surface of leaf primordia. Subcellular localization analysis showed that TaYAB1 and TaCRC fusion protein were localized in the nucleus, and they might play a key role in transcription and transcriptional regulation during the process of wheat dorsoventral polarity (Zhao et al., 2009). Evolutional study in plants showed that YABBY genes have not been found in bryophytes and lycopodialeo genomes, suggesting the YABBY is a unique transcription factor for seed plants (Finet et al., 2016).

Pomegranate is an important commercial fruit tree, which is widely grown in tropics and subtropics areas. There are bisexual flowers (vase shape) and functional male flowers (bell shape) in pomegranate. Ovule sterility causes functional male flowers which fall off after flowering and seriously affect the fruit yield (Wetzstein et al., 2013; Chen et al., 2017). The genome data provide critical support for the function research of pomegranate gene (Qin et al., 2017; Yuan et al., 2018). In this study, the YABBY genes were identified in pomegranate. PgYABBY genes were computationally and experimentally characterized. A comparison of expression patterns with flowers in development stages were studied in the bisexual and functional male flowers of pomegranate. Our results have potential significance for the function research of PgYABBY genes.

2. Materials and Methods

2.1 Plant Material and Data Collection for Expression Analysis

From April to June 2018, the bisexual flowers and functional male flowers of pomegranate were collected at Baima Base for Teaching and Scientific Research of Nanjing Forestry University. Flowers were categorized into eight stages according to their vertical diameter: 3.0–5.0 mm (P1), 5.1–8.0 mm (P2), 8.1–10.0 mm (P3), 10.1–12.0 mm (P4), 12.1–14.0 mm (P5), 14.1–16.0 mm (P6), 16.1–18.0 mm (P7), and 18.1–20.0 mm (P8). Three biological repeats were set up at each stage. Samples were frozen with liquid nitrogen and stored in –78°C refrigerator.

The transcription data of the whole genome in pomegranate was obtained from NCBI database (http://www.ncbi.nlm.nih.gov/), including leaves, roots, bisexual flowers, functional male flowers, exocarp, pericarp, and other tissues. YABBY protein sequences of *Arabidopsis thaliana* were downloaded from the Arabidopsis Information Source (https://www.arabidopsis.org/), and that of tomato was downloaded from PlantTFDB database (http://planttfdb.cbi.pku.edu.cn/index.php).

2.2 Identification and Sequence Analysis of YABBY Transcription Factors

YABBY protein sequences in PlantTFDB database (http://planttfdb.cbi.pku.edu.cn/download_seq.php?Fam=YABBY) were used as queries to perform BLAST against the pomegranate genome database (E-value<$1e^{-5}$, identity>50%), duplication was removed and YABBY protein sequences were screened and selected. The HMM profile of the Pfam YABBY domain (PF04690) was performed against pomegranate protein databases using HMMER 3.0 software package (E-value<$1e^{-5}$). Moreover, all obtained YABBY protein sequences were further analyzed on CDD website (https://www.ncbi.nlm.nih.gov/cdd; Marchler-Bauer et al., 2017) to verify the presence of the C2C2 domain at N terminal and YABBY domain at C terminal.

ExPASy Proteomics Server (https://web.expasy.org/protparam/) was used to predict the physicochemical properties of YABBY protein, including amino acid sequence length, molecular weight and isoelectric point.

2.3 Phylogenetic Analysis

Multiple sequences alignment of pomegranate, *Arabidopsis* and tomato YABBY proteins were carried out using ClustalX 2.1 software, and artificially corrected them. An unrooted Neighbor-Joining (NJ) phylogenetic tree was constructed with all of the YABBY protein sequences from pomegranate, Arabidopsis, tomato using MEGA 7.0 (Kumar et al., 2016). Bootstrap analysis was performed using 1,000 repetitions with the following parameters: Complete deletion, Poisson model.

2.4 Gene Structure Analysis and Motif Identification

The exon-intron structure of the YABBY genes were determined based on alignment of the corresponding coding sequences and full-length sequences, then the results were displayed using Gene Structure Display Server (GSDS 2.0: http://gsds.cbi.pku.edu.cn) online tools.

Motifs of the YABBY proteins were identified using MEME online tools (http://meme-suite.org/tools/meme) with default parameter, and the motif characteristics of the YABBY proteins were obtained.

2.5 Protein Structure Analysis and Protein Interaction Network

In order to identify three-dimensional structure of the YABBY proteins, Phyre 2.0 was used to predicted structures of 6 YABBY proteins (Kelley and Sternberg, 2009). By searching, three-dimensional protein models of six proteins were constructed.

Ensemblplants (http://plants.ensembl.org/index.html) website and GO tools (http://www.geneontology.org/) were used to predict their functions. The protein-protein interaction network of *YABBY* family was analyzed through String (https://string-db.org/).

2.6 Prediction Promoter Elements

To identify putative *cis*-acting elements in promoters, 2000 bp gene sequences upstream of the initiation codon were obtained from the pomegranate genome sequence by TBtools (Chen et al., 2018). The *cis*-acting elements on the promoter region were analyzed by PlantCARE (http://bioinformatics.psb.ugent.be/webtools/plantcare/html/).

2.7 Expression Patterns by Transcriptome Data

To examine expression patterns of YABBY genes in different pomegranate tissues and organs, the published transcriptome data were download for expression analysis. Firstly, all RNA-Seq data were qualitatively controlled by fastp (Chen et al., 2018) to obtain cleaned reads. Then the sequence was indexed with pomegranate transcriptome data. Through Kallisto 0.44.0 software (Nicolas et al., 2016), and the gene expression was further calculated and analyzed. The corresponding expression level (TPM value) of *YABBY* genes was thus obtained. TPM value was transformed into Log_2 (TPM+1). Finally, the thermal map of *PgYABBY* was drawn with R programmed pheatmap.

2.8 Expression Patterns by Quantitative Real-time PCR (qRT-PCR)

Total RNA was extracted using the BioTeke plant total RNA extraction kit (centrifugal column type), and first-strand cDNA was prepared by reverse transcription kit (PrimeScript™RT reagent Kit with gDNA Eraser, TaKaRa). Then the cDNA was used as template for expression analysis on *YABBY* family genes at different developmental stages of pomegranate.

According to DNA and CDS sequences, specific quantitative primers (Table 1) were designed. Pomegranate *PgActin* was used as internal reference gene. Each 20 μL of reaction mixture contained 10 μL of TB Green *Premix Ex Taq*, 0.4 μL of ROX Reference Dye II, 0.4 μL of upstream/downstream primers, 2 μL of cDNA template, and 6.8 μL of ddH_2O. The PCRs were performed on the Applied Biosystems 7500 and the thermal cycler was set as follows: pre-denaturation at 95°C for 30 s, denaturation at 95°C for 5 s, and denaturation at 60°C for 34 s for 40 cycles, and fluorescence was acquired at the second step of each cycle; dissolution curve was gained as follows: 95°C for 15 s, 60°C for 60 s, and 95°C for 15 s. The data were quantitatively analyzed by $2^{-\Delta\Delta CT}$ method (Livak and Schmittgen, 2001).

The experimental data represent at least three independent biological repeats. Data were analyzed using SPSS software (22.0, USA).

Table 1 Primers used in this study.

Gene	*Primer Sequence*
PgYAB1	F: ACGCATCAAAGCAGGAAA; R: GCTGTGGGAGGGAAGAAC
PgINO	F: GCAAAGGCAGAACATCCA; R: CAGGTCCTCATCTTCATTGTCG
PgCRC	F: CTCTGTTCCGCCATCTTC; R: ACCTCAGTGCCATAACCC
PgYAB2	F: TTCGGAGCGAGTCTGTTA; R: CGATTAGGTGCTGTTGTTT
PgYAB5	F: GAGAAGAGGCAGCGAGTG; R: GCCGAAATGGATGTGAGG
PgYAB3	F: GCAGTAGCCTGTTCAAGACC; R: TGGCGGCACATTTCGTAT
PgActin	F: AGTCCTCTTCCAGCCATCTC; R: CACTGAGCACAATGTTTCCA

3. Results

3.1 Identification and Sequence Analysis of YABBY Transcription Factor

A total of six *PgYABBY* genes were identified in pomegranate, using two methods of local blast and hmmsearch model. All candidate proteins were confirmed to belong to YABBY family.

The physicochemical properties of PgYABBY amino acid sequences were analyzed by ExPASy Proteomics Server. The results showed that the length of coding region of the six *PgYABBY* genes ranged from 507 bp (*PgCRC*) to 675 bp (*PgYAB3*) (Table 2). The numbers of amino acids translated by YABBY varied from 168 aa –224 aa, and the molecular weight of the protein ranged from 18228.52 Da–24736.28 Da. The isoelectric point ranged from 8.46 (*PgINO*) to 9.02 (*PgCRC* and *PgYAB5*), indicating their property of basic proteins.

Table 2 Information of YABBY gene in pomegranate.

Gene name	*Gene ID*	*CDS (bp)*	*Exon No.*	*Amino Acid Residues*	*MW(Da)*	*PI*	*AT Homologous Gene*
PgCRC	Pg016200	507	6	168	18228.52	9.02	AT1G69180 *AtCRC*
PgINO	Pg016102	660	7	219	24447.56	8.46	AT1G23420 *AtINO*
PgYAB1	Pg014578	606	7	201	22481.82	8.84	AT2G45190 *AtYAB1*
PgYAB2	Pg016743	630	6	209	23370.49	8.94	AT1G08465 *AtYAB2*
PgYAB5	Pg026102	588	7	195	21760.74	9.02	AT2G26580 *AtYAB5*
PgYAB3	Pg030843	675	7	224	24736.28	8.58	AT4G00180 *AtYAB3*

3.2 Phylogenetic Analysis

To analyze the evolutionary relationships in YABBY proteins, six PgYABBYs, six AtYABBYs, and nine SlYABBYs were used to construct the phylogenetic tree. According to the previous phylogenetic analysis on YABBYs in Arabidopsis and tomato, PgYABBYs were classified into five distinct groups, YAB1/YAB3, YAB2, INO, CRC, and YAB5. Among them, the YAB1/YAB3 group contains two members, PgYAB1 and PgYAB3 were respectively clustered into YAB1 and YAB3 group. The group of INO, CRC, YAB2, and YAB5 had one member, PgINO, PgCRC, PgYAB2, and PgYAB5, respectively.

3.3 Gene Structures of PgYABBYs

The gene structures were investigated on gene structure display server (GSDS). The length and splicing varied among the six *PgYABBYs*. The number of exons was different. *PgYAB2* and *PgCRC* had five introns, and *PgYAB1*, *PgINO*, *PgYAB5*, and *PgYAB3* all contained six introns, which indicated that the gene structure with six introns was the main form of *PgYABBY*s. We speculated that the length of *PgYABBYs* was closely related to the change of introns length.

3.4 Motif Identification of PgYABBYs

The multiple sequence alignment of six YABBY proteins showed that PgYABBY proteins contain C2C2 zinc finger conserved domain near the N terminus (Cys-22-Cys-53) and the YABBY conserved domain which was located toward the C-terminal end of the protein (Figure 1A) (Bartholmes et al., 2012; Zhao et al., 2017). Four conservative motifs of PgYABBY members were searched and identified by MEME online tools (Figure 1B/C). The subfamily members showed similar gene structure and protein conservative motif distribution, such as four members of YAB1/YAB3 subfamily with seven exons, two members with six exons (PgYAB2 and PgCRC). All the PgYABBY members contained Motif 1 and Motif 2 encoded YABBY domain by three to five exons. These results proved that the protein domain and motif of PgYABBYs were highly conservative.

3.5 Protein Structure Analysis and Protein Interaction Network

The secondary structure of YABBY family proteins in pomegranate was mainly composed of irregular curls, and the α-helices and β-corner, which were distributed throughout these proteins. The tertiary structure of PgYABBYs was predicted by Phyre2. These six proteins represented diverse conformations of YABBY proteins.

Protein function prediction revealed that all family members had multicellular developmental function. PgYAB1 and PgYAB3 are involved in regulating the fate of paraxial/abaxial polar cells during embryogenesis and organogenesis. They serve as regulators of floral organ formation and development. In addition, PgYAB5 and PgYAB2 are identity genes of paraxial cells, which can regulate the development of apical meristem of embryo. PgINO regulate the development of ovule outer integument and the polar growth of integument. PgCRC regulate the development of nectary, inhibit the transverse development of early pistil, and promote the fusion of carpel.

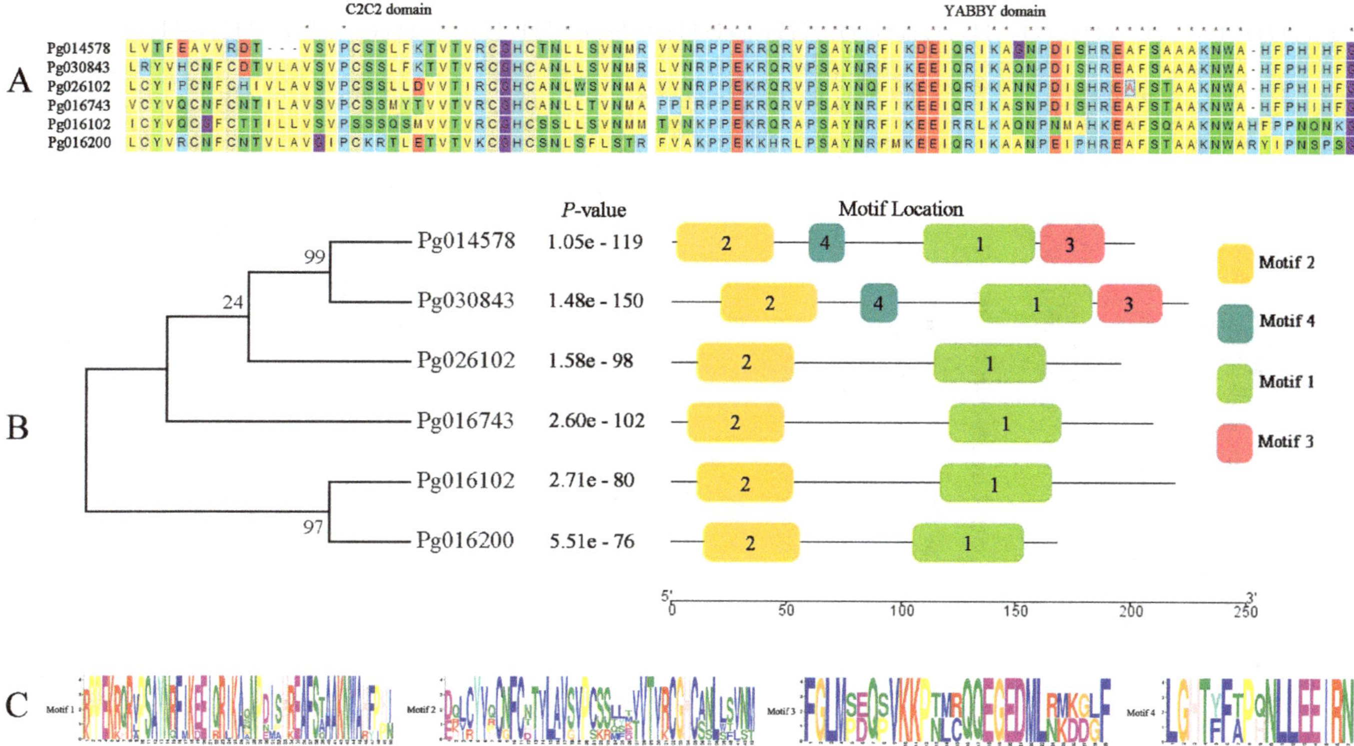

Fig. 1 Conservative protein motif of YABBY family in pomegranate.

The protein interaction between PgYABBYs was analyzed by String software to know potential functions and signal transduction or metabolic pathways. The result showed that the regulatory network of PgYAB1 and PgYAB3 was similar, PgYAB5 and PgYAB2 was similar, PgINO and PgCRC had independent regulatory networks in protein-protein interaction database. PgYAB1 and PgYAB3 interact with KAN, WUS and SPL proteins to regulate the differentiation of floral organs at the first and second cycles. PgYAB5 and PgYAB2 interact with unspecified proteins such as AS, ETT, BOB, and ELO3. PgINO, BEL1, KAN, ANT, SPL, and SUP coordinate to regulate ovule development, the same spatial pattern of INO expression was established by Villanueva et al. (1999). PgCRC, AG and SHP proteins coordinate to regulate the differentiation of carpel.

3.6 Cis-acting Elements Located in YABBY Gene Promoters

In this study, *Cis*-element analysis showed that each of *PgYABBY* members contained a wide variety of *cis*-acting elements, ranging from 36 *cis*-acting elements in *PgCRC* to 28 *cis*-acting elements in *PgYAB2*. *PgYABBY* family genes respond to different plant hormone and abiotic stress, including abscisic acid response element (ABRE), antioxidant response element (ARE), methyl jasmonate response element (CGTCA-motif), LTR, hydrophilic salicylic acid response element (TCA-motif), auxin response element (TGA-motif and AuxRR-core), TGACG-motif, and gibberellin-responsive element (GARE-motif and P-box). Among the six *PgYABBY* genes, four members contained ABRE, six family genes contained ARE and four genes contain CGTCA-motif. All these genes were supposed to be related to plant responses to stress. Pomegranate had two PgYABBY genes with TGA-motif (*PgCRC*, *PgYAB1*). *PgYAB1*, *PgINO*, *PgCRC*, *PgYAB5*, and *PgYAB3* had cold stress response elements LTR. *PgYAB1*, *PgYAB3*, *PgCRC*, and *PgINO* had MeJA-responsiveness elements. *PgYAB1*, *PgYAB2* and *PgYAB3* had AuxRR-core elements. Only *PgCRC* contained gibberellin element P-box, *PgYAB5* had a TCA-motif.

3.7 Expression Patterns of YABBY Genes in Different Organs

According to published transcriptome data of pomegranate, *PgYABBY* genes are expressed in leaf, root, flower, fruit, endocarp, and pericarp. The analysis revealed the different expression levels of *YABBY* genes. For example, *PgCRC* is expressed during the development of both bisexual and functional male flowers, suggesting that *PgCRC* may be involved in the differentiation of female and male organs in pomegranate flowers; *PgINO* is expressed in the middle and late developmental stages in bisexual flowers and exocarps. *PgYAB3* and *PgYAB1* showed similar expression patterns, namely, they expressed in the early stage of differentiation and formation of bisexual and functional male flower buds. This may indicate that they are both involved in regulating the development of leaves, male and female organs. *PgYAB2* is expressed in the early and middle stages of the development of bisexual and functional male flower buds of pomegranate, as well as in leaves. The gene may be related to the differentiation and development of male and female organs in pomegranate flowers. *PgYAB5* is greatly expressed in leaves, indicating that it may play an important role in regulating leaf development.

3.8. qRT-PCR Analysis of PgYABBYs During Flower Development in Pomegranate

qRT-PCR was used to analyze expression patterns of *PgYABBY* genes during flowers development. The expression levels of *PgYAB1* and *PgYAB3* are significantly decreased at the early flower bud stage, but they are the lowest at the late flowering stage. The mRNA levels of *PgYAB5* and *PgCRC* in bisexual flowers are higher than those in functional male flowers. In pomegranate functional male flower, *PgYAB5* transcript level seems to have no obvious changes during P2–P8. The levels of *PgYAB1* and *PgYAB5* in bisexual flowers are higher than those in functional male flowers at P2. *PgYAB2* is increased gradually after P2, and then suddenly decreased at P4–P7, but increased again at P8. The expression level of *PgINO* is significantly increased at the early flower bud stages, and then decreased after P5. The mRNA level of *PgINO* in bisexual flowers is higher than that in functional male flowers, especially at P3, P4 and P7.

4. Discussion

The YABBY family is a plant-specific transcription factor, with the typical zinc-finger and YABBY conserved domain. YABBY family genes play important regulatory roles in the development of lateral organs and abaxial-adaxial polarity. Many YABBY genes are identified in plants, with a large variable among species. For example, six *AtYABBY* genes have been identified in Arabidopsis, eight *OsYABBY* genes in rice (Toriba et al., 2007), 23 *GhYABBY* genes in cotton (Xu et al., 2015), 13 *ZmYABBY* genes in maize (Gemin et al., 2014) and 11 *BrYABBY* genes in *Brassica pekinensis* (Li Xiaobo et al., 2012). In this study, six *PgYABBY*s encoding 168–224 amino acids were identified in pomegranate. The six YABBY genes were identified in pomegranate and Arabidopsis, indicating that this family could reflect a lack of expansion during evolution. *PgYABBY* genes contained 5–6 introns, and seven exons were the main structure of *PgYABBYs*. The genetic structure of *PgYABBYs* was consistent with that of *Arabidopsis* and Chinese cabbage (Li et al., 2012). The YABBY gene structure is relatively conservation.

To understand the PgYABBY evolutionary relationships, we constructed a phylogenetic tree including YABBY proteins from three species. According to the phylogenetic relationship of YABBY genes between Arabidopsis and tomato, PgYABBY genes were divided into five distinct clades, which was in line with the classification of YABBY in tomatoes and apples (Huang et al., 2013; Shao Hongxia et al., 2017). In YAB1/YAB3 clades, *PgYAB1* and *PgYAB3* were identified as parallel homologous gene pairs, indicating that *PgYABBYs* expand in a species-specific manner (Chen et al., 2017). This apparent recent evolution implied that the gene pairs might perform similar functions. Each group of *PgYABBYs* had homologous genes with Arabidopsis. It was speculated that the members of YABBY gene family of pomegranate and Arabidopsis may have similar biological functions.

In Arabidopsis, YABBY genes are related to formation of adaxial-abaxial polarity and floral organ development. Our result found that PgYABBY were widely expressed in many structures: stem, leaf, flower, nectary, carpel, and ovule. *PgYAB1* and *PgYAB3*, or *PgYAB2* and *PgYAB5* showed similar protein interaction networks. This implied that these genes might perform similar functions in pomegranate. *PgINO* and *PgCRC* specifically regulated ovule, carpel, and nectary development. The results were identical to results in the previous reports (Stahle et al., 2009; Sarojam et al., 2010).

Cis-acting elements are located at the gene promoter site and have specific binding with transcription factors to regulate gene transcription (Riechmann et al., 2000). In this study, we found that the promoter sequence of *PgYABBYs* contained multiple *cis*-acting elements related to hormone response and abiotic stress, suggesting that *PgYABBY* might be closely related to hormone signal transduction and stress response. In particular, the discovery of hormone-inducing related elements, such as auxin and gibberellin, might indicate that *PgYABBY* genes were associated with growth and development of pomegranate.

Based on the published transcriptome data of pomegranate, *PgYABBYs* were extremely low expressed in root and pericarp. As an ortholog of the *INO*, the *PgINO* were expressed specifically in reproductive organs (Eckardt, 2010). The homolog of *AtYAB3* and *AtYAB5*, *PgYAB3* and *PgYAB5* were highly expressed in leaves. The observation that *PgYAB1* and *PgYAB3* in the same subfamily showed similar pattern of expressions, suggested that they performed similar functions. *PgYAB5* was expressed in vegetative tissues, while *PgCRC* expression was restricted in the flower. *PgINO*, *PgCRC* and *PgYAB5* were expressed as reported in Arabidopsis (Eckardt, 2010). These results provided preliminarily evidences that *PgYABBY*s were related to the development of pomegranate flowers and fruits. The results of qRT-PCR showed that the expression patterns of *PgYABBYs* in the two types of flowers were relatively similar. Interestingly, the expression *PgINO* was significantly higher than other members during the development of flowers, and the expression in bisexual flowers were also significantly higher than that in functional male flowers. The evidence suggested that *INO* is involved in polarity of the ovule primordium, the anlage to the outer integument (Villanueva et al., 1999). *INO* is necessary for the formation and asymmetric growth of the ovule outer integument in *Arabidopsis*. We speculated that *PgINO* might be a key gene in regulating the differentiation and development of ovule in pomegranate.

In this study, six YABBY genes were identified in pomegranate. PgYABBY genes were computationally and experimentally characterized through systematical analysis of gene structure, function and evolution. Our results illustrated the dynamic transcription patterns of PgYABBYs during flower development, indicating that *PgINO* might participate in the differentiation of flower organs in pomegranate.

References

Bartholmes, C., O. Hidalgo and S. Gleissberg. 2012. Evolution of the YABBY gene family with emphasis on the basal eudicot *Eschscholzia californica* (Papaveraceae). *Plant Biology*, *14*: 11–23.

Bartley, G.E. and B.K. Ishida. 2002. Digital fruit ripening: Data mining in the TIGR tomato gene index. *Plant Molecular Biology Reporter*, *20*(2): 115–30.

Bartley, G.E. and B.K. Ishida. 2003. Developmental gene regulation during tomato fruit ripening and *in vitro* sepal morphogenesis. *BMC Plant Biology*, *3*(1): 4.

Bartley, G.E. and B.K. Ishida. 2007. Ethylene-sensitive and insensitive regulation of transcription factor expression during *in vitro* tomato sepal ripening. *Journal of Experimental Botany*, *58*(8): 2043–51.

Bowman, J.L. 2000. The YABBY gene family and abaxial cell fate. *Current Opinion in Plant Biology*, *3*(1): 17–22.

Bowman, J.L. and D.R. Smyth. 1999. *CRABS CLAW*: A gene that regulates carpel and nectary development in Arabidopsis, encodes a novel protein with zinc finger and helix-loop-helix domains. *Development*, *126*(11): 2387–96.

Chen, L.N., J. Zhang, H.X. Li, J. Niu, H. Xue, B.B. Liu, Q. Wang, X. Luo, F.H. Zhang, D.G. Zhao and S.Y. Cao. 2017. Transcriptomic analysis reveals candidate genes for female sterility in pomegranate flowers. *Front. Plant Science*, *8*: 1430.

Chen, X.L., S.Y. Liu, T. Sun, Z.C. Liu, J.B. Tao, Y.X. Zhao and A.X. Wang. 2017. Bioinformatics analysis on *YABBY* gene family in tomato. *Journal of Northeast Agricultural University*, *48*(10): 11–19.

Chen, C.J., R. Xia, H. Chen and Y.H. He. 2018. TBtools: A Toolkit for Biologists integrating various HTS-data org handling tools with a user-friendly interface. *BioRxiv*. doi: https://doi. /10.1101/289660.

Chen, S., Y. Zhou, Y. Chen and J. Gu. 2018. fastp: An ultra-fast all-in-one FASTQ preprocessor. *Bioinformatics*, *34*(17): i884–i890.

Cong, B., L.S. Barrero and S.D. Tanksley. 2008. Regulatory change in YABBY-like transcription factor led to evolution of extreme fruit size during tomato domestication. *Nature Genetics*, *40*(6): 800–804.

Dai, M.Q., Y.F. Hu, Y. Zhao, H.F. Liu and D.X. Zhou. 2007. A *WUSCHEL-LIKE HOMEOBOX* gene represses a YABBY gene expression required for rice leaf development. *Plant Physiology*, *144*(1): 380–90.

Eckardtn, N.A. 2010. YABBY genes and the development and origin of seed plant leaves. *The Plant Cell*, *22*(7): 2103.

Eshed, Y., A. Izhaki, S.F. Baum, S.K. Floyd and J.L. Bowman. 2004. Asymmetric leaf development and blade expansion in *Arabidopsis* are mediated by KANADI and YABBY activities. *Development*, *131*(12): 2997–3006.

Finet, C., S.K. Floyd, S.J. Conway, B. Zhong, C.P. Scutt and J.L. Bowman. 2016. Evolution of the YABBY gene family in seed plants. *Evolution and Development*, *18*(2): 116–26.

Fourquin, C., A. Primo, I. Martinez-Fernandez, E. Huet-Trujillo and C. Ferrandiz. 2014. The *CRC* orthologue from Pisum sativum shows conserved functions in carpel morphogenesis and vascular development. *Annals of Botany*, *114*(7): 1535–44.

Ge, M., Y.D. Lu, T.F. Zhang, T. Li, X.L. Zhang and H. Zhao. 2014. Genome-wide identification and analysis of YABBY gene family in maize. *Jiangsu Journal of Agricultural Science*, *30*(6): 1267–27.

Golz, J.F., M. Roccaro, R. Kuzoff and A. Hudson. 2004. GRAMINIFOLIZ promotes growth and polarity of Antirrhinum leaves. *Development*, *131*(15): 3661–70.

Ha, C.M., J.H. Jun and J.C. Fletcher. 2010. Control of Arabidopsis leaf morphogenesis through regulation of the YABBY and KNOX families of transcription factors. *Genetics*, *186*(1): 197–206.

Han, H.Q., Y. Liu, M.M. Jiang, H.Y. Ge and H.Y. Chen. 2015. Identification and expression analysis of YABBY family genes associated with fruit shape in tomato (*Solanum lycopersicum* L.). *Genetics and Molecular Research*, *14*(2): 7079–91.

Han, X.X. 2014. *Function Analysis of YABBY5 and CRC Gene in Fruit Shaping of Fingered Citron*. Jinhua, Zhejiang Normal University.

Huang, Z.J., H.J. Van, G. Gonzalez, H. Xiao and K.E. Vander. 2013. Genome-wide identification, phylogeny and expression analysis of *SUN*, *OFP*, and *YABBY* gene family in tomato. *Molecular Genetics and Genomics*, *288*(3-4): 111–29.

Juarez, M.T., R.W. Twigg and M.C. Timmermans. 2004. Specification of adaxial cell fate during maize leaf development. *Development, 131*(18): 4533–44.

Kelley, L.A. and M.J.E. Sternberg. 2009. Protein structure prediction on the Web: A case study using the Phyre server. *Nature Protocols, 4*: 363–71.

Kim, M., T. Pham, A. Hamidi, S. McCormick, R.K. Kuzoff and N. Sinha. 2003. Reduced leaf complexity in tomato wiry mutants suggests a role for *PHAN* and *KNOX* genes in generating compound leaves. *Development, 130*(18): 4405–15.

Kumar, S., G. Stecher and K. Tamura. 2016. MEGA7: Molecular Evolutionary Genetics Analysis Version 7.0 for Bigger Datasets. *Molecular Biology and Evolution, 33*(7): 1870–74.

Li, X.B., C.C. Yang and N.W. Qiu. 2012. Bioinformatic analysis of YABBY protein family in Arabidopsis and Chinese Cabbage. *Shandong Agricultural Sciences, 44*(12): 1–6.

Liao, F.L., X.X. Han, W.R. Chen, Y. Guo, C.X. Zhang, Z.Y. Chen, Y.Y. Zhou and W.D. Guo. 2016. Fruit morphogenesis observation and expression analysis of fruit-shaping-related genes in fingered citron. *Acta Horticulturae Sinica, 43*(11): 2141–50.

Liu, H.L., Y.Y. Xu, Z.H. Xu and K. Chong. 2007. A rice YABBY gene, *OsYABBY4*, preferentially expresses in developing vascular tissue. *Development Genes and Evolution, 217*(9): 629–37.

Livak, K.J. and T.D. Schmittgen. 2001. Analysis of relative gene expression data using real-time quantitative PCR and the 2-ΔΔCT method. *Methods, 25*(4): 402–08.

Marchler-Bauer, A., Y. Bo, L. Han, J. He, C.J. Lanczycki, S. Lu, F. Chitsaz, M.K. Derbyshire, R.C. Geer, N.R. Gonzales, M. Gwadz, D.I. Hurwitz, F. Lu, G.H. Marchler, J.S. Song, N. Thanki, Z. Wang, R.A. Yamashita, D. Zhang, C. Zheng, L.Y. Geer and S.H. Bryant. 2017. CDD/SPARCLE: Functional classification of proteins via subfamily domain architectures. *Nucleic Acids Research, 45*(D1): D200–D203.

Nicolas, L.B., P. Harold, M. Páll and P. Lior. 2016. Near-optimal probabilistic RNA-seq quantification. *Nature Biotechnology, 34*: 525–27.

Ono, N.N., M.T. Britton, J.N. Fass, C.M. Nicolet, D. Lin, L. Tian. 2011. Exploring the transcriptome landscape of pomegranate fruit peel for natural product biosynthetic gene and SSR marker discovery. *Journal of Integrative Plant Biology, 53*(10): 800–813.

Ophir, R., A. Sherman, M. Rubinstein, R. Eshed, M.S. Schwager, R. Harel-Beja, I. Bar-Ya'akov and D. Holland. 2014. Single-nucleotide polymorphism markers from *de novo* assembly of the pomegranate transcriptome reveal germplasm genetic diversity. *PLoS One, 9*(2): e88998.

Qin, G.H., C.Y. Xu, R. Ming, H.B. Tang, G. Romain, M.K. Elena, Y.D. Hu, X.K. Yi, Y.J. Qi, X.Y. Xu, Z.H. Gao, H.F. Pan, J.B. Jian, Y.P. Tian, Z. Yue and Y.L. Xu. 2017. The pomegranate (*Punica granatum* L.) genome and the genomics of punicalagin biosynthesis. *Plant Journal, 91*(6): 1108–28.

Riechmann, J.L., J. Heard, G. Martin, L. Reuber, C. Jiang, J. Keddie, L. Adam, O. Pineda, O.J. Ratcliffe, R.R. Samaha, R. Creelman, M. Pilgrim, P. Broun, J.Z. Zhang, D. Ghandehari, B.K. Sherman and G. Yu. 2000. *Arabidopsis* transcription factors : Genome wide comparative analysis among eukaryotes. *Science, 290*(5499): 2105–10.

Sarojam, R., P.G. Sappl, A. Goldshmidt, I. Efroni, S.K. Floyd, Y. Eshed and J.L. Bowman. 2010. Differentiating Arabidopsis shoots from leaves by combined YABBY activities. *The Plant Cell, 22*(7): 2113–30.

Sawa, S., K. Watanabe, K. Goto, E. Kanaya, E.H. Morita and K. Okada. 1999. *FILAMENTOUS FLOWER*, a meristem and organ identity gene of Arabidopsis, encodes a protein with a zinc finger and HMG-related domains. *Genes and Development, 13*(9): 1079–88.

Shao, H.X., H.F. Chen, D. Zhang, H.Q. Wu, C.P. Zhao and M.Y. Han. 2017. Identification, evolution and expression analysis of the YABBY gene family in apple (*Malus* × *domestica* Borkh.). *Acta Agriculturae Zhejiangensis, 29*(7): 1129–38.

Sieber, P., M. Petrascheck, A. Barberis and K. Schneitz. 2004. Organ polarity in Arabidopsis *NOZZLE* physically interacts with members of the YABBY family. *Plant Physiology, 135*(4): 2172–85.

Siegfried, K.R., Y. Eshed, S.F. Baum, D. Otsuga, G.N. Drews and J.L. Bowman. 1999. Members of the YABBY gene family specify abaxial cell fate in Arabidopsis. *Development, 126*(18): 4117–28.

Stahle, M.I., J. Kuehlich, L. Staron, A.G. Arnim and J.F. Golz. 2009. YABBYs and the transcriptional corepressors LEUNIG and LEUNIG_HOMOLOG maintain leaf polarity and meristem activity in Arabidopsis. *The Plant Cell, 21*(10): 3105–18.

Tanaka, W., T. Toriba, Y. Ohmori, A. Yoshida, A. Kawai, T. Mayama-Tsuchida, H. Ichikawa, N. Mitsuda, M. Ohme-Takagi and H.Y. Hirano. 2012. The YABBY gene *TONGARI-BOUSHI1* is involved in lateral organ development and maintenance of meristem organization in the rice spikelet. *The Plant Cell, 24*(1): 80–95.

Tanaka, W., T. Toriba and H.K. Hirano. 2017. Three TOB1-related YABBY genes are required to maintain proper function of the spikelet and branch meristems in rice. *New Phytologist, 215*(2): 825–39.

Toriba, T., K. Harada, A. Takamura, H. Nakamura, H. Ichikawa, T. Suzaki and H.Y. Hirano. 2007. Molecular characterization the YABBY gene family in *Oryza sativa* and expression analysis of *OsYABBY1*. *Molecular Genetics and Genomics, 277*(5): 457–68.

Villanueva, J.M., J. Broadhvest, B.A. Hauser, R.J. Meister, K. Schneitz and C.S. Gasser. 1999. *INNER NO OUTER* regulates abaxial-adaxial patterning in *Arabidopsis* ovules. *Genes and Development, 13*: 3160–69.

Wetzstein, H.Y., W. Yi and J.A. Porter. 2013. Flower position and size impact ovule number per flower, fruit set, and fruit size in Pomegranate. *Journal of the American Society for Horticultural Science, 138*(3): 159–66.

Xu, Z.Z., W.C. Ni, X.G. Zhang, Q. Guo, P. Xu and X.L. Shen. 2015. Genome-wide analysis of the YABBY gene family in Cotton. *Biotechnology Bulletin, 31*(11): 146–52.

Yamada, T., S. Yokota, Y. Hirayama, R. Imaichi, M. Kato and C.S. Gasser. 2011. Ancestral expression patterns and evolutionary diversification of YABBY genes in angiosperms. *Plant Journal, 67*(1): 26–36.

Yamaguchi, T., N. Nagasawa, S. Kawasaki, M. Matsuoka, Y. Nagato and H.Y. Hirano. 2004. The YABBY gene *DROOPING LEAF* regulates carpel specification and midrib development in *Oryza sativa*. *The Plant Cell, 16*(2): 500–509.

Yuan, Z.H., Y.M. Fang, T.K. Zhang, Z.J. Fei, F.M. Han, C.Y. Liu, M. Liu, W. Xiao, W.J. Zhang, M.W. Zhang, Y.H. Ju, H.L. Xu, H. Dai, Y.J. Liu, Y.H. Chen, L.L. Wang, J.Q. Zhou, D. Guan, M. Yan, Y.H. Xia, X.B. Huang, D.Y. Liu, H.M. Wei and H.K. Zheng. 2018. The pomegranate (*Punica granatum* L.) genome provides insights into fruit quality and ovule developmental biology. *Plant Biotechnology Journal, 16*: 1363–74.

Zhang, X.L., Z.P. Yang, J. Zhang and L.G. Zhang. 2013. Ectopic expression of *BraYAB1-702*, a member of YABBY gene family in Chinese cabbage, causes leaf curling, inhibition of development of shoot apical meristem, and flowering stage delaying in *Arabidopsis thaliana*. *International Journal of Molecular Sciences, 14*(7): 14872–91.

Zhao, W., H.Y. Su, J. Song, X.Y. Zhao and X.S. Zhang. 2006. Ectopic expression of *TaYAB1*, a member of YABBY gene family in wheat, causes the partial abaxialization of the adaxial epidermises of leaves and arrests the development of shoot apical meristem in Arabidopsis. *Plant Science, 170*(2): 364–71.

Zhao, W., H.Y. Su and L. Wang. 2009. Bioinformatics analysis of YABBY genes from wheat. *Modern Agricultural Science and Technology, 7*: 139–41.

Zhao, S.P., D. Lu, T.F. Yu, Y.J. Ji, W.J. Zheng, S.X. Zhang, S.C. Chai, Z.Y. Chen and X.Y. Cui. 2017. Genome-wide analysis of the YABBY family in soybean and functional identification of *GmYABBY10* involvement in high salt and drought stresses. *Plant Physiology Biochemistry, 119*: 132–46.

15

Exploring Genomic Horizons

A Comprehensive Study on Identification and Evolutionary Analysis of the AOMT Gene Family in Pomegranate

Xinhui Zhang[1] and *Zhaohe Yuan*[1*]

Anthocyanin *O*-methyltransferases (AOMTs) have been recognized as being capable of anthocyanin methylation, which increases anthocyanin diversity and stability and improves the protection of plants from environmental stress. Meanwhile, no detailed identification or genome-wide analysis of the AOMT gene family members in pomegranate (*Punica granatum*) have been reported. Three published pomegranate genome sequences offer substantial resources with which to explore gene evolution based on the whole genome. Altogether, 58 identified OMTs from pomegranate and five other species were divided into the AOMT and the OMT groups, according to their phylogenetic tree, and the AOMTs derived from the OMTs. AOMTs in the same subclade have a similar gene structure and protein conserved motifs. The PgAOMT family evolved and expanded primarily via whole-genome duplication and tandem duplication. The PgAOMTs expression pattern in peel and aril development by qRT-PCR verification indicated that the PgAOMTs had tissue-specific patterns. The main fate of AOMTs were neo- or non-functionalization after duplication events. High expression genes of *PgOMT04* and *PgOMT09* were speculated to contribute to 'Taishanhong' pomegranate's bright red peel color. Finally, this chapter integrated the above analysis to infer the evolutionary scenario of the AOMT family.

[1] College of Forestry, Nanjing Forestry University, Nanjing 210037, China.
* Corresponding author: Zhaohe Yuan, zhyuan88@hotmail.com

1. Introduction

Gene duplication and loss events constitute the main factor of gene family evolution (Demuth and Hahn, 2009). Whole genome duplication (WGD) and small-scale duplication events are the major sources of most duplicates (Panchy, Lehti-Shiu, and Shiu, 2016). After duplication events, the identical copies are a predominant feature of individual genes and whole genomes, while a large fraction of duplications was lost (Elisabeth, Leng, and Alexander, 2018). Some of these duplicates' retention contributes to neo-functionalization, such as increasing metabolic activity, evolving new structure, and novel adaptive traits (Elisabeth et al., 2018; Panchy et al., 2016). After WGD, the strong-biased retention of regulatory and developmental genes has ramifications for evolution in the longer term (Van de Peer, Mizrachi, and Marchal, 2017). These regulatory and developmental genes control multiple reactions in land plants biochemical pathways. Thus, it is significant to elucidate the evolution of the regulatory and developmental gene families after WGD.

Anthocyanins, which are the largest group of flavonoids, are significant pigments to plants. By conferring to the fruit pulp, pericarp, flower, and leaf, a red, orange, or purple color, anthocyanins contribute to their appealing appearance (Chiu et al., 2010; Du et al., 2015; Montefiori et al., 2011; Y.N. Yang et al., 2015). Through these bright colors, plants attract the coevolving birds or insects to spread seeds and pollens (Y. Gao et al., 2019). Anthocyanins have potential health benefits. Some research suggests that they possess antioxidant, antitumor, anti-inflammatory, and immunomodulation activity (He and Giusti, 2010). In addition, anthocyanins play crucial roles in enhancing tolerance in response to multiple abiotic stresses (Kovinich, Kayanja, Chanoca, Otegui, and Grotewold, 2015). Like other phenolic compounds, anthocyanins are usually modified by diverse moieties, including acyl, hydroxyl, glycosyl, and methyl groups (Du et al., 2015; J. Luo et al., 2007; Noda et al., 2017; Offen et al., 2006). These modifications can increase anthocyanin categories and improve its stability.

The methylation of anthocyanin is mediated by *O*-methyltransferases (OMT). The OMT family can be divided into three major types based on sequence homology and substrate variance (Ferrer, 2003). Type 1 OMTs mediate the methylation of flavonoids, type 2 OMTs mediate the methylation of caffeoyl CoA, and type 3 OMTs mediate the methylation of small phenolic compounds. To date, a few of the OMTs that were responsible for methylation of anthocyanins were characterized. Anthocyanin *O*-methyltransferases (AOMTs) named FAOMT and VvAOMT were isolated from two grape cultivars and were verified that catalyzed anthocyanin methylation *in vitro* and planta (Hugueney et al., 2009; Lücker, Martens, and Lund, 2010). In the tomato genome, 57 candidates OMT genes were identified, while only one gene (*AnthOMT*) exhibited a strong correlation with the accumulation of methylated anthocyanin using gene silencing tests (Gomez Roldan et al., 2014). Through following *PsAOMT* and *PtAOMT* from purple-flowered and red-flowered Paeonia plants, respectively, and using site-directed mutagenesis, the pair of homologous genes were characterized as being involved in anthocyanin methylation and one key amino acid substitution led to a vast difference in catalytic activity between them (Du et al., 2015).

Pomegranate (*Punica granatum*) has achieved considerable attention due to its high antioxidant activity, soft seed characteristic, rich color in peel and aril (the edible part of pomegranate), and active pharmaceutical ingredients (Barathikannan et al., 2016; Yuan et al., 2018; X. Zhang, Zhao, Ren, Wang, and Yuan, 2020). In addition, part of the important stress relevant and nutritionally beneficial secondary metabolites of pomegranate were found in its seed oil (Barathikannan et al., 2016; Costa et al., 2020). Indeed, lipid droplets that make up the seed oils are an ancient response mechanism and these droplets can accumulate secondary metabolism (de Vries and Ischebeck, 2020). Pomegranate peel and aril color, as a consequence of the anthocyanin accumulation, is a crucial feature in determining its fruit commodity value. Moreover, the high anthocyanin content and biological activities in fruit indicate that their consumption would be of great benefit to health, including such afflictions as the effect of colon, breast, lung, and gastric cancer and that they could be used to produce functional foods (Bowen-Forbes, Zhang, and Nair, 2010; Kalaycıoğlu and Erim, 2017). While phenylpropanoid metabolisms and anthocyanin production is not limited to pomegranate, indeed, the molecular mechanisms that regulate this process are found across land plants, and probably even in the algal relatives of land plants (Albert et al., 2018; de Vries, de Vries, Slamovits, Rose, and Archibald, 2017; Fürst-Jansen, de Vries, and de Vries, 2020).

Previous studies found that pomegranate has experienced a paleohexaploidy event shared with eudicots and a paleotetraploidy event shared with *Eucalyptus grandis* (Yuan et al., 2018). These two duplication events resulted in varying degrees' expansion of the pomegranate multiple gene families. Yuan et al. (2018) constructed a phylogenetic tree with six species of AOMT genes and put forward the assumption that high copy number of AOMTs were responsible for producing anthocyanin. In addition, nine AOMT genes were identified in pomegranate. The diverse AOMTs may have a link with pomegranate distinct colors (Yuan et al., 2018).

In this study, we carry out an analysis of AOMT genes, a gene family involved in anthocyanin methylation, to explore the evolution scenario of the gene family after duplication events. A total of 58 putative OMT genes were identified from pomegranate, *Durio zibethinus*, *Citrus sinensis*, *Arabidopsis thaliana*, *Malus domestica*, and *Vitis vinifera,* and were divided into two classes (the OMT group and AOMT group) through phylogenetic analysis. We present detailed analyses of OMTs, including gene structure, conserved motif distribution, expression pattern, and gene duplication.

2. Materials and Methods

2.1 *Sample Collection*

The pomegranate 'Taishanhong' ($2n = 2x = 18$) has been widely cultivated in China as an important commercial cultivar because of its large body, bright red peel color, and sweet taste (Yuan et al., 2018). In our test, pomegranate materials were obtained from the pomegranate germplasm resource nursery of Nanjing Forestry University.

To verify the reliability of the AOMT candidate gene sequences, the results of RT-PCR and gene cloning were compared with the predicted genes. Fresh young leaves of pomegranate were selected as materials and stored at − 80°C after liquid nitrogen freezing.

To study the differential expression of AOMT genes in peel and aril at different stages during development, the fruits of four different development stages (July, August, September, and October) were selected at random. The peel and aril were separated and stored at −80°C after liquid nitrogen freezing.

2.2 Data Sources and Gene Identification of AOMTs

Pomegranate genome (ASM220158v1) was downloaded from NCBI (https://www.ncbi.nlm.nih.gov) (Yuan et al., 2018). *A. thaliana* AOMT protein sequences were downloaded from the *A. thaliana* database (http://www.arabidopsis.org)., *M. domestica*, *C. sinensis*, and *V. vinifera* AOMT protein sequences were downloaded from Phytozome, and *D. zibethinus* was downloaded from NCBI.

HMMER v3.1b1 was used to search for sequences with AOMT domains according to the downloaded PFAM (the protein families' database) file (PF01596, e-value $\leq 10^{-10}$) (Clements, Eddy, and Finn, 2011). The protein domains were detected by CDD (Marchler-Bauer et al., 2011) (https://www.ncbi.nlm.nih.bov/cdd) and SMART (Ponting, 1998) (http://smart.embl-heidelberg.de). The sequences with no AOMT domain or incomplete domain were removed. If there was a redundant sequence, the longest one was retained.

2.3 Phylogenetic Analysis

To explore the phylogenetic relationship of the AOMT genes among six species, including 10 pomegranate OMT genes, 16 OMT *D. zibethinus* genes, seven *C. sinensis* OMT genes, six *A. thaliana* OMT genes, nine *M. domestica* OMT genes, and 10 *V. vinifera* OMT genes, these identified AOMT gene sequences were utilized to perform multiple sequence alignment analysis with MAFFT (a multiple sequence alignment program) software (Katoh and Standley, 2016). Then, a maximum likelihood tree was established using PhyML (Guindon et al., 2010) with the JTT model, SPR topology search, 1,000 bootstrap replicates, and aLRT statistics. The tree was visualized using Figtree v1.4.3 (http://tree.bio.ed.ac.uk/software/figtree/).

2.4 Exon-Intron Structure

Gene Structure Displayer server (GSDS) website (G. Gao, 2014) (http://gsds.cbi.pku.edu.cn) was used to analyze the AOMT gene structure. Each plant species GFF annotation was downloaded corresponding to the genome. The required annotation contents were filtered out using Perl script and were uploaded to the GSDS website for the AOMT exon-intron structure.

2.5 Motif Analysis

To further identify and analyze the protein conserved motifs of the PgAOMT family members, the online tool Multiple Em for Motif Elicitation (MEME: http://meme.nbcr.net/ meme/intro.html) (Bailey et al., 2009) was employed with the following parameters: the motif width was 20~100 aa, and the number of motifs was six.

2.6 RNA Extraction, cDNA Synthesis, and RT-PCR Analysis

Total RNA of pomegranate young leaves was extracted using RNAprep Pure total RNA extraction kit. RNA was purified using DNase I digestion kit (Vazyme-innovation in enzyme technology, Nanjing, China). The obtained RNA was reverse transcribed into cDNA by HiScript®II 1st Strand cDNA Synthesis Kit (Vazyme-innovation in enzyme technology, Nanjing, China). The pomegranate target genes were screened according to the verified *A. thaliana*, *M. domestica*, *D. zibethinus*, *C. sinensis*, and *V. vinifera* AOMT homologous gene. Then, the primers were designed depending on the candidate genes with Primer5.0 for PCR amplification based on the predicted genes through ePCR. Finally, the GBclonart seamless cloning kit (GBI, Suzhou, China) was used to recombine gene as Zhang et al. (2019) described and the cloned genes were sequenced.

2.7 Expression Analyses of PgAOMT Gene Family Members

The public transcriptomes pomegranate peel (SRR5279392, SRR5279393, SRR5279394) and aril (SRR5279386, SRR5279387, SRR5279388) during development were downloaded from NCBI. Quality controls for these transcriptomes were examined by the NGS QC Toolkit v2.3.3. Then, the gene expression was quantified by the kallisto v0.43.1. The data was calculated by reads per kilobase per million mapped reads (RPKM) as transcript abundance and the RPKM values were transformed in log2 fold change. The differently expressed PgAOMT genes in peel and aril across development periods were analyzed with DESeq2 software (Love, Huber, and Anders, 2014). According to the threshold p-value adjust <0.01 and log2(fold-change) ≥ 2, differentially expressed genes (DEGs) were screened out. Upregulated DEGs and downregulated DEGs were classified. The volcano map generation was exhibited by R software. To validate the result of the RNA-seq data, three samples for each tissue (peel and aril) were collected for the expression levels of PgAOMT genes.

RNA samples of aril and peel during development (July, August, September, and October) were used for qRT-PCR experiments. The operation method of RNA extraction and cDNA reserve transcription is the same as 2.7 mentioned. qRT amplification was performed in triplicate using the Luna® Universal qPCR Master

Mix (NEB, Ipswich, MA, USA). The expression levels of these genes were assessed using the method described by Liu et al. (2018). The correlative expression data was calculated according to the $2^{-(\Delta\Delta CT)}$ method (Livak and Schmittgen, 2001). Finally, the expression spectrum was drawn out using the R software.

3. Results

3.1 Identification of the AOMT Genes in Pomegranate

To identify the AOMT genes of pomegranate, we used SelectHMM to (e-value ≤ 10) screen and CDD and SMART to identify all possible AOMT members which contain complete domains. Altogether 10 OMT candidate genes were identified from the pomegranate genome.

According to the on-line analysis software ExPASY (Panu et al., 2012), we provided characteristics of PgOMT genes including protein length (PL), molecular weight (MW), isoelectric point (PI), instability index, and mean value of hydrophilicity (GRAVY) (Table 1). The 10 predicted PgOMT proteins ranged from 93 (*PgOMT08*) to 414 amino acids (aa) (*PgOMT10*), with an average of 232 aa. The proteins MW ranged from 10,153.52 (*PgOMT08*) to 45,306.63 (*PgOMT10*) Ku, with an average of 25,669.56 Ku. The PI varied from 4.79 (*PgOMT07*) to 9.26 (*PgOMT10*), with an average of 5.83. The instability index varied from 23.63 (*PgOMT06*) to 56.67 (*PgOMT05*). All proteins were hydrophilic except PgOMT10 and PgOMT07 based on the GRAVY.

Table 1 The basic information of PgOMT gene family.

Species	*Gene ID*	*Gene Name*	*Protein Length/aa*	*Molecular Weight/ku*	*PI*	*Instability Index*	*GRAVY*
Pomegranate	*Pg002344.1*	*PgOMT01*	250	27,320.50	5.45	46.31	−0.088
	Pg002346.1	*PgOMT02*	250	27,519.67	5.05	45.46	−0.052
	Pg002348.1	*PgOMT03*	239	26,378.53	5.35	31.20	−0.079
	Pg002351.1	*PgOMT04*	242	27,172.27	5.44	36.49	−0.160
	Pg002849.1	*PgOMT05*	183	20,421.47	6.21	56.67	−0.152
	Pg003086.1	*PgOMT06*	109	11,927.73	5.89	23.63	−0.110
	Pg006183.1	*PgOMT07*	286	31,785.38	4.79	39.20	0.009
	Pg017894.1	*PgOMT08*	93	10,153.52	5.18	45.87	−0.200
	Pg021629.1	*PgOMT09*	258	28,709.93	5.72	37.09	−0.246
	Pg026019.1	*PgOMT10*	414	45,306.63	9.26	34.14	0.093

3.2 Phylogenetic Analysis of PgAOMT Gene Family

A comprehensive and robust phylogenetic tree is the basis of studying phylogenetic genomics. To study the AOMT gene family evolutionary relationship, a total of

58 candidate OMT genes were identified from pomegranate (10 OMT genes), *D. zibethinus* (16 OMT genes), *C. sinensis* (7 OMT genes), *A. thaliana* (6 OMT genes), *M. domestica* (9 OMT genes), and *V. vinifera* (10 OMT genes) and a maximum likelihood (ML) tree (Figure 1) was constructed using PhyML (Guindon et al., 2010). This ML tree has robust bootstrap-supported values in general.

Based on the phylogenetic tree branches and bootstrap-supported values, the OMT genes can be divided into Class I and Class II. In Class I, six species OMT genes were evenly distributed, while in Class II, just *D. zibethinus*, *M. domestica*, and *V. vinifera* were the main species. We defined Class I and Class II as AOMT group and OMT group, respectively (Figure 1). Meanwhile, it was speculated that the AOMT group evolved from the OMT group.

It is apparent that each of the four species (*A. thaliana*, *V. vinifera*, *M. domestica,* and *D. zibethinus*) constituted the gene cluster respectively in the basal part of the ML tree (Figure 1), indicating that OMT gene family has undergone species-specific expansion during evolution. In Class I, the branches were made up of *GSVIVT01020603001*, *GSVIVT01015243001*, *GSVIVT01015246001*, *GSVIVT01015245001*, *GSVIVT01015244001*, *PgOMT06*, *PgOMT05*, *PgOMT08*, *PgOMT04*, *PgOMT03*, PgOMT01, and *PgOMT02*, which represented a ratio of 5:7 of grape to pomegranate (Figure 1). It was indicated that pomegranate experienced one WGD event after paleohexaploidy shared with *V. vinifera* and considerable fraction duplicates were lost during the subsequent gene-loss event (Jiao et al., 2012; Kaltenegger, Eich, and Ober, 2013; Lee and Irish, 2011; Yuan et al., 2018).

3.3 Gene Structure and Protein Conserved Motifs of PgAOMT Genes Family

To study the AOMT gene structure, we analyzed the distribution and amount of the exons and introns among six species OMT genes on the basis of phylogenetic similarities. The results showed that the majority of OMT genes contained introns and the number of introns varied from 0 to 8, only *AT1G24733*, *MD02G1230800*, and *PgOMT08* had no intron. In the same subclade, OMTs seem to have similar intron/exons distribution, while there were some special cases such as *GSVIVT01010469001*, *MD16G1118200*, and *PgOMT10*, that were obviously different from the genes of the same subclade in length and number of intron. In addition, the introns of early copy genes tended to be longer. These supported the assumption that the length and number of introns facilitate gene family differentiation. Combined with previous phylogenetic analysis, it was speculated that gene duplication events with gene loss events influenced PgOMTs gene structure.

To research the distribution of AOMT proteins conserved motifs, six species protein sequences were analyzed using MEME. Most OMT proteins have six conserved motifs. However, some proteins such as *AT1G24733* only had motif 1 and 2. Combined with the phylogenetic tree, it explained why *AT1G24733* and *AT1G24735* that formed an individual branch. It was observed that four AOMTs

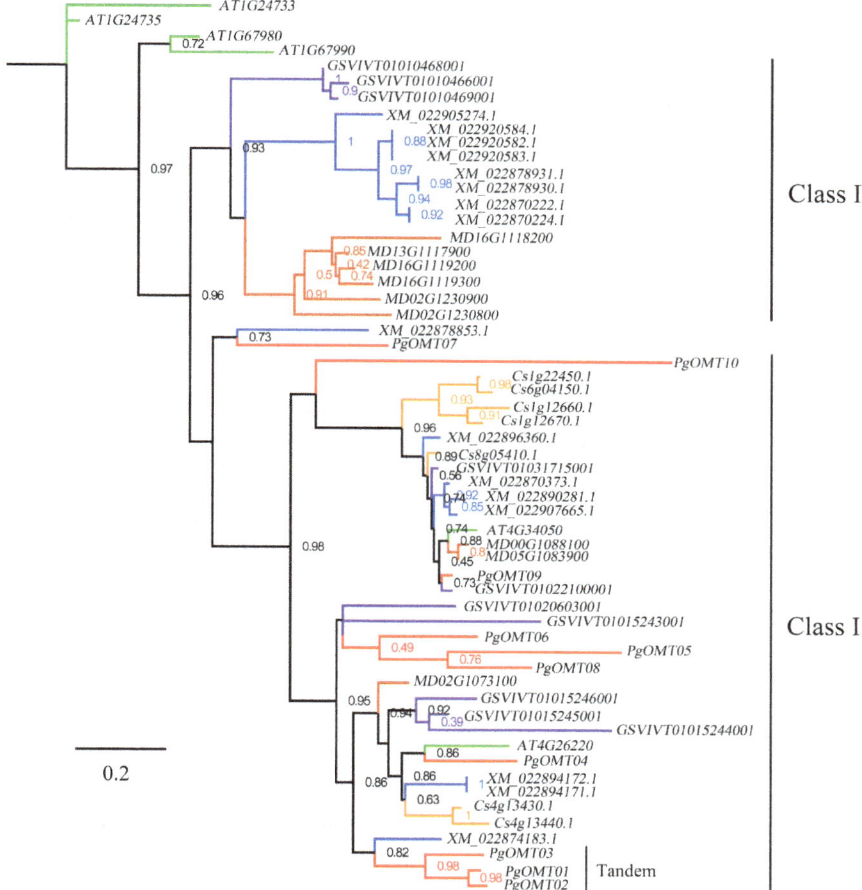

Fig. 1 OMT (*O*-methyltransferases) gene family phylogenetic tree of six species including pomegranate, *A. thaliana*, *V. vinifera*, *M. domestica*, *C. sinensis*, and *D. zibethinus* presented by various branch colors. A total of 58 sequences were aligned with MAFFT (a multiple sequence alignment program) and the phylogenetic tree was constructed using JTT (Jones-Taylor-Thornton) model. According to the phylogenetic relationship, 58 OMTs were divided into Class I (AOMT (anthocyanin *O*-methyltransferases) group) and Class II (OMT group). Node values beside branches were bootstrap-supported values generated from 1,000 replicates. Pg: pomegranate; AT: *A. thaliana*; Cs: *C. sinensis*; GSVIVT: *V. vinifera*; MD: *M. domestica*; XM: *D. zibethinus*.

including *PgOMT01*, *PgOMT02*, *PgOMT03*, and *PgOMT04* belonging to the same subclade contained all six motifs. While motif distribution of three other AOMTs from the same subclade were different, as follows, *PgOMT10* possesses motif 1, 2, and 6; *PgOMT06* possesses motif 1 and 4; *PgOMT05* contains motif 1, 2, 5, and 6; and *PgOMT08* contains motif 3 and 6. These illustrated that there were obvious differences among PgOMT proteins after the WGD event. The loss and retention of motifs may be related to the evolution of subclade.

3.4 RT-PCR Analysis of PgAOMT Candidate Gene Sequences

The RT-PCR technique was used to detect the reliability of five OMT genes. Five target sequences were obtained after reverse transcription. The sequencing reads of five OMT genes were aligned with predicted PgAOMTs downloaded from the database using MEGA-X software (Kumar, Stecher, Li, Knyaz, and Tamura, 2018). Through alignment, it was found that although there were differences among several base pairs in partial regions, the structure of the five sequencing OMT genes was consistent with the predicted sequences on the whole.

3.5 Expression Patterns of PgAOMT in Peel and Aril during Different Development Stages

Some research has revealed that AOMT genes play roles in anthocyanin methylation that influence the plant coloration and improve self-protection from environmental stress (Du et al., 2015; Gomez Roldan et al., 2014). To gain insight into the functions of the PgAOMT genes in the development of aril and peel, the expression patterns of PgOMT genes in aril and peel were analyzed using public RNA-Seq data from NCBI. The volcano plot analysis showed that PgOMT09 in aril was significantly downregulated. The DEG analyses demonstrated that PgOMT09 played pivotal roles in aril development.

To further understand the physiological role of the PgAOMTs, we determined the PgAOMT gene family members' expression in peel and aril during different developmental stages (July, August, September, and October). The histograms revealed that the transcript abundance of seven PgAOMT genes was significantly different in aril and peel, suggesting that there were variations in their functions of PgAOMT in aril and peel development. Notably, *PgOMT04* exhibited high expression both in peel and aril development. *PgOMT09* expression was downregulated significantly at the early stage of aril development. *PgOMT01*, *PgOMT02*, and *PgOMT03* belonging to the same subclade exhibited low or no expression. These phenomena indicated that during the genome duplication events, some genes (such as *PgOMT09*) evolved new functions and some genes (such as *PgOMT01*, *PgOMT02*, and *PgOMT03*) became nonfunctional pseudogenes due to segmental large-scale duplication events that arose from tandem duplication (Panchy et al., 2016). Additionally, PgAOMT genes had aril/peel-specific expression patterns in pomegranate. For instance, *PgOMT01* and *PgOMT03* only expressed in aril. *PgOMT05*, *PgOMT07*, and *PgOMT10* were more highly expressed in aril than in peel.

4. Discussion

AOMT enzymes are known to play roles in anthocyanin methylation. They can increase diversity and improve the stability of anthocyanin (Du et al., 2015; Gomez Roldan et al., 2014). Few AOMTs were identified and characterized such as *VvAOMT* in grape, *CkmOMT2* in cyclamen, *AnthOMT* in tomato, *DIFe1*, *DIFe2a*, and *DIFe2b* in *Petunia* spp. and *PsAOMT* and *PtAOMT* in purple-flowered and red-flower *Paeonia* plants, respectively (Du et al., 2015; Gomez Roldan et al., 2014; Hugueney et al., 2009; Kondo et al., 2009; Provenzano et al., 2014). So far, few articles have described the evolution of the OMT gene family. Three published pomegranate genomes have been sequenced and these provide valuable resources to explore functional genes in genome-wide data (X. Luo et al., 2019; Qin et al., 2017; Yuan et al., 2018). In addition, pomegranate possesses diverse anthocyanins (Ben-Simhon et al., 2011; Ben-Simhon et al., 2015) and AOMTs expansion occurred after duplication events (Yuan et al., 2018). Here, we identified and analyzed the expression patterns of AOMTs based on phylogenomic analyses to speculate on the evolution of PgAOMT gene family.

It has been reported that pomegranate shared the paleohexaploidy event with all eudicots and the paleotetraploidy event with *E. grandis* (Yuan et al., 2018). Gene duplication has been deemed as a predominant mechanism to increase functional diversification and the enhanced expression divergence of duplicated genes can substantially contribute to physiological diversification (Carocha et al., 2015). Hence, exploring the duplication events to illuminate gene functional diversification is significant. A total of 58 OMTs were identified from the selected six species. They were divided into Class I and Class II by constructing an ML tree that analyzed through bootstrap-supported values (Figure 1). Of the 10 candidate PgOMT genes, nine PgOMTs belong to Class I and only one belonged to Class II (Figure 1). We defined Class I as AOMT group and Class II as OMT group. It was speculated that AOMT group evolved from OMT group. In the subclade of Class I, there was a ratio of five grape to seven pomegranate, suggesting that there might be pomegranate genome duplication and loss events (Jaillon et al., 2007; Yuan et al., 2018). In contrast with the number of AOMTs in other species, PgAOMTs copy number was more than that of *E. grandis* (6) and *C. sinensis* (2) (Myburg et al., 2014; Xu et al., 2013). In our study, the number of AOMTs from pomegranate was more than that from other species (Figure 1), suggesting a lineage-specific expansion event occurred in pomegranate. High copy number of PgAOMTs may contribute to the diverse anthocyanins forming distinct colors of pomegranate (Yuan et al., 2018).

By analyzing AOMT genes structure combined with the phylogenetic tree, it was obvious that the gene structure of AOMTs in Class I have some differences, but members in the same subclade seem to be similar. In the clade of *PgOMT06*, *PgOMT05*, and *PgOMT08*, the three PgAOMTs have fewer introns than others. The absence of introns of the three PgAOMTs proved the presumption that there might be gene loss events in PgAOMT family due to neo-, sub- or non-functionalization

(Hartmann, Tesch, Nothwang, and Bininda-Emonds, 2013; M. Liu, Wen, Sun, Ma, and Chen, 2019). Furthermore, we found that the motifs encoding PgAOMTs in earlier evolved clades were absent in varying degrees in Class I. For example, PgOMT10 only contained motif 1, 2, and 6; PgOMT06 only contained motif 1 and 4; PgOMT05 possessed motif 1, 2, and 5; PgOMT08 possessed motif 3 and 6, whereas the PgAOMTs that later evolved contained all six motifs. In general, the AOMTs phylogenetic analysis is consistent with the gene structure.

Some gene families evolved and expanded through tandem duplication or segmental duplication along with high rates of birth and death. Gene family's expansion in these ways is of important significance to diversify their functions (Cannon, Mitra, Baumgarten, Young, and May, 2004). Vice versa, gene function may feedback on copy number and genome organization and thus give rise to the widely varying patterns of tandem or segmental duplication (Cannon et al., 2004). In brief, genes exist in the form of clusters after gene tandem duplication (Tremblay Savard, Bertrand, and El-Mabrouk, 2011). Deciphering gene clusters evolution is vital to offer new insight into gene family evolutionary scenario. By analyzing the gene mapping in pomegranate, it presented the existence of tandem duplication (Figure 1: *PgOMT01*, *PgOMT02*, and *PgOMT03*), which are identical to those confirmed by Yuan et al. (2018). Pomegranate genome large-scale duplication gave rise to AOMT tandem duplication (Yuan et al., 2018). Gene tandem duplication events drove the evolution of PgAOMT family to some extent.

Pseudogenes tend to have significantly lower expression compared with annotated genes (Zou, Lehti-Shiu, Thibaud-Nissen, Prakash, and Shiu, 2009). They are considered as nonfunctional genes that have similar structure to functional genes in a domain family. Hence, numerous pseudogenes were misidentified as functional genes during the genome annotation process (Zou et al., 2009). Three tandem duplicated genes of *PgOMT01*, *PgOMT02*, and *PgOMT03* have low or no gene expression levels in peel or aril development, suggesting they were pseudogenes and genes may experience rapid birth and death during AOMT family evolution (Nei and Rooney, 2005). In addition, the expression of *PgOMT04* and *PgOMT10* demonstrated significantly higher than others both in peel and aril development. The high expression supported the assumption that the PgAOMTs evolved new functions after genome duplication (T. Zhang et al., 2019).

Anthocyanin accumulation is known to be induced for resisting abiotic stresses (Guo and Wang, 2010). AOMTs have been identified that are capable of anthocyanin methylation, particularly endows plant with an enhanced protection, have experienced genome duplication, presumably due to selection imposed by rapid changes in abiotic environments (Gomez Roldan et al., 2014; Yuan et al., 2018). As we all know, methylation modification plays roles in increasing anthocyanin diversity and stability. In addition, the anthocyanin methylation level is significantly correlated with the hue towards red or purple of plant tissues (Du et al., 2015; Kondo et al., 2009; R.Z. Yang et al., 2009). Zhao et al. (2015) proposed that AOMT could be responsible for pomegranate peel and aril color transition from white to red. It has been proposed that the methylation of B-ring hydroxyl groups generates the shift towards red (Tanaka, Sasaki, and Ohmiya, 2008). In 'Taishanhong' pomegranate, Cyanidin-3Glucoside and Delphinidin-3-glucoside were detected as the major anthocyanins, but methylated anthocyanin contents have not been reported (Feng, Yuan, Zhao, Yin, and Feng, 2015).

Subsequently, qRT-PCR results showed that AOMT genes were expressed in peel and aril of pomegranate fruit. *PgOMT04* and *PgOMT09* in peel and *PgOMT04* in aril with higher expression throughout the development period, indicating that these O-methylations of anthocyanin and others may be closely related with fruit development and the ripening process. *PgOMT09* was downregulated continuously in aril during the early development process, suggesting that it played roles in early development. *PgOMT03* and *PgOMT10* in peel and aril and *PgOMT10* in aril showed fluctuating expression pattern, suggesting that they may play roles in specific stages of fruit development. In contrast, *PgOMT09* expressed both in peel and aril, with distinct expression pattern, indicating that there was different accumulation of O-methylation anthocyanin between them. Interestingly, *PgOMT01* was specifically expressed in pomegranate peel, suggesting that its specific formation of O-methylated anthocyanin or others would occur in fruit peel.

Conclusion

In this study, we identified 10 OMTs from pomegranate genome and they were divided into two groups (AOMT and OMT group) by phylogenetic analysis. Pomegranates have undergone WGD, tandem duplication, and gene loss events, presumably due to selection for improving their behavioral response to rapid changes of environment. The expression patterns of PgAOMTs indicated that some members function in anthocyanin methylation which enhances diversity and stability of anthocyanins and improves protection against stresses such as ultraviolet. This study also provides fundamental information on AOMTs for the coloration of pomegranate.

References

Albert, N.W., Thrimawithana, A.H., McGhie, T.K., Clayton, W.A., Deroles, S.C., Schwinn, K.E., Bowman, J.L., Jordan, B.R. and Davies, K.M. 2018. Genetic analysis of the liverwort *Marchantia polymorpha* reveals that R2R3MYB activation of flavonoid production in response to abiotic stress is an ancient character in land plants. *New Phytologist, 218*: 554–66.

Bailey, T.L., Boden, M., Buske, F.A., Frith, M., Grant, C.E., Clementi, L., Ren, J., Li, W. W. and Noble, W.S. 2009. MEME Suite: Tools for motif discovery and searching. *Nucleic Acids Research, 37*: W202–W208.

Barathikannan, K., Venkatadri, B., Khusro, A., Al-Dhabi, N.A., Agastian, P., Arasu, M.V., Choi, H.S. and Kim, Y.O. 2016. Chemical analysis of *Punica granatum* fruit peel and its *in vitro* and *in vivo* biological properties. *BMC Complement Altern. Med., 16*: 264.

Ben-Simhon, Z., Judeinstein, S., Nadler-Hassar, T., Trainin, T., Bar-Ya'akov, I., Borochov-Neori, H. and Holland, D. 2011. A pomegranate (*Punica granatum* L.) WD40-repeat gene is a functional homologue of *Arabidopsis TTG1* and is involved in the regulation of anthocyanin biosynthesis during pomegranate fruit development. *Planta, 234*: 865–81.

Ben-Simhon, Z., Judeinstein, S., Trainin, T., Harel-Beja, R., Bar-Ya'akov, I., Borochov-Neori, H. and Holland, D. 2015. A 'White' anthocyanin-less pomegranate (*Punica granatum* L.) caused by an insertion in the coding region of the leucoanthocyanidin dioxygenase (LDOX; ANS) gene. *PLoS One*, *10*: e0142777.

Bowen-Forbes, C.S., Zhang, Y. and Nair, M.G. 2010. Anthocyanin content, antioxidant, anti-inflammatory, and anticancer properties of blackberry and raspberry fruits. *Journal of Food Composition and Analysis*, *23*: 554–60.

Cannon, S.B., Mitra, A., Baumgarten, A., Young, N.D. and May, G. 2004. The roles of segmental and tandem gene duplication in the evolution of large gene families in *Arabidopsis thaliana*. *BMC Plant Biology*, *4*: 1–21.

Carocha, V., Soler, M., Hefer, C., Cassan-Wang, H., Fevereiro, P., Myburg, A.A., Paiva, J.A.P. and Grima-Pettenati, J. 2015. Genome-wide analysis of the lignin toolbox of *Eucalyptus grandis*. *New Phytologist*, *206*: 1297–1313.

Chiu, L.-W., Zhou, X., Burke, S., Wu, X., Prior, R.L. and Li, L. 2010. The purple cauliflower arises from activation of an MYB transcription factor. *Plant Physiology*, *154*: 1470–80.

Clements, J., Eddy, S.R. and Finn, R.D. 2011. HMMER web server: Interactive sequence similarity searching. *Nucleic Acids Research*, *39*: W29–W37.

Costa, A. M.M., Moretti, L.K., Simões, G., Silva, K.A., Calado, V., Tonon, R.V. and Torres, A.G. 2020. Microencapsulation of pomegranate (*Punica granatum* L.) seed oil by complex coacervation: Development of a potential functional ingredient for food application. *LWT*, *131*: 109519.

de Vries, J., de Vries, S., Slamovits, C.H., Rose, L.E. and Archibald, J.M. 2017. How embryophytic is the biosynthesis of phenylpropanoids and their derivatives in streptophyte algae? *Plant and Cell Physiology*, *58*: 934–45.

de Vries, J. and Ischebeck, T. 2020. Ties between stress and lipid droplets pre-date seeds. *Trends in Plant Science*, *25*: 1203–14.

Demuth, J.P. and Hahn, M.W. 2009. The life and death of gene families. *Bioessays*, *31*: 29–39.

Du, H., Wu, J., Ji, K.-X., Zeng, Q.-Y., Bhuiya, M.-W., Su, S., Shu, Q.-Y., Ren, H.-X., Liu, Z.-A. and Wang, L.-S. 2015. Methylation mediated by an anthocyanin, O-methyltransferase, is involved in purple flower coloration in Paeonia. *Journal of Experimental Botany*, *66*: 6563–77.

Elisabeth, K., Leng, S. and Alexander, H. 2018. The effects of repeated whole genome duplication events on the evolution of cytokinin signaling pathway. *BMC Evolutionary Biology*, *18*(1): 76.

Feng, Z., Yuan, Z., Zhao, X., Yin, Y. and Feng, L. 2015. Composition and contents of anthocyanins in different pomegranate cultivars. *Acta Horticulturae*, 35–41.

Ferrer, J.L. 2003. Chapter two Structural, functional, and evolutionary basis for methylation of plant small molecules. *Recent Advances in Phytochemistry*, *37*: 37–58.

Fürst-Jansen, J.M. R., de Vries, S. and de Vries, J. 2020. Evo-physio: On stress responses and the earliest land plants. *Journal of Experimental Botany*, *71*: 3254–69.

Gao, G. 2014. GSDS 2.0: An upgraded gene feature visualization server. *Bioinformatics*, *31*: 1296.

Gao, Y., Zhu, N., Zhu, X., Wu, M., Jiang, C.-Z., Grierson, D., Luo, Y., Shen, W., Zhong, S., Fu, D.-Q. and Qu, G. 2019. Diversity and redundancy of the ripening regulatory networks revealed by the fruitENCODE and the new CRISPR/Cas9 CNR and NOR mutants. *Horticulture Research*, *6*: 39.

Gomez Roldan, M.V., Outchkourov, N., van Houwelingen, A., Lammers, M., Romero de la Fuente, I., Ziklo, N., Aharoni, A., Hall, R.D. and Beekwilder, J. 2014. An O-methyltransferase modifies accumulation of methylated anthocyanins in seedlings of tomato. *The Plant Journal*, *80*: 695–708.

Guindon, S., Dufayard, J.F., Lefort, V., Anisimova, M., Hordijk, W. and Gascuel, O. 2010. New algorithms and methods to estimate maximum-likelihood phylogenies: Assessing the performance of PhyML 3.0. *Syst Biol.*, *59*: 307–21.

Guo, J. and Wang, M.-H. 2010. Ultraviolet A-specific induction of anthocyanin biosynthesis and PAL expression in tomato (*Solanum lycopersicum* L.). *Plant Growth Regulation*, *62*: 18.

Hartmann, A.-M., Tesch, D., Nothwang, H.G. and Bininda-Emonds, O.R.P. 2013. Evolution of the cation chloride cotransporter family: Ancient origins, gene losses, and subfunctionalization through duplication. *Molecular Biology and Evolution*, *31*: 434–47.

He, J. and Giusti, M.M. 2010. Anthocyanins: Natural colorants with health-promoting properties. *Annu. Rev. Food Sci. Technol., 1*: 163–87.

Hugueney, P., Provenzano, S., Verriès, C., Ferrandino, A., Meudec, E., Batelli, G., Merdinoglu, D., Cheynier, V., Schubert, A. and Ageorges, A. 2009. A novel cation-dependent O-Methyltransferase involved in anthocyanin methylation in grapevine. *Plant Physiology, 150*: 2057–70.

Jaillon, O., Aury, J.M., Noel, B., Policriti, A., Clepet, C., Casagrande, A., Choisne, N., Aubourg, S., Vitulo, N. and Jubin, C. 2007. The grapevine genome sequence suggests ancestral hexaploidization in major angiosperm phyla. *Nature, 449*: 463–67.

Jiao, Y., Leebens-Mack, J., Ayyampalayam, S., Bowers, J.E., McKain, M.R., McNeal, J., Rolf, M., Ruzicka, D.R., Wafula, E., Wickett, N.J., Wu, X., Zhang, Y., Wang, J., Zhang, Y., Carpenter, E.J., Deyholos, M.K., Kutchan, T.M., Chanderbali, A.S., Soltis, P.S., Stevenson, D.W., McCombie, R., Pires, J.C., Wong, G.K.-S., Soltis, D.E. and Depamphilis, C.W. 2012. A genome triplication associated with early diversification of the core eudicots. *Genome Biology, 13*: R3–R3.

Kalaycıoğlu, Z. and Erim, F. B. 2017. Total phenolic contents, antioxidant activities, and bioactive ingredients of juices from pomegranate cultivars worldwide. *Food Chemistry, 221*: 496–507.

Kaltenegger, E., Eich, E. and Ober, D. 2013. Evolution of homospermidine synthase in the convolvulaceae: A story of gene duplication, gene loss, and periods of various selection pressures. *The Plant Cell, 25*: 1213–27.

Katoh, K. and Standley, D.M. 2016. A simple method to control over-alignment in the MAFFT multiple sequence alignment program. *Bioinformatics, 32*: 1933–42.

Kondo, E., Nakayama, M., Kameari, N., Tanikawa, N., Morita, Y., Akita, Y., Hase, Y., Tanaka, A. and Ishizaka, H. 2009. Red-purple flower due to delphinidin 3,5-diglucoside, a novel pigment for *Cyclamen* spp., generated by ion-beam irradiation. *Plant Biotechnology, 26*: 565–69.

Kovinich, N., Kayanja, G., Chanoca, A., Otegui, M.S. and Grotewold, E. 2015. Abiotic stresses induce different localizations of anthocyanins in *Arabidopsis. Plant Signal Behav., 10*: e1027850–e1027850.

Kumar, S., Stecher, G., Li, M., Knyaz, C. and Tamura, K. 2018. MEGA X: Molecular evolutionary genetics analysis across computing platforms. *Molecular Biology Evolution, 35*: 1547–49.

Lee, H.-L. and Irish, V.F. 2011. Gene duplication and loss in a MADS box gene transcription factor circuit. *Molecular Biology and Evolution, 28*: 3367–80.

Liu, M., Wen, Y., Sun, W., Ma, Z. and Chen, H. 2019. Genome-wide identification, phylogeny, evolutionary expansion, and expression analyses of *bZIP* transcription factor family in tartaty buckwheat. *BMC Genomics, 20*: 1–18.

Liu, X., Liu, X., Zhang, Z., Sang, M., Sun, X., He, C., Xin, P. and Zhang, H. 2018. Functional analysis of the *FZF1* genes of *Saccharomyces uvarum. Front Microbiology, 9*: 96.

Livak, K.J. and Schmittgen, T.D. 2001. Analysis of relative gene expression data using real-time quantitative PCR and the $2^{-\Delta\Delta CT}$ method. *Methods, 25*: 402–408.

Love, M. I., Huber, W. and Anders, S. 2014. Moderated estimation of fold change and dispersion for RNA-seq data with DESeq2. *Genome Biology, 15*: 550.

Lücker, J., Martens, S. and Lund, S.T. 2010. Characterization of a *Vitis vinifera* cv. Cabernet Sauvignon 3′,5′-O-methyltransferase showing strong preference for anthocyanins and glycosylated flavonols. *Phytochemistry, 71*: 1474–84.

Luo, J., Nishiyama, Y., Fuell, C., Taguchi, G., Elliott, K.A., Hill, L., Tanaka, Y., Kitayama, M., Yamazaki, M. and Bailey, P. 2007. Convergent evolution in the BAHD family of Acyl transferases: Identification and characterization of anthocyanin acyl transferases from *Arabidopsis thaliana. Plant Journal, 50*: 678–95.

Luo, X., Li, H., Wu, Z., Yao, W., Zhao, P., Cao, D., Yu, H., Li, K., Poudel, K. and Zhao, D. 2020. The pomegranate (*Punica granatum* L.) draft genome dissects genetic divergence between soft- and hard-seeded cultivars. *Plant Biotechnology Journal, 18*: 955–68.

Marchler-Bauer, A., Lu, S., Anderson, J.B., Chitsaz, F., Derbyshire, M.K., DeWeese-Scott, C., Fong, J.H., Geer, L.Y., Geer, R.C., Gonzales, N.R., Gwadz, M., Hurwitz, D.I., Jackson, J.D., Ke, Z., Lanczycki, C.J., Lu, F., Marchler, G.H., Mullokandov, M., Omelchenko, M.V., Robertson, C.L.,

Song, J.S., Thanki, N., Yamashita, R.A., Zhang, D., Zhang, N., Zheng, C. and Bryant, S.H. 2011. CDD: A Conserved Domain Database for the functional annotation of proteins. *Nucleic Acids Res.*, *39*: 225–29.

Montefiori, M., Espley, R.V., Stevenson, D., Cooney, J., Datson, P.M., Saiz, A., Atkinson, R. G., Hellens, R.P. and Allan, A.C. 2011. Identification and characterization of *F3GT1* and *F3GGT1*, two glycosyltransferases responsible for anthocyanin biosynthesis in red-fleshed kiwifruit (*Actinidia chinensis*). *Plant Journal*, *65*: 106–18.

Myburg, A.A., Grattapaglia, D., Tuskan, G.A., Hellsten, U., Hayes, R.D., Grimwood, J., Jenkins, J., Lindquist, E., Tice, H. and Bauer, D. 2014. The genome of *Eucalyptus grandis*. *Nature*, *510*: 356.

Nei, M. and Rooney, A.P. (2005). Concerted and birth-and-death evolution of multigene families. *Annual Review of Genetics*, *39*: 121–52.

Noda, N., Yoshioka, S., Kishimoto, S., Nakayama, M., Douzono, M., Tanaka, Y. and Aida, R. 2017. Generation of blue chrysanthemums by anthocyanin B-ring hydroxylation and glucosylation and its coloration mechanism. *Science Advances*, *3*: e1602785.

Offen, W., Martinez-Fleites, C., Yang, M., Kiat-Lim, E., Davis, B.G., Tarling, C.A., Ford, C.M., Bowles, D.J. and Davies, G.J. 2006. Structure of a flavonoid glucosyltransferase reveals the basis for plant natural product modification. *The EMBO Journal*, *25*: 1396–1405.

Panchy, N., Lehti-Shiu, M.D. and Shiu, S.H. 2016. Evolution of gene duplication in plants. *Plant Physiology*, *171*: 2294–2316.

Panu, A., Manohar, J., Konstantin, A., Delphine, B., Gabor, C., Edouard, d. C., Séverine, D., Volker, F., Arnaud, F. and Elisabeth, G. 2012. ExPASy: SIB bioinformatics resource portal. *Nucleic Acids Research*, *40*: W597–W603.

Ponting, C.P. 1998. SMART, a simple modular architecture research tool: Identification of signaling domains. *Proceedings of the National Academy of Sciences*, *95*: 5857–64.

Provenzano, S., Spelt, C., Hosokawa, S., Nakamura, N., Brugliera, F., Demelis, L., Geerke, D.P., Schubert, A., Tanaka, Y., Quattrocchio, F. and Koes, R. 2014. Genetic control and evolution of anthocyanin methylation. *Plant Physiology*, *165*: 962–77.

Qin, G., Xu, C., Ming, R., Tang, H., Guyot, R., Kramer, E.M., Hu, Y., Yi, X., Qi, Y. and Xu, X. 2017. The pomegranate (*Punica granatum* L.) genome and the genomics of punicalagin biosynthesis. *The Plant Journal*, *91*: 1108–28.

Tanaka, Y., Sasaki, N. and Ohmiya, A. 2008. Biosynthesis of plant pigments: Anthocyanins, betalains, and carotenoids. *The Plant Journal*, *54*: 733–49.

Tremblay Savard, O., Bertrand, D. and El-Mabrouk, N. 2011. Evolution of orthologous tandemly arrayed gene clusters. *BMC Bioinformatics*, *12*: S2.

Van de Peer, Y., Mizrachi, E. and Marchal, K. 2017. The evolutionary significance of polyploidy. *Nature Review Genetics*, *18*: 411–24.

Xu, Q., Chen, L.L., Ruan, X., Chen, D., Zhu, A., Chen, C., Bertrand, D., Jiao, W.B., Hao, B.H. and Lyon, M.P. 2013. The draft genome of sweet orange (*Citrus sinensis*). *Nature Genetics*, *45*: 59.

Yang, R.Z., Wei, X.L., Gao, F.F., Wang, L.S., Zhang, H.J., Xu, Y.J., Li, C.H., Ge, Y.X., Zhang, J.J. and Zhang, J. 2009. Simultaneous analysis of anthocyanins and flavonols in petals of lotus (*Nelumbo*) cultivars by high-performance liquid chromatography-photodiode array detection/electrospray ionization mass spectrometry. *Journal of Chromatography A.*, *1216*: 106–12.

Yang, Y.N., Yao, G.F., Zheng, D., Zhang, S.L., Wang, C., Zhang, M.Y. and Wu, J. 2015. Expression differences of anthocyanin biosynthesis genes reveal regulation patterns for red pear coloration. *Plant Cell Rep.*, *34*: 189–98.

Yuan, Z., Fang, Y., Zhang, T., Fei, Z., Han, F., Liu, C., Liu, M., Xiao, W., Zhang, W., Wu, S., Zhang, M., Ju, Y., Xu, H., Dai, H., Liu, Y., Chen, Y., Wang, L., Zhou, J., Guan, D., Yan, M., Xia, Y., Huang, X., Liu, D., Wei, H. and Zheng, H. 2018. The pomegranate (*Punica granatum* L.) genome provides insights into fruit quality and ovule developmental biology. *Plant Biotechnology Journal*, *16*: 1363–74.

Zhang, T., Liu, C., Huang, X., Zhang, H. and Yuan, Z. 2019. Land-plant phylogenomic and pomegranate transcriptomic analyses reveal an evolutionary scenario of CYP75 genes Subsequent to whole genome duplications. *Journal of Plant Biology*, *62*: 48–60.

Zhang, X., Zhao, Y., Ren, Y., Wang, Y. and Yuan, Z. 2020. Fruit breeding in regard to color and seed hardness: A Genomic View from Pomegranate. *Agronomy*, *10*: 991.

Zhao, X., Yuan, Z., Yin, Y. and Feng, L. 2015. Patterns of pigment changes in pomegranate (*Punica granatum* l.) peel during fruit ripening. *Acta Horticulturae*, *1089*: *83*–89.

Zou, C., Lehti-Shiu, M.D., Thibaud-Nissen, F.O., Prakash, T. and Shiu, B.S.H. 2009. Evolutionary and expression signatures of pseudogenes in *Arabidopsis* and Rice. *Plant Physiology*, *151*: 3–15.

16

Genome-wide Identification and Expression Analysis of the CLC Gene Family in Pomegranate Reveals Its Role in Salt Resistance

Cuiyu Liu[1] and *Zhaohe Yuan*[2*]

Pomegranate is an important commercial fruit tree, with moderate tolerance to salinity. The balance of Cl^- and other anions in pomegranate tissues are affected by salinity; however, the accumulation patterns of anions are poorly understood. The chloride channel (CLC) gene family is involved in conducting Cl^-, NO_3^-, HCO_3^-, and I^-, but its characteristics have not been reported on pomegranate. This chapter illustrates characteristics of pomegranate CLC gene family, including sequence analysis, evolutionary traits, conserved motifs and residues, and the expression patterns under salt stress. The findings suggested that the *PgCLC* genes play important roles in balancing of Cl^- and NO_3^- in pomegranate tissues under NaCl stress. This chapter offers a theoretical foundation for the further functional characterization of the *CLC* genes in pomegranate.

1. Introduction

Chlorine (Cl^-) is an essential micronutrient for plants, predominantly occurring in the form of Cl^- (Marschner, 2012; White and Broadley, 2001). It is mainly involved in plant physiological activities, such as photosynthesis, regulation of stomatal opening and closing, stabilization of the membrane potential, regulation

[1] Research Institute of Subtropical Forestry, Chinese Academy of Forestry. Hangzhou, 311400, China.

[2] College of Forestry, Nanjing Forestry University. Nanjing, 210037, China.

* Corresponding author: zhyuan88@hotmail.com

of intracellular pH gradients, and electrical excitability (White and Broadley, 2001). Excess and/or deficiency of Cl^- leads to weak plant growth, low yield, and poor quality (Karaivazoglou et al., 2005; Teakle and Tyerman, 2010). In a salinized environment, mostly caused by high NaCl, the foliar salt damage of some plants was mainly caused by Na^+ (Munns and Tester, 2008), while that of other plants, such as tobacco (*Nicotiana Tabacum*) (Karaivazoglou et al., 2005), grape (*Vitis Vinifera*) (Tregeagle et al., 2010), citrus (*Citrus aurantium*) (Moya et al., 2003), and soybean (*Glycine max*) (Luo et al., 2002) was mainly caused by Cl^-. The accumulation patterns of anions, such as Cl^-, NO_3^-, HCO_3^-, and SO_4^{2-} in plant tissues were associated with the plant salt tolerance (Teakle and Tyerman, 2010).

Chlorine channel (CLC) proteins are highly associated with uptake and transport of these anions, like Cl^-, NO_3^-, HCO_3^-, I^-, and Br^- (Amtmann et al., 2009; Gojon et al., 2009;, Lu et al., 2020; Miller et al., 2007).The first CLC family gene (*CLC-0*) was identified from the electric organ of marine ray (*Torpedo marmorata*) (Jentsch et al., 1990), and since then, some new members have been found in bacteria, yeast, mammals, and plants (Jentsch and Thomas, 2008). In land plants, the first CLC gene, *CLC-Nt1*, was cloned in tobacco (Lurin et al., 1996). Subsequently, numerous *CLC* gene homologues were isolated from *Arabidopsis* (De Angeli et al., 2006), rice (*Oryza sativa*) (Nakamura et al., 2006), soybean (*Glycine max*), and trifoliate orange (*Poncirus trifoliata*) (Wei et al., 2015), etc. All the CLC proteins have a highly conserved voltage-gated chloride channel (Voltage-gate CLC) domain and two CBS (cystathionine beta synthase) domains of putative regulatory function (Lu et al., 2020). Also, the CLC gene family members contain three highly conserved regions related to anion selectivity: GxGIPE (I), GKxGPxxH (II), and PxxGxLF (III) (Xing et al., 2020). If the x residue in the conserved region (I) is P [proline, Pro], NO_3^- is preferentially transported, whereas if it is substituted by S [serine, Ser], Cl^- is preferentially transported (Zifarelli and Pusch, 2010). The first x residue in conserved region II and the next fourth residue of the conserved region III can both be E (Glu) residue, which are signatures for CLC antiporters (Accardi and Picollo, 2010). However, if any other amino acids are found at these positions, such as in AtCLCe, AtCLCf, and AtCLCg, these proteins may exert CLC channels' activity (Accardi and Picollo, 2010). Therefore, CLC proteins may act as Cl^- channels or as Cl^-/H^+-exchangers (antiporters) (Jentsch and Thomas, 2008). The Cl^- channels mediate passive transport by dissipating pre-existing electrochemical gradients, while the antiporters mediate active transport by coupling with energy consumption to move the substrate against an electrochemical gradient (Accardi and Picollo, 2010). In higher plants, CLC proteins play vital roles in the control of electrical excitability, turgor maintenance, stomatal movement, ion homeostasis, and in responses to biotic and/or abiotic stress (Jentsch and Michael, 2018; Jossier et al., 2010).

In *Arabidopsis*, there are seven reported CLC genes, *AtCLCa*~*AtCLCg*, which play different roles in diverse cell organelles (Guo et al., 2014; Jossier et al., 2010). Barbierbrygoo et al. (2011) and Marmagne et al. (2007) suggested that AtCLCa ~ AtCLCd and AtCLCg were clustered into a distinct branch, belonging to eukaryotic

CLCs, while AtCLCe and AtCLCf are closely related to prokaryotic CLC channels. *AtCLCa* codes for NO_3^-/H^+ exchangers localized in the vacuolar membrane, which is critically involved in this nitrate accumulation in the vacuole (De Angeli et al., 2006). *AtCLCb*, coding for a vacuolar antiporter, shares 80% identity with *AtCLCa*, it is highly expressed in young roots, hypocotyl, and cotyledons (Fechtbartenbach et al., 2010). *AtCLCc* is essential for the detoxification of cytosol by sequestrating Cl^- into the vacuoles under salt stress, and it is strongly expressed in guard cells, pollen, and roots (Jossier et al., 2010). AtCLCd and AtCLCf, both localized in Golgi membranes, may play a role in the acidification of the *trans*-Golgi vesicles network (Guo et al., 2014, Marmagne et al., 2007), while AtCLCe is targeted to the thylakoid membranes in chloroplasts (Marmagne et al., 2007). *AtCLCg*, the closest homolog to *AtCLCc* (62% identity), plays a physiological role in the Cl^- homeostasis during NaCl stress (Tam et al., 2016). In other plants, many *CLC* genes are involved in anions transport and in the response to salt stress. For instance, the expression level of *OsCLC-1* is upregulated in rice under NaCl stress (Nakamura et al., 2006); *PtrCLC* genes are profoundly induced in orange by salt stress (Wei et al., 2015); *GmCLC1* has been found to possess enhanced salt tolerance in transgenic *Arabidopsis* seedlings by reducing the Cl^- accumulation in shoots (Wei et al., 2016); and *GsCLC-c2* over-expression contributes to Cl^- and NO_3^- homeostasis, and therefore confers the salt tolerance on wild soybean (Wei et al., 2019).

In Pomegranate, it was found that the Cl^- content was double the Na^+ content in pomegranate tissues, and uptake of other anions was also affected by various concentration of salinity (Liu et al., 2020). Researchers reported that the accumulation patterns of anions, such as Cl^-, NO_3^-, HCO_3^-, and SO_4^{2-} in plant tissues were associated with the plant salt tolerance (Teakle and Tyerman, 2010). Also, the NO_3^-/Cl^- was even equal to the K^+/Na^+, which was confirmed as one of the critical determinants of plant salt resistance (Apse and Blumwald, 2007; Munns and Tester, 2008). Therefore, the study on the CLC gene family is contributed to elucidate the pomegranate salt tolerance and their roles of *PgCLC* genes in balancing of anions in pomegranate tissues under NaCl stress.

2. Identification of CLCs in Pomegranate

An HMM profile was used to identify the putative *CLC* genes in pomegranate genome. All seven putative *CLC* genes were named *PgCLC-B* to *PgCLC-G* according to the homologous *AtCLCs* (Table 1). CLC transporters and channels have regulatory functions when ATP, ADP, AMP, or adenosine are bound at the CBS domains (Elgebali et al., 2019). In pomegranate, each one of the CLCs contains a highly conserved Volgate_CLC domain near the N-terminus and two CBS domains at the C-terminus. The specific effect implies that individual CLC transporters and channels are sensitive to the cell's metabolic state (Accardi and Picollo, 2010; Lu et al., 2020). PgCLCs contain 698 ~ 797 amino acids and have molecular weights of 75.7 ~ 87.9 kDa. The predicted isoelectric points (pI) of all the PgCLC proteins range from 5.86– 8.44. The grand average of the hydrophobicity (GRAVY) values are all positive values, indicating that the PgCLCs were hydrophobic proteins. There are a few transmembrane helices (TMHs) in the PgCLCs, ranging from 9–11, which are associated with the ion transport.

Table 1 Characteristics of the CLC genes in pomegranate.

Gene ID	*Name*	*Length*	*Mw (kDa)*	*pI*	*GRAVY*	*Orthologs*	*TMHs*
CDL15_Pgr005627	*PgCLC-B*	797	87.9	6.49	0.259	*AtCLC-B*	9
CDL15_Pgr027626	*PgCLC-C1*	698	75.7	7.53	0.364	*AtCLC-C*	10
CDL15_Pgr013895	*PgCLC-C2*	717	78.1	5.92	0.325	*AtCLC-C*	10
CDL15_Pgr008552	*PgCLC-D*	788	86.9	8.57	0.175	*AtCLC-D*	11
CDL15_Pgr019810	*PgCLC-E*	764	81.3	5.86	0.188	*AtCLC-E*	10
CDL15_Pgr012201	*PgCLC-F*	765	81.7	6.54	0.035	*AtCLC-F*	11
CDL15_Pgr015371	*PgCLC-G*	709	77.3	8.44	0.468	*AtCLC-G*	10

***Note*:** Molecular weight (Mw), isoelectric points (pI), grand average of the hydrophobicity (GRAVY), transmembrane helices (TMHs).

3. Phylogenetic Analysis of the CLC Gene Family in Pomegranate

The CLC gene family is an evolutionarily well-conserved family, which has been found in prokaryotes and eukaryotes (Lu et al., 2020). To elucidate the evolutionary traits of the CLC gene family in land plants, 15 interesting species that had available reference genome sequences were investigated. The results showed two obvious clades of the CLC gene tree, clade I was the major group bearing a moderate support (BS = 61%) and clade II contained two subgroups (Figure 1). PgCLC-E and PgCLC-F belonged to clade II and other PgCLCs belonged to clade I. The divergence of clades I and II might have occurred before the origin of land plants due to each clade consisting of taxa from embryophytes (Figure 1). The CLCs topology was consistent with that of *Arabidopsis* (Barbierbrygoo et al., 2011; Marmagne et al., 2007), tobacco (Hui et al., 2018), tea (*Camellia sinensis*) (Xing et al., 2020), and trifoliate orange (Wei et al., 2015).

Phylogenetic analyses also indicated multiple rounds of ancient gene expansion (Figure 1). Numerous early whole-genome duplication (WGD) events in plants, including the gamma event shared by core-eudicots (Jaillon et al., 2007), the WGD event shared by angiosperms (Genome, 2013; Initiative, 2019; Jiao et al., 2011), and the seed-plant WGD event (Initiative, 2019; Jiao et al., 2011), contribute to gene duplications. The diversity of gene copy numbers from different lineages (Figure 1A) might be related to the rounds of WGD events shared with the taxon (Zhang et al., 2019). The gene duplication between PgCLC-C1 and PgCLC-C2 (the red star in Figure 1B) was supported by the duplication burst shared by core eudicots (Ren et al., 2018). The gene duplication between the CLC-C and CLC-G subfamilies was due to one duplicate shared with angiosperms (the purple star in Figure 1B) (Ren et al., 2018). In the CLC-A/B subfamily, only one member, PgCLC-B, was identified in pomegranate, while there were two members from *Arabidopsis* and *Eutrema* due to a specific gene duplication shared by plants of Brassicaceae (Barker et al., 2009). The phylogenetic analyses also found a gene expansion in seed plants, with a gene birth from an ancient gene duplication (the green star in Figure 1B) and a subsequent gene death. The CLC-A/B/C/G subfamily (Figure 1) exhibited a gene loss event in gymnosperms after experiencing the seed-plant WGD event (Initiative, 2019; Jiao et al., 2011) despite the fact that the absence of the gene might have resulted from the putative incompleteness of the genome assembly and annotation. Recent phylogenetic studies have also found land plant-scale gene birth and expansion, such as in the CYP75 gene family (Zhang et al., 2019) and GH28 gene family (Initiative, 2019).

Here, the phylogenetic results showed that seven putative *PgCLC* genes originated before the divergence of land plants and were retained after experiencing six duplications, including at least one ancient core eudicots-specific duplication (PgCLC-C1 and PgCLC-C2) and one angiosperm-specific expansion (PgCLC-C1/C2 and PgCLC-G) (Figure 1).

Additionally, multiple sequence alignment was performed to meticulously analyze the conserved regions of CLC proteins. Members of the CLC-A/B subfamily had a P [proline, Pro] residue in the conserved region GxGIPE (I), while other proteins of the CLC-C, CLC-G, and CLC-D subfamilies in clade I had an S [serine, Ser] residue in the conserved region I. These critical residues were recognized to have a close relation with anion selectivity. The P [proline, Pro] preferentially transported NO_3^- (Zifarelli and Pusch, 2010), whereas the S [serine, Ser] preferentially transported Cl^- (Zifarelli and Pusch, 2010). Thus, PgCLC-B was likely to be an NO_3^-/H^+ exchanger that mainly transported NO_3^- rather than Cl^- (De Angeli et al., 2006; Fechtbartenbach et al., 2010), while PgCLC-C, PgCLC-D, and PgCLC-G might have high affinity for Cl^- (Figure 1C) (De Angeli et al., 2006; Fechtbartenbach et al., 2010). The presence of the conserved gating glutamate (E) in the conserved region (II) and the proton glutamate (E) residues in the next fourth residue of the conserved region (III) were signatures for CLC antiporters (Accardi and Picollo, 2010). Otherwise, the conserved gating glutamate (E) of the CLC-G subfamily and the proton glutamate (E) residue of the CLC-E and CLC-F subfamilies were substituted by other amino acids, which suggested that the members of these three subfamilies might be CLC ion channels (Figure 1D). Based on these results, four PgCLC proteins (PgCLC-B, PgCLC-C1, PgCLC-C2, and PgCLC-D) were assumed as CLC antiporters, while the other three PgCLCs (PgCLC-E, PgCLC-F, and PgCLC-G) were likely CLC channels. The results were in line with the findings in Arabidopsis (Accardi and Picollo, 2010).

4. Expression Patterns of PgCLCs under NaCl Stress

To further investigate the expression patterns of the *PgCLC* genes, a qRT-PCR analysis of pomegranate roots and leaves was performed under different NaCl

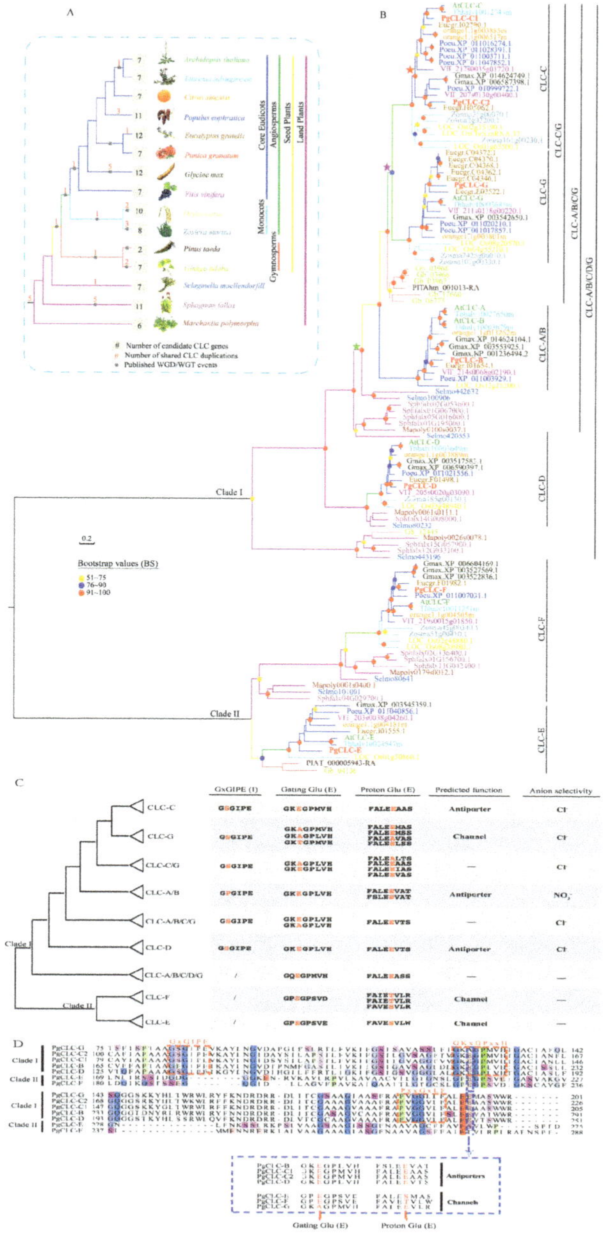

Fig. 1 Phylogenetic analysis of the CLC gene family in land plants. (A) Species tree with different branch colors show distinct species. (B) Phylogenetic tree of the CLC gene family in land plants present in various colors. (C) CLC gene tree with collapsed branches and the conserved residues shared by each subfamily. (D) Partial sequence alignment of the seven CLC proteins in pomegranate.

stress conditions. The results showed that all the *PgCLC* genes had tissue-specific expression patterns, with high expression levels in leaves and low expression levels in roots. The tissue-specific expression of seven *PgCLCs* indicated different mechanisms of transporting anions in pomegranate roots and leaves. In leaves, the high expression levels of *PgCLCs* suggested the inclusion of Cl^- into leaf cells or organelles. However, the low expression levels of *PgCLCs* in roots suggested the exclusion of Cl^- from root cells. The recovery of *PgCLC-C1*, *PgCLC-C2*, and *PgCLC-D* in roots was contributed to the sequestration of Cl^- into the root vacuoles at high salinity levels. Therefore, *PgCLCs* were supposed to alleviate the deleterious effects of Cl^- via excluding the Cl^- from root cells and sequestrating Cl^- into the leaf vacuoles (Jossier et al., 2010; Zhang et al., 2017). On the other hand, under moderate salinity (≤200 mM NaCl), the low expression levels of three Cl^- channels, *PgCLC-E*, *PgCLC-F*, and *PgCLC-G* in leaves, suggested the capacity for pomegranate to inhibit the Cl^- influx into cells or organelles (Liu et al., 2020; Marmagne et al., 2007).

The expression level of *PgCLC-B* (a NO_3^-/H^+ exchanger) suggested that the decreased uptake of NO_3^- in roots might be due to the inhibition of *PgCLC-B* activity under salt stress (De Angeli et al., 2006; Fechtbartenbach et al., 2010). The inhibition of nitrogen uptake was also associated with nitrate transporter (NRTs) (Kiba et al., 2012; Wang et al., 2012). Meanwhile, the increased expression level of *PgCLC-B* in leaves indicated an acceleration of transporting NO_3^- into leaves to mitigate the nitrogen deficiency (Fechtbartenbach et al., 2010).

In a ward, these findings suggested that the *PgCLC* genes played important roles in uptake and transport of Cl^- and NO_3^- in pomegranate tissues under salt stress (Amtmann et al., 2009; Gojon et al., 2009; Jossier et al., 2010; Miller et al., 2007). Researchers found that *PtrCLC* genes were dramatically induced in response to NaCl stress, and *PtrCLC6* showed a leaf-specific expression pattern in trifoliate orange (Wei et al., 2015). Also, all the expressed *NtCLC* genes had a low expression level in tobacco roots under salt stress (Hui et al., 2018). In addition, the functional characterization of each *PgCLC* genes need further study.

This study made a comprehensive, genome-wide inventory of the CLC gene family in pomegranate. The *PgCLC* genes displayed a tissue-specific expression pattern under salt stress. *PgCLCs* were supposed to play important roles in balancing of Cl^- and NO_3^- in pomegranate tissues under salt stress. It not only reveals the roles of *PgCLCs* in uptake and transport of these anions under different NaCl concentration, but also provides a reference for the further study on functions of the *CLC* gene.

References

Accardi, A., and Picollo, A. 2010. CLC channels and transporters: Proteins with borderline personalities. *Biochim. Biophys. Acta.*, *1798*: 1457–64.

Amtmann, A., Armengaud, P., Salt, D.E., and Williams, L. 2009. Effects of N, P, K, and S on metabolism: New knowledge gained from multi-level analysis. *Curr. Opin. Plant Biol.*, *12*: 275–83.

Apse, M.P., and Blumwald, E. 2007. Na^+ transport in plants. *FEBS Lett.*, *581*: 2247– 54.

Barbierbrygoo, H.L.N., Angeli, A.D., Filleur, S., Frachisse, J.M., Gambale, F., Thomine, S.B., and Wege, S. 2011. Anion Channels/Transporters in Plants: From Molecular Bases to Regulatory Networks. *Annu. Rev. Plant Biol., 62*: 25–51.

Barker, M.S., Vogel, H., and Schranz, M.E. 2009. Paleopolyploidy in the Brassicales: Analyses of the Cleome Transcriptome Elucidate the History of Genome Duplications in Arabidopsis and Other Brassicales. *Genome Biol. Evol., 1*: 391–99.

De Angeli, A., Monachello, D., Ephritikhine, G., Frachisse, J., Thomine, S., Gambale, F., and Barbier-Brygoo, H. 2006. The nitrate/proton antiporter *AtCLCa* mediates nitrate accumulation in plant vacuoles. *Nature, 442*: 939–42.

Elgebali, S., Mistry, J., Bateman, A., Eddy, S.R., Luciani, A., Potter, S.C., Qureshi, M., Richardson, L., Salazar, G.A., and Smart, A. 2019. The Pfam protein families database in 2019. *Nucleic Acids Res., 47*: D427– D432.

Fechtbartenbach, J.V.D., Bogner, M., Dynowski, M., and Ludewig, U. 2010. CLC-b-mediated NO_3^-/H^+ exchange across the tonoplast of *Arabidopsis* vacuoles. *Plant Cell Physiol., 51*: 960–68.

Genome, A. 2013. The Amborella Genome and the Evolution of Flowering Plants. *Science, 342*: 1241089.

Gojon, A., Nacry, P., and Davidian, J.C. 2009. Root uptake regulation: A central process for NPS homeostasis in plants. *Curr. Opin. Plant Biol., 12*: 328.

Guo, W., Zuo, Z., Cheng, X., Sun, J., Li, H., Li, L., and Qiu, J. 2014. The chloride channel family gene CLCd negatively regulates pathogen-associated molecular pattern (PAMP)-triggered immunity in Arabidopsis. *J. Exp. Bot., 65*: 1205–15.

Hui, Z., Jin, J., Jin, L., Li, Z., Xu, G., Ran, W., Zhang, J., Niu, Z., Chen, Q., and Liu, P. 2018. Identification and analysis of the chloride channel gene family members in tobacco (*Nicotiana tabacum*). *Gene, 676*: 56–64.

Initiative, O.T.P.T. 2019. One thousand plant transcriptomes and the phylogenomics of green plants. *Nature, 574*: 679–85.

Jaillon, O., Aury, J., Noel, B., Policriti, A., Clepet, C., Casagrande, A., Choisne, N., Aubourg, S., Vitulo, N., and Jubin, C. 2007. The grapevine genome sequence suggests ancestral hexaploidization in major angiosperm phyla. *Nature, 449*: 463–67.

Jentsch, and Thomas, J. 2008. CLC Chloride Channels and Transporters: From Genes to Protein Structure, Pathology, and Physiology. *Crit. Rev. Biochem. Mol. Biol., 43*: 3–36.

Jentsch, T.J., and Michael, P. 2018. CLC Chloride Channels and Transporters: Structure, Function, Physiology, and Disease. *Physiol. Rev., 98*: 1493–90.

Jentsch, T.J., Steinmeyer, K., and Schwarz, G. 1990. Primary structure of Torpedo marmorata chloride channel isolated by expression cloning in Xenopus oocytes. *Nature, 348*: 510–14.

Jiao, Y., Wickett, N.J., Ayyampalayam, S., Chanderbali, A.S., Landherr, L., Ralph, P.E., Tomsho, L.P., Hu, Y., Liang, H., and Soltis, P.S. 2011. Ancestral polyploidy in seed plants and angiosperms. *Nature, 473*: 97–100.

Jossier, M., Kroniewicz, L., Dalmas, F., Le, T.D., Ephritikhine, G., Thomine, S., Barbierbrygoo, H., Vavasseur, A., Filleur, S., and Leonhardt, N. 2010. The Arabidopsis vacuolar anion transporter, AtCLCc, is involved in the regulation of stomatal movements and contributes to salt tolerance. *Plant J. Cell Mol. Biol., 64*: 563–76.

Karaivazoglou, N.A., Papakosta, D.K., and Divanidis, S. 2005. Effect of chloride in irrigation water and form of nitrogen fertilizer on Virginia (flue-cured) tobacco. *Field Crops Res., 92*: 61–74.

Kiba, T., Feria-Bourrellier, A.B., Lafouge, F., Lezhneva, L., Boutet-Mercey, S., Orsel, M., Bréhaut, V., Miller, A., Daniel-Vedele, F., Sakakibara, H., and Krapp, A. 2012. The *Arabidopsis* nitrate transporter NRT2.4 plays a double role in roots and shoots of nitrogen-starved plants. *The Plant Cell, 24*: 245–58.

Liu, C., Yan, M., Huang, X. and Yuan, Z. 2020. Effects of NaCl stress on growth and ion homeostasis in pomegranate tissues. *Eur. J. Hortic. Sci., 85*: 42–50.

Lu, S.N., Wang, J.Y., Chitsaz, F., Derbyshire, M.K., Geer, R.C., Gonzales, N.R., Gwadz, M., Hurwitz, D.I., Marchler, G.H., and Song, J.S. 2020. CDD/SPARCLE: The conserved domain database in 2020. *Nucleic Acids Res., 48*: D265– D268.

Luo, Q.Y., Yu, B.J., and Liu, Y.L. 2002. Stress of Cl^- is Stronger than That of Na^+ on *Glycine max* Seedlings under NaCl Stress. *J. Integr. Agr.*, *1*: 1404–1409.

Lurin, C., Geelen, D., Barbierbrygoo, H., Guern, J., and Maurel, C. 1996. Cloning and functional expression of a plant voltage-dependent chloride channel. *The Plant Cell*, *8*: 701–11.

Marmagne, A., Vinaugerdouard, M., Monachello, D., De Longevialle, A.F., Charon, C., Allot, M., Rappaport, F., Wollman, F., Barbierbrygoo, H., and Ephritikhine, G. 2007. Two members of the Arabidopsis CLC (chloride channel) family, AtCLCe and AtCLCf, are associated with thylakoid and Golgi membranes, respectively. *J. Exp. Bot.*, *58*: 3385–93.

Marschner, P. 2012. *Marschner's Mineral Nutrition of Higher Plants*. Academic Press, Australia.

Miller, A.J., Fan, X., Orsel, M., Smith, S.J., and Wells, D.M. 2007. Nitrate transport and signaling. *J. Exp. Bot.*, *58*: 2297.

Moya, J.L., Gómez-Cadenas, A., Primo-Millo, E., and Talon, M. 2003. Chloride absorption in salt-sensitive Carrizo citrange and salt-tolerant Cleopatra mandarin citrus rootstocks is linked to water use. *J. Exp. Bot.*, *54*: 825–33.

Munns, R., and Tester, M. 2008. Mechanisms of salinity tolerance. *Annu. Rev. Plant Biol.*, *59*: 651–81.

Nakamura, A., Fukuda, A., Sakai, S., and Tanaka, Y. 2006. Molecular cloning, functional expression and subcellular localization of two putative vacuolar voltage-gated chloride channels in rice (*Oryza sativa* L.). *Plant Cell Physiol.*, *47*: 32–42.

Ren, R., Wang, H.F., Guo, C.C., Zhang, N., Zeng, L., Chen, Y.M., Ma, H., and Qi, J. 2018. Wide-spread whole genome duplications contribute to genome complexity and species diversity in angiosperms. *Molecular Plant*, *11*: 414–28.

Tam, N.C., Astrid, A., Mathieu, J., Sylvain, D., Sébastien, T., and Sophie, F. 2016. Characterization of the chloride channel-like, *AtCLCg*, involved in chloride tolerance in *Arabidopsis thaliana*. *Plant Cell Physiol.*, *57*: 764–75.

Teakle, N.L., and Tyerman, S.D. 2010. Mechanisms of Cl^- transport contributing to salt tolerance. *Plant Cell Environ.*, *33*: 566–89.

Tregeagle, J.M., Tisdall, J.M., Tester, M., Walker, R.R., Barrettlennard, E.G. and Setter, T.L. 2010. Cl^- uptake, transport, and accumulation in grapevine rootstocks of differing capacity for Cl^- exclusion. *Funct. Plant Biol.*, *37*: 665–73.

Wang, H., Zhang, M., Guo, R., Shi, D., Liu, B., Lin, X. and Yang, C. 2012. Effects of salt stress on ion balance and nitrogen metabolism of old and young leaves in rice (*Oryza sativa* L.). *BMC Plant Biol.*, *12*: 194.

Wei, P., Che, B., Shen, L., Cui, Y., Wu, S., Cheng, C., Liu, F., Li, M., Yu, B., and Lam, H. 2019. Identification and functional characterization of the chloride channel gene, GsCLC-c2 from wild soybean. *BMC Plant Biol.*, *19*: 1–15.

Wei, P., Wang, L., Liu, A., Yu, B., and Lam, H.M. 2016. *GmCLC1* confers enhanced salt tolerance through regulating chloride accumulation in soybean. *Front. Plant Sci.*, *7*: 1082.

Wei, Q., Gu, Q.Q., Wang, N., Yang, C., and Peng, S. 2015. Molecular cloning and characterization of the chloride channel gene family in trifoliate orange. *Biol. Plantarum*, *59*: 645–53.

White, P.J., and Broadley, M.R. 2001. Chloride in Soils and Its Uptake and Movement within the Plant: A Review. *Ann. Bot.*, *88*: 967–88.

Xing, A., Ma, Y., Wu, Z., Nong, S., Zhu, J., Sun, H., Tao, J., Wen, B., Zhu, X., and Fang, W. 2020. Genome-wide identification and expression analysis of the CLC superfamily genes in tea plants (*Camellia sinensis*). *Funct. Integr. Genomics*, 1–12.

Zhang, H., Zhao, F., Tang, R., Yu, Y., Song, J., Wang, Y., Li, L. and Luan, S. 2017. Two tonoplast MATE proteins function as turgor-regulating chloride channels in *Arabidopsis*. *Proc. Natl. Acad. Sci.*, *114*: E2036– E2045.

Zhang, T., Liu, C., Huang, X., Zhang, H. and Yuan, Z. 2019. Land-plant phylogenomic and pomegranate transcriptomic analyses reveal an evolutionary scenario of CYP75 genes subsequent to whole genome duplications. *J. Plant Bio.*, *62*: 48–60.

Zifarelli, G., and Pusch, M. 2010. CLC transport proteins in plants. *Febs Letters*, *584*: 2122–27.

17

Decoding Anthocyanin Biosynthesis

bZIP Gene Family Study in Pomegranate

Sha Wang[1] and *Zhaohe Yuan*[1*]

In this chapter, 65 PgbZIPs were identified and analyzed from the genome of 'Taishanhong' pomegranate by bioinformatics analysis. We divided them into 13 groups (A, B, C, D, E, F, G, H, I, J, K, M, and S) according to the phylogenetic relationship with those of *Arabidopsis*, each containing a different number of genes. The regularity of exon/intron number and distribution was consistent with the classification of groups in the evolutionary tree. We selected *PgbZIP16* and *PgbZIP34* as candidate genes which affect anthocyanin accumulation. The full-length CDS region of *PgbZIP16* and *PgbZIP34* was cloned from pomegranate petals by homologous cloning technique, encoding 170 and 174 amino acids, which were 510 bp and 522 bp, respectively. Subcellular localization assays suggested that both *PgbZIP16* and *PgbZIP34* were nucleus-localized. Real-time quantitative PCR (qPCR) results demonstrated that the expression of *PgbZIP16* in red petals was 5.83 times of that in white petals, while *PgbZIP34* was 3.9 times. The results of transient expression in tobacco showed that consistent trends were observed in anthocyanin concentration and expression levels of related genes, which both increased and then decreased. Both *PgbZIP16* and *PgbZIP34* could promote anthocyanin accumulation in tobacco leaves.

1. Background

Anthocyanins belonging to flavonoids are water-soluble pigments providing various color for plants, especially in fruits and flowers (V. Bendokas et al., 2020).

[1] College of Forestry, Nanjing Forestry University, Nanjing 210037, China.
[*] Corresponding author: zhyuan88@hotmail.com

Anthocyanin accumulation in plants is usually to resist abiotic stress, such as drought, ultraviolet radiation, hormone, and low temperature (Landi et al., 2015; Luo et al., 2020; Naing et al., 2021; Kristine et al., 2009). Recently, anthocyanins have deserved increasing attention for their potential health benefits. It has been presented that daily intake of natural foods rich in anthocyanins has a forceful protective effect on the human body. Moreover, it can play a role in preventing cardiovascular disease and obesity (Kelly et al., 2017; de Pascual-Teresa et al., 2014; Li et al., 2016; Cassidy et al., 2018). In many plants, it has been proposed that the anthocyanin biosynthesis pathway is mainly regulated by R2R3-MYB, bHLH, and WD40 repeats factors to control their downstream structural genes (Zhang et al., 2019). While one other transcription factor *bZIP* is found to promote anthocyanin synthesis in combination with MYBs to increase their expression (Gangappa et al., 2016).

Transcription factors (TFs) are the most important part of the plant growth and development regulatory network, which activate or inhibit genes expression by combining with specific promoter sequences. *bZIP* transcription factors regulate many plant processes through the interaction of DNA-binding motifs, transcriptional- activation motifs, nuclear localization signals, and oligomerization sites (Landi et al., 2015). Many TFs can be classified into different gene families according to their conserved domains. Presently, at least 64 families of transcription factors have been identified in the plants (Perez-Rodriguez et al., 2010). As one of the most abundant and conserved gene families in eukaryotes, the basic leucine zipper motif (*bZIP*) gene family has an integral role in growth development and abiotic stress responses in plants (Nijhawan et al., 2008). The conserved *bZIP* domain has about 40~80 amino acid residues, which includes two parts, a highly conserved DNA-binding basic composed of 20 amino acids and a relatively diversified leucine zipper region (Talanian et al., 1990). The basic amino acid region is located at the C-terminal region and through a fixed N-x7-R/K structure for sequence-specific DNA binding. The leucine zipper region located at the N-terminal region consists of several heptapeptide repeats or hydrophobic amino acid residues, such as methionine, isoleucine, valine, etc. This domain's main function is by forming dimers through the leucine zipper domain (Landschulz et al., 1998; Ellenberger et al., 1992).

The *bZIP* transcription factor family has been comprehensively identified in several plants, such as 78, 58, 69, 55, 89, 114, 45, 45 in *Arabidopsis* (Jakoby et al., 2002; Dröge-Laser et al., 2018), maize (*Zea mays*) (Wei et al., 2012), tomato (*Solanum lycopersicum*) (Li et al., 2015), grape (*Vitis vinifera*) (Liu et al., 2014), rice (*Oryza sativa*) (Nijhawan et al., 2008), apple (*Malus domestica*) (Li et al., 2016), poplar (*Populus simonii*) (Zhao et al., 2021), and Chinese jujube (*Ziziphus jujuba*) (Zhang et al., 2020), respectively. The *Arabidopsis bZIP* gene family consists of 78 members divided into 13 groups (groups A–K, M, and S) (Dröge-Laser et al., 2018). Currently, a large number of *bZIP* genes have been found to play important roles in the processes of plant growth and development, such as seed maturation and germination (Toh et al., 2012), flower development (Strathmann et al., 2001), vascular development (Gibalová et al., 2017), and embryogenesis (Guan et al., 2009). For example, the *AtbZIP11* affects plant root development by linking low-energy signals to auxin-mediated control of primary root growth (Weiste et al., 2017). Overexpression of the *ZmbZIP4* in maize can also lead to an increase in the number of lateral roots, longer primary roots, and improved plant roots (Ma et al., 2018). In addition, the *bZIP* genes also play an important role in plant biotic and abiotic stress (Yoshida et al., 2010; Wu et al., 2018; Gai et al., 2020). In wheat, *TabZIP15* promotes the combination of ABF/AREB and ABRE (ABA response element) cis-acting elements through the expression of ABF/AREB to induce downstream target gene expression responding to plant salt and drought stress (Bi et al., 2021). Similar results were observed for the *GsbZIP67* gene in Alfalfa (*Medicago sativa*), overexpression of *GsbZIP67* promoted the growth of plant roots and shoots and changed the physiological indicators of transgenic plants under bicarbonate salt-alkali stress (Wu et al., 2018).

Interestingly, a large number of studies have shown that some *bZIP* genes are involved in plant anthocyanin biosynthetic pathway (Schepens et al., 2004). ELONGATED HYPOCOTYL5 (HY5), one member of the *bZIP* gene family, was activated in a light-dependent manner to promote pigment accumulation. *HY5* could directly bind to G-box or ACE-box of MYB factors, including PRODUCTION OF ANTHOCYANIN PIGMENT1 (*PAP1*), PRODUCTION OF FLAVONOL GLYCOSIDES (*MYB12* and *MYB111*), and *MYB-like Domain* (*MYBD*) to promote their gene expression (Shin et al., 2007; Stracke et al., 2010; Shin et al., 2013; Nguyen et al., 2015). Besides *MYBs*, *HY5* co-regulate with *PIF3* the expression of anthocyanin biosynthesis structure genes (Schepens et al., 2004). Moreover, it has been proposed that the overexpression *bZIP* gene *MdHY5* in apple callus induces anthocyanin accumulation by upregulating *MdMYB10* expression and its downstream genes (An et al., 2017). Overexpression of *CRY1a* could increase accumulation of anthocyanin in tomato and *SlHY5* silencing could decrease *CRY1a*-induced anthocyanin accumulation (Liu et al., 2018). In addition, *HY5* positively regulates the cold responses through activation of anthocyanin biosynthesis genes such as chalcone synthase (*CHI*) and chalcone isomerase (*CHS*) (Catalá et al., 2011). Under the induction of abscisic acid (ABA), *MdbZIP44* (*HY5*) positively regulates the anthocyanin accumulation by enhancing the interaction between *MdMYB1* and its downstream target genes (An et al., 2018).

Pomegranate belongs to the Lythraceae family. It is one of the important characteristic economic forest species in the world. It has achieved considerable attention due to its high antioxidant activity, rich color in peel and aril (the edible part of pomegranate), nutritious value, active pharmaceutical ingredients, and anthocyanin (Yuan et al., 2018; Ben et al., 2017). Pomegranate has fruit and flower pomegranate, its attractive appearance, the long forefronts of flowering, and it is gradually being used in landscaping. A comprehensive analysis shows that pomegranate has strong health functions, high ornamental value and ecological and economic profitability, and great development prospects (Chater et al., 2018). In recent years, most scholars have sequenced and assembled the genomes of different pomegranate varieties, and obtained high-quality genome maps, such as 'Dabenzi' (Qin et al., 2017), 'Taishanhong' (Yuan et al., 2018), and 'Tunisia' soft-seed pomegranate (Luo et al., 2019), which provide an important molecular biology basis for pomegranate genetic improvement research.

In this study, we use bioinformatics methods to identify the members of the pomegranate *bZIP* transcription factor family members, and analyze the physical and chemical properties, conserved domains, evolutionary relationships, cis-acting elements, and tissue and organ expression of transcription factors. At the same time, two candidate genes related to anthocyanin synthesis were identified in pomegranate for the first time, and the gene cloning, subcellular location, and differential expression of flower with different colors were analyzed. These results provide a reference for studying the expression of pomegranate *bZIP* gene family during the growth and development of pomegranate and biotic and abiotic stress, and provide a basis for further elucidating the formation mechanism of pomegranate flower color.

2. Materials and Methods

2.1 *Identification and Characterization of bZIP Gene Family Members of Pomegranate*

Sequences with E-value < e^{-5} were identified by HMMER v3.2.1 software based on the hidden Markov models (HMM) profile of the *bZIP* gene family domain (PF00170) downloaded from the Pfam database (http://pfam.xfam.org/). We used the hmmsearch (http://www.hmmer.org/) with *bZIP* to search the 'Taishanhong' pomegranate amino acid sequences, with a threshold of E-value ≤ $1e^{-5}$, and manually removed redundancy (Finn et al., 2011; Chen et al., 2020). At the same time, the *bZIP* proteins of other species were downloaded from Plant Transcription Factor Database (http://planttfdb.gao-lab.org/index.php) as seed files (EI-Gebali et al., 2019). Sequence similarity searches to genes in the whole genome sequence of pomegranate were conducted using the BlastP program on a local NCBI database (E-value < $1e^{-10}$, identity > 50%), and duplicates were removed.

Combing the comparison results of HMMER and BlastP, using the online software SMART (http://smart.embl-heidelberg.de/), CDD (https://www.ncbi.nlm.nih.gov/cdd), and pfam (http://pfam.xfam.org/) databases were used to confirm the integrity of the conserved *bZIP* domains, and the sequences that did not contain *bZIP* conserved structural domains were removed (Marchler-Bauer et al., 2017; Letunic et al., 2018). Isoelectric point (PI), molecular weight (MV), and instability indices of the identified *bZIP* protein were obtained using the online software ExPaSy-Protparam (https://web.expasy.org/protparam/). Subcellular localization of PgbZIPs was predicted by Cell-PLoc-2 (http://www.csbio.sjtu.edu.cn/bioinf/Cell-PLoc-2/) (Shen et al., 2010).

2.2 *Phylogenetic Analysis*

To explore the phylogenetic relationships of the pomegranate *bZIP* gene family, all the *Arabidopsis bZIP* protein sequences were obtained from TAIR database (https://www.arabidopsis.org/). The amino acid sequences of pomegranate and *Arabidopsis bZIPs* were imported into MEGA 7.0 and multiple sequence comparisons were performed using MUSCLE (Dröge-Laser et al., 2018). Thereafter, we used the maximum likelihood method of MEGA 7.0 to construct phylogenetic trees (Kumar et al., 2016; Lefort et al., 2017). The classification of the pomegranate *bZIP* protein family was referenced from previous studies in Arabidopsis. Finally, the phylogenetic tree was visualized using the EvolView website (https://www.evolgenius.info/evolview/) (Balakrishnan et al., 2019).

2.3 *Gene Structure and Conservative Motif Analysis*

Information on 65 PgbZIPs was obtained from the genome annotation GFF files, and the gene structure of PgbZIP gene family was analyzed by the online Gene Structure Display Serve (GSDS, http://gsds.cbi.pku.edu.cn/). We used ClustalW and WebLogo for multiple sequence comparison and visualization analysis, respectively (Thompson et al., 1997; Crooks et al., 2004). Finally, the conserved patterns of the *bZIP* gene family were identified through the online website MEME (http://meme-suite.org/) (Balakrishnan et al., 2019).

2.4 *Analysis of Cis-acting Elements of Pomegranate bZIP Gene Family*

The 1.5 kb promoter sequence upstream of the transcription start site of each PgbZIP gene was extracted from the pomegranate genome sequence and predicted by PlantCARE for cis-regulatory elements (http://bioinformatics.psb.ugent.be/webtools/plantcare/html/) (Magali et al., 2020). The results were manually deleted and saved as a file in bed format, and the file was submitted to the online website GSDS 2.0 for visualization.

2.5 *Expression Analysis of Pomegranate bZIP Gene Family*

To study the expression of PgbZIPs in different tissues and organs, we used the NCBI database (http://www.ncbi.nlm.nih.gov/) illumina sequencing platform to obtain seven tissue transcript data of pomegranate hermaphrodite: functional male flower, leaf, root, endocarp, ectocarp, and pericarp. Their accession numbers were 'Dabenzi' SRR5279388, SRR5279391, SRR5279394–SRR5279397; 'Tunisia' SRR5446592, SRR5446595, SRR5446598, SRR5446601, SRR5446604, SRR5446607, and SRR5678820; 'Baiyushizi' SRR5678819, 'Black127' SRR1054190, 'nana' SRR1055290, and 'Wonderful' SRR080723. The RNA-Seq were quality filtered using the fastp software (Chen et al., 2018). To quantify annotated transcript abundance, we used Kallisto version 0.44.0 to obtain transcriptome data, and the transformed TPM value $\log_2{}^{(TPM+1)}$ were visualized using the TBtools.

2.6 Experiment Material

Research conducted at Baima Base for Teaching and Scientific Research of Nanjing Forestry University, the test materials were 'Liuhuahong', 'Liuhuafen', and 'Liuhuabai' white pomegranate. No permission is required for sample collection. Pomegranate samples were collected in June 2020, and samples were frozen in liquid nitrogen and stored in a –80°C refrigerator for backup.

2.7 Gene Cloning and Subcellular Localization

BioTeke Plant Total RNA Extraction Kit (spin column type) was used to extract RNA from three flower colors of pomegranate petals. cDNA was obtained using reverse transcription kit (PrimeScriptTM RT reagent Kit with gDNAEraser, TaKaRa), and stored at –20°C. Oligo 7.0 software was used to design the primers and the primer sequence was synthesized by Shanghai Bioengineering Co., Ltd.

The PCR reaction system was as follows: 2×Taq Plus Master Mix: 25 µl; F: 1 µl; R: 1 µl; DNA template: 2 µl; Nuclease-free ddH2O: 21 µl. The PCR reaction procedure was as follows: 95°C: 3 min; 95°C: 15 s, 58°C: 45 s, 72°C: 1 min, a total of 35 cycles; 72°C: 5 min; 4°C: storage. The PCR products were separated by 1% agarose gel electrophoresis, and the gel was cut to recover the target fragments. After recovery, ligation, transformation, and sequencing, the CDS sequences of the pomegranate *PgbZIP16* and *PgbZIP34* genes were finally obtained. ExPaSy-Translate online tool (https://web.expasy.org/translate/) was used to translate it into an amino acid sequence.

The correct recombinant plasmid obtained by sequencing was transferred into Agrobacterium GV3101 by freeze-thaw method, and then the tobacco leaves were infected by Agrobacterium tumefaciens-mediated method for 3–4 weeks. The empty vector pBI121 with GFP tag was transformed into Agrobacterium tumefaciens GV3101 in this study. *pBI121-GFP* was used as a control. After 24 hours of dark culture and light culture, respectively, the fluorescence signal was observed under a confocal microscope and photographed.

2.8 Expression Specificity of Pomegranate PgbZIP16 and PgbZIP34

Real-time fluorescent quantitative PCR was used to study the expression patterns of *PgbZIP16* and *PgbZIP34* genes in the petals of three different flower colors of pomegranate. The BioEasy Master Mix Plus (SYBR Green) was used as a fluorescent dye, and the reaction program was as follows: 95°C: 3 min; followed by 40 cycles of 95 °C for 30 s and 60 °C for 15 s. Specific primers were designed for qPCR. The pomegranate actin was used as the internal reference gene, and was performed on three biological and three technical replicates for each treatment. Comparison of relative gene expression data of flowers was done using the $2^{-\Delta\Delta Ct}$ method (Xu et al., 2012; Livak et al., 2001).

2.9 *Agrobacterium Infiltration*

PgbZIP16 and *PgbZIP34* recombinant plasmids with GUS tags were transferred into Agrobacterium tumefaciens GV3101, and the activated Agrobacterium tumefaciens was inoculated into 50 mL of LB liquid medium at a ratio of 1:100, and incubated at 28°C for 16 h with shaking at 210 r/min. Then centrifuged at 4000 rpm/min for 10 min to collect the bacteria and the bacteria were re-suspended and used in permeate (10 mmol/L MES + 10 mmol/L MgCl2.6H2O + 100 mmol/L AS, pH 5.6) to re-suspend the bacteria. Then the injection solution was prepared proportionally, placed at room temperature for 2–4 h, injected into the abaxial surface of tobacco leaves, sampled daily after infestation, stored in a refrigerator at –80°C, and tested seven days later.

2.10 *Overexpression of PgbZIP16 in Arabidopsis and GUS Activity Assay*

Wild-type *Arabidopsis* plants were grown in the incubator (*Arabidopsis* seeds were kept in our laboratory). We constructed the pBI121-*PgbZIP16* overexpression vector and transformed it into Agrobacterium GV3101, inoculated it in 50 mL LB liquid medium, incubated it at 210 r/min at 28°C for 48 h, and then collected the bacteria by centrifugation at 4000 rpm/min within 10 min. After re-suspension of bacteria with an osmotic agent (0.05% sliwet77 + 5% sucrose + 1/2MS liquid medium), *Arabidopsis* plants were subsequently transformed according to the flower dip method and incubated in the dark for 48 h. After three infestations, *Arabidopsis* seeds were collected and screened for *PgbZIP16* transgene-positive plants (Clough et al., 1998).

Seeds of the identified positive plants were planted in 1/2 MS Petri dishes containing 25 μg/L. Positive seedlings with two true leaves were photographed after two weeks, and selected plants were selected and placed in a prepared GUS staining solution and incubated overnight at 37°C. Positive stained plants were decolorized with anhydrous ethanol and photographed with a stereomicroscope after all the green color faded (Jefferson et al., 1987).

2.11 *Statistical Analysis*

The data are shown as the means ± standard errors (SEs) of three or six independent biological replicates. Statistical differences between samples were analyzed by LSD and Duncan (D) ($p < 0.05$). Data analysis and visualization were processed using SPSS 20.0 and Origin 2018.

3. Results

3.1 *Identification and Characterization of the bZIP Transcription Factor Family in Pomegranate*

In this study, we identified 65 gene family members from the whole genome of ‘Taishanhong’ pomegranate. For subsequent analysis, they renamed them according to scaffold (Table 1). The physicochemical properties of *bZIP* transcription factors were analyzed by ExPASy online-tool. The results showed that the molecular weight of pomegranate *bZIP* family proteins ranged from 14803.81–380571.64 Da, the theoretical isoelectric point ranged from 4.74–9.91, and the protein lengths ranged from 128–1543 aa, with the shortest being 128 aa (PgbZIP11) and the longest being 1543 aa (PgbZIP8). These results provide a theoretical basis for further purification, activity and function studies of PgbZIP protiens. The subcellular location prediction of each member indicated that all members of the *bZIP* gene family are expressed in the nucleus.

3.2 Phylogenetic Tree and bZIP Conservative Domain Analysis

The results of the multi-sequence alignment of protein sequences of 65 *bZIP* family members in pomegranate showed that the core conserved structural domain of the PgbZIPs protein had an average length of 50 aa. The *bZIP* structural domain consists of a basic region and a leucine zipper. The basic region was located at the C-terminus and contains a fixed N-X7-R/K motif bound to a specific DNA sequence, while the leucine zipper region was located at the N-terminal end and consists of several repetitive heptapeptide or hydrophobic amino acid residues. The highly conserved leucine residues were sometimes replaced by isoleucine, methionine, valine, etc. Our results are consistent with a previous study in *Arabidopsis* (Dröge-Laser et al., 2018).

To explore the homologous evolutionary relationships and classification of the *bZIP* family, we constructed a phylogenetic tree using the *bZIP* members of pomegranate, *Eucalyptus megacephalus* and *Arabidopsis*. Referring to the evolutionary relationship and naming rules of *Arabidopsis bZIP* genes, a deep cluster analysis was carried out on the whole phylogenetic tree. The pomegranate *bZIP* gene family was divided into 13 groups (A, B, C, D, E, F, G, H, I, J, K, M, and S). These 13 groups differed greatly in size. Two of the groups have only one member, namely, group B and M. The largest group has 16 members (group S). Throughout the evolutionary tree, the *bZIP* genes of the three species were distributed in almost all these 13 subgroups, indicating that the *bZIP* genes showed different divergence in gene function in pomegranate, *Eucalyptus megacephalus*, and *Arabidopsis*. Meanwhile, some *bZIP* genes of pomegranate, *Eucalyptus megacephalus* and *Arabidopsis* each clustered together in a small clade, suggesting that a co-speciation event and species-specific duplication events occurred during the *bZIP* family divergence. Similar to the evolutionary relationships in *Arabidopsis*, our further analysis revealed that two pairs of homologous genes, PgbZIP16/AtHY5 and PgbZIP34/AtHYH in group H, were able to influence anthocyanin accumulation.

3.3 Gene Structure and Protein Conserved Motifs of PgbZIP Genes Family

As the composition of introns/exons and types and numbers of introns were typical marks of evolution within certain gene families, we explored the gene structures

Table 1 The identified PgbZIP genes and their related information.

Gene Name	*Gene ID*	***Location***	***Group***	***CDS***	***AA***	***MW(Da)***	***pI***	***Subcelluar Localization***
PgbZIP1	Pg000486.1	scaffold1:692075:692518	S	443	147	16488.75	7.86	Nucleus
PgbZIP2	Pg000487.1	scaffold1:695206:695652	S	447	149	36053.89	5.18	Nucleus
PgbZIP3	Pg000951.1	scaffold1:2591499:2593093	I	729	243	59393.95	5.12	Nucleus
PgbZIP4	Pg011031.1	scaffold2:1782851:1795679	D	2265	755	184985.25	4.88	Nucleus
PgbZIP5	Pg011379.1	scaffold2:3151711:3152349	S	639	213	54055.25	5.09	Nucleus
PgbZIP6	Pg010908.1	scaffold2:3418553:3421108	C	1320	440	47485.28	7.75	Nucleus
PgbZIP7	Pg016213.1	scaffold3:1034859:1035461	S	600	200	22636.22	5.59	Nucleus
PgbZIP8	Pg019746.1	scaffold4:1634806:1637852	I	1041	347	87468.71	4.97	Nucleus
PgbZIP9	Pg019532.1	scaffold4:1723642:1725546	A	834	278	30788.86	9.29	Nucleus
PgbZIP10	Pg019474.1	scaffold4:2417973:2420458	A	834	278	29691.17	5.91	Nucleus
PgbZIP11	Pg019929.1	scaffold4:3861901:3862287	S	384	128	14981.66	6.91	Nucleus
PgbZIP12	Pg022634.1	scaffold5:1188977:1190472	A	780	260	28329.43	8.35	Nucleus
PgbZIP13	Pg022422.1	scaffold5:1764422:1764892	S	471	157	38400.71	5.17	Nucleus
PgbZIP14	Pg022742.1	scaffold5:2775118:2777326	A	1311	437	46593.41	9.68	Nucleus
PgbZIP15	Pg022303.1	scaffold5:3712021:3712449	S	426	142	16587.97	9.42	Nucleus
PgbZIP16	Pg024592.1	scaffold6:2540285:2543069	H	507	169	18466.50	9.91	Nucleus
PgbZIP17	Pg026584.1	scaffold7:690047:692762	A	1038	346	38062.41	8.42	Nucleus
PgbZIP18	Pg026564.1	scaffold7:944817:948069	D	1164	388	95110.83	5.03	Nucleus
PgbZIP19	Pg026477.1	scaffold7:2126264:2129463	J	1554	518	125872.84	4.98	Nucleus
PgbZIP20	Pg026888.1	scaffold7:2982554:2984794	A	1245	415	44938.52	9.49	Nucleus
PgbZIP21	Pg001581.1	scaffold10:3315235:3317319	E	1125	375	92957.96	4.98	Nucleus
PgbZIP22	Pg002913.1	scaffold11:2030336:2032657	A	1266	422	45882.25	8.87	Nucleus

Contd.

Table 1 contd.

Gene Name	*Gene ID*	***Location***	***Group***	***CDS***	***AA***	***MW(Da)***	***pI***	***Subcelluar Localization***
PgbZIP23	Pg003837.1	scaffold12:3213559:3215308	E	891	297	32957.73	5.79	Nucleus
PgbZIP24	Pg005395.1	scaffold13:113807:114247	S	438	146	16629.94	6.37	Nucleus
PgbZIP25	Pg005194.1	scaffold13:3081012:3081581	S	567	189	21419.06	8.93	Nucleus
PgbZIP26	Pg008236.1	scaffold16:613569:614306	F	738	246	60324.09	5.10	Nucleus
PgbZIP27	Pg008260.1	scaffold16:920825:924942	D	1296	432	106682.40	5.00	Nucleus
PgbZIP28	Pg008051.1	scaffold16:1996563:1999431	D	1254	418	102298.97	5.01	Nucleus
PgbZIP29	Pg008941.1	scaffold17:183002:184828	A	813	271	28884.20	9.44	Nucleus
PgbZIP30	Pg008855.1	scaffold17:988420:990879	G	1035	345	36244.38	5.39	Nucleus
PgbZIP31	Pg009560.1	scaffold18:76414:78344	E	927	309	76112.58	5.08	Nucleus
PgbZIP32	Pg009379.1	scaffold18:2446066:2448483	I	924	308	75878.49	5.04	Nucleus
PgbZIP33	Pg010362.1	scaffold19:3431036:3432620	A	1128	376	41024.31	8.95	Nucleus
PgbZIP34	Pg012358.1	scaffold21:2286741:2288172	H	519	173	19322.86	9.89	Nucleus
PgbZIP35	Pg013506.1	scaffold22:1812013:1819851	B	4629	1543	380571.64	4.74	Nucleus
PgbZIP36	Pg013360.1	scaffold22:5624189:5627104	I	1074	358	88826.51	4.98	Nucleus
PgbZIP37	Pg013743.1	scaffold23:1468781:1470885	I	1299	433	107204.34	4.98	Nucleus
PgbZIP38	Pg014054.1	scaffold24:1165026:1166900	D	705	235	57216.90	5.14	Nucleus
PgbZIP39	Pg014221.1	scaffold24:1205470:1206418	A	807	269	30034.36	9.23	Nucleus
PgbZIP40	Pg015819.1	scaffold29:475717:476172	S	453	151	16520.56	9.50	Nucleus
PgbZIP41	Pg015844.1	scaffold29:999707:1002603	I	1806	602	148814.05	4.92	Nucleus
PgbZIP42	Pg015870.1	scaffold29:1365294:1366040	F	747	249	59751.28	5.14	Nucleus
PgbZIP43	Pg016758.1	scaffold30:1040570:1041935	A	798	266	29406.84	9.23	Nucleus
PgbZIP44	Pg017636.1	scaffold33:350792:351202	S	411	137	32817.67	5.23	Nucleus

Contd.

Table 1 contd.

Gene Name	*Gene ID*	*Location*	*Group*	*CDS*	*AA*	*MW(Da)*	*pI*	*Subcelluar Localization*
PgbZIP45	Pg017872.1	scaffold34:149123:150849	F	870	290	68888.04	5.12	Nucleus
PgbZIP46	Pg017805.1	scaffold34:1334693:1335175	S	480	160	17740.69	6.31	Nucleus
PgbZIP47	Pg018175.1	scaffold35:221629:223223	K	822	274	66700.39	5.10	Nucleus
PgbZIP48	Pg018974.1	scaffold38:374554:377270	C	1221	407	43936.74	5.58	Nucleus
PgbZIP49	Pg019276.1	scaffold39:1591489:1593020	K	909	303	74386.34	5.07	Nucleus
PgbZIP50	Pg020198.1	scaffold40:873396:873791	S	393	131	14803.81	9.35	Nucleus
PgbZIP51	Pg020064.1	scaffold40:1476711:1477762	A	657	219	24176.26	9.47	Nucleus
PgbZIP52	Pg021901.1	scaffold48:629216:639124	D	1371	457	112445.22	5.00	Nucleus
PgbZIP53	Pg023016.1	scaffold50:634939:635538	S	597	199	22700.04	5.60	Nucleus
PgbZIP54	Pg023400.1	scaffold52:433306:436854	G	1194	398	42055.23	6.26	Nucleus
PgbZIP55	Pg023806.1	scaffold54:1143344:1149748	M	888	296	73776.05	5.05	Nucleus
PgbZIP56	Pg023869.1	scaffold55:84392:86720	G	1053	351	37924.99	5.56	Nucleus
PgbZIP57	Pg025024.1	scaffold60:329234:333156	D	1542	514	125729.20	4.97	Nucleus
PgbZIP58	Pg028547.1	scaffold80:538801:545326	D	1446	482	119119.95	4.94	Nucleus
PgbZIP59	Pg028644.1	scaffold80:650755:652351	J	894	298	73357.03	5.07	Nucleus
PgbZIP60	Pg029370.1	scaffold87:425556:427793	E	975	325	80691.55	5.01	Nucleus
PgbZIP61	Pg002089.1	scaffold105:550189:557112	D	1461	487	119232.03	4.96	Nucleus
PgbZIP62	Pg003440.1	scaffold114:422966:423556	S	591	197	49650.00	5.11	Nucleus
PgbZIP63	Pg004996.1	scaffold128:119987:121887	A	978	326	35872.31	8.98	Nucleus
PgbZIP64	Pg011906.1	scaffold201:109649:114841	C	1230	410	44481.01	5.41	Nucleus
PgbZIP65	Pg012452.1	scaffold211:113511:113990	S	477	159	17460.66	6.60	Nucleus

of PgbZIP genes structures to further understand their evolutionary trajectory. We analyzed the intron/exon and motif structure of each member. As expected, members of the different groups had different gene structures, conserved domains, and numbers of introns/exons, with the number of introns ranging from 0–11. For example, there were no introns in the S group, while *PgbZIP54* and *PgbZIP61* had the largest number of introns with 11 introns.

To investigate the distribution of conserved patterns *bZIP* proteins, 65 PgbZIP protein sequences were analyzed by MEME. The number and type of conserved motifs contained in each protein sequence varied. The distribution of different conserved motifs was revealed with different functions of different genes. Group D has the most motifs, with PgbZIP4 containing 16 motifs, and Group S has the least motifs, with PgbZIP15 containing two motifs.

3.4 Cis-acting Elements of Pomegranate bZIP Gene Family

To further explore the potential mechanism of *bZIP* gene in biotic and abiotic stress, we submitted the 1500 bp upstream sequence of the PgbZIP translation start site to Plant CARE for detection of cis-acting elements. The PgbZIP gene family cis-acting elements were mapped using the online website GSDS2.0. Meanwhile, we analyzed and screened 12 cis-acting elements, mainly including ABA-responsive element ABRE, drought-inducible response element MBS, low-temperature response element LTR, defense and stress-responsive element TC-rich repeats, trauma-responsive element WUN-motif, gibberellin-responsive element P-box, anaerobic-inducible cis-regulatory element ARE, meristematic tissue expression-related cis-regulatory element CAT-box, regulatory element MYB of secondary metabolic pathways, and common cis-acting element CAAT-box of promoter and enhancer regions. Pomegranate had 65 PgbZIPs, each with one or more cis-acting elements, suggesting that expression of PgbZIPs may be associated with these abiotic stresses. In total, 65 genes had one or more ABA response elements and 40 PgbZIPs had one or more LTR response elements, indicating that PgbZIPs may be significantly responsive to ABA and low temperature stresses. Eighteen PgbZIPs had TC-rich repeats response elements, and 17 PgbZIPs had WUN-motif response elements. In conclusion, the analysis of cis-acting elements suggested that the PgbZIP genes may respond to different abiotic stresses.

3.5 Tissue-differential Gene Expression Patterns of Pomegranate bZIP Genes

To explore the expression patterns of the pomegranate *bZIPs* gene family in different tissues, we analyzed the expression of *bZIP* genes in pomegranate roots, stems, flowers, endocarp, exocarp, and leaves using transcriptome analysis data by RNA-Seq. The analysis showed that the expression pattern of *PgbZIP* genes was distinctly tissue-organ specific. The expression of *PgbZIP1*, *PgbZIP13*, *PgbZIP18*, *PgbZIP24*, *PgbZIP41*, *PgbZIP43*, *PgbZIP46*, and *PgbZIP64* was higher than other members, indicating that this gene may be involved in the transcriptional regulation of various physiological and biochemical processes during pomegranate development. In contrast, the expression of *PgbZIP3*, *PgbZIP29*, *PgbZIP55*, and *PgbZIP59* was generally lower. Compared with other tissues, *PgbZIP44* and *PgbZIP47* were highly expressed in roots, *PgbZIP11* in buds and pericarp, and *PgbZIP34* and *PgbZIP46* in young leaves. *PgbZIP16* and *PgbZIP34* were somewhat expressed in pericarp, leaves, flowers, and fruits, but relatively low in root development.

3.6 Cloning and Analysis of PgbZIP16 and PgbZIP34

The 510 bp and 522 bp open reading frame (ORF) of the *PgbZIP16* and *PgbZIP34* genes were amplified from the mixed-sample cDNA. The ORF encodes 170 and 174 aa, respectively. The predicted protein molecular weights were 39591.74 and 41604.78 Da, and the theoretical isoelectric points were 5.22 and 5.18, respectively. The amino acid sequence analysis of the proteins of these two genes contains a *bZIP* domain (BRLZ Domain) located at sites 87 ~ 151 and 96 ~ 160, respectively. Evolutionary tree analysis showed that *PgbZIP16* and *PgbZIP34* genes belong to *HY5* and *HYH* type transcription factors in the *bZIP* gene family, respectively, and play an important role in the transcriptional regulation of anthocyanin synthesis. Therefore, *PgbZIP16* and *PgbZIP34* were selected for cloning and functional analysis in this paper.

Reverse transcription quantitative PCR (RT-qPCR) revealed that *PgbZIP16* and *PgbZIP34* were highly expressed in 'Liuhuahong'. Among the three different colors of pomegranate flowers, the expression level *PgbZIP16* in 'Liuhuahong' was 5.83 times that of 'Liuhuabai' and 5 times that of 'Liuhuafen'. *PgbZIP34* had a similar expression trend, and its expression level was 3.9 times that of 'Liuhuabai' and 2.3 times that of 'Liuhuafen'.

To determine the subcellular localization of PgbZIP16 and PgbZIP34, the construct encoding PgbZIP16 and PgbZIP34 fused to green fluorescent protein (GFP) were transformed into tobacco leaves. Intense fluorescence from 35S:GFP-PgbZIP16 and 35S:GFP-PgbZIP34 were detected in the nucleus, indicating that PgbZIP16 and PgbZIP34 localize to nucleus. These results suggested that PgbZIP16 and PgbZIP34 might function as a transcription factor in regulating anthocyanin biosynthesis.

3.7 Functional Studies in Tobacco

To investigate the functions of *PgbZIP16* and *PgbZIP34* genes, we constructed pBI121-*PgbZIP16* and pBI121-*PgbZIP34* overexpression vectors and transferred them into tobacco leaves by injection method. The results showed consistent trends in anthocyanin accumulation and gene expression levels, both of which increased and then decreased. The anthocyanin in *PgbZIP16* transgenic tobacco leaves started to increase significantly on the 3rd day, and reached the highest level on the 5th day, which was 0.067 $mg \cdot g^{-1}$ FW. It was 5.58 times higher than that in non-infested leaves and 2.79 times higher than that in pBI121 null leaves. Gene expression was consistent with the level of anthocyanin content, with the highest at day 5, which was 9.72-fold higher than that of un-infested leaves and 3.74-fold higher than that of pBI121 null leaves. The anthocyanin in *PgbZIP34* transgenic tobacco leaves started to increase significantly on the 3rd day, and reached the highest level on the 5th day, which was 0.047 $mg \cdot g^{-1}$ FW. It was 3.92-fold higher than that of un-infested leaves and 1.96 times more than that of pBI121 nulled leaves. Gene expression was consistent with anthocyanin content levels, with the highest being on day 5, which was 6.56 times higher than that of un-infested leaves and 2.54 times higher than that of pBI121 unloaded leaves. In summary, *PgbZIP16* had an important role in anthocyanin accumulation; but because they are homologous sequences, therefore, we further explored the genetic regulation of anthocyanin accumulation by *PgbZIP16* in *Arabidopsis*.

3.8 Genetic Regulation of PgbZIP16 on Arabidopsis Anthocyanin Accumulation

It has been previously reported that *HY5* acts as a transcriptional activator that positively regulates the biosynthesis of anthocyanins in plants (Shin et al., 2013; Nguyen et al., 2015; An et al., 2017). *HY5* regulated the accumulation of anthocyanins through directly binding to the promoters of *CHS*, *CHI*, *F3H*, *F3'H*, *DFR*, and *ANS* (Chattopadhyay et al., 1998; Jungeun et al., 2007). To investigate whether *PgbZIP16* plays a role in promoting anthocyanin biosynthesis in *Arabidopsis*, we constructed a pBI121-*PgbZIP16* overexpression vector and transformed *Arabidopsis* using the flower-dip method. The *PgbZIP16* overexpression *Arabidopsis* (*PgbZIP16*-6, *PgbZIP16*-16, *PgbZIP16*-21) were compared with the control *Arabidopsis* carrying only the empty vector of the 35S:pBI121 vector. It was confirmed that *Arabidopsis* overexpressing *PgbZIP16* expressed *PgbZIP16* at significantly higher levels than the control *Arabidopsis*. The *PgbZIP16* was constructed into the pBI121 vector to drive stable expression of the GUS reporter gene. Histochemical staining for GUS activity showed that *Arabidopsis* seedlings overexpressing *PgbZIP16* were expressed in stem segments. Anthocyanin content was measured in control and *PgbZIP16* overexpressing *Arabidopsis* leaves. It was found to be significantly higher in the leaves of *PgbZIP16* overexpressing *Arabidopsis* than in the control. In transgenic *Arabidopsis*, there was a concordance between the expression of most structural genes on the anthocyanin biosynthetic pathway (*FLS*, *4CL*, *CHI*, *CHS*, *F3H*, *F3'H*, *DFR*, *UF3GT*, *UGT1*, *UGT2*, and *ANS*) and *PgbZIP16* genes, which were significantly upregulated compared with the control. Moreover, *PgbZIP16* significantly upregulated the expression levels of *UF3GT*, *ANS*, and *DFR* genes in transgenic *Arabidopsis* and enhanced anthocyanin accumulation. Therefore, the *PgbZIP16* gene has a genetic regulatory effect on anthocyanin accumulation in *Arabidopsis*.

4. Discussion

The *bZIP* transcription factor plays a crucial role in plant growth, development, and abiotic stress responses, such as seed maturation, flower development, stress response, etc. (Toh et al., 2012; Strathmann et al., 2001; Gai et al., 2020). Currently, the *bZIP* gene family has been studied in the model plant, especially for *Arabidopsis* and crops, such as *AtbZIP11/18* in *Arabidopsis* (Gibalová et al., 2017; Weiste et al., 2017), *GsbZIP67* in soybeans (Wu et al., 2018), *CabZIP25* in pepper (Gai et al., 2020), *TabZIP15* in wheat (Bi et al., 2021), and *MdHY5*/*MdbZIP44* in apple (An et al., 2017; An et al., 2018). Till date, despite the sequencing of the whole pomegranate gene has been completed, little is known about the *bZIP* gene family in pomegranate (Ben et al., 2017; Qin et al., 2017; Luo et al., 2019). So, we identified and analyzed the expression pattern of *bZIP* based on phylogenetic analysis to speculate on the evolution of the PgbZIP gene family.

In this study, a total of 65 PgbZIP genes were identified in pomegranate using bioinformatics methods (Table 1). The *bZIP* members in pomegranate were similar to those in *Arabidopsis* (Dröge-Laser et al., 2018), apple (Li et al., 2016), poplar (Zhao et al., 2021), and jujube (Zhang et al., 2020), which may have been caused by ancient polyploid events. To further understand the evolutionary relationships between *Arabidopsis* and pomegranate, we constructed a phylogenetic tree of *bZIP* genes following a clustering approach (Dröge-Laser et al., 2018). Due to the highly conserved *bZIPs* sequences, genes with the same function belong to the same group, which provides a reference for studying this gene family. Similar to the *Arabidopsis* grouping, the pomegranate *bZIP* genes were divided into 13 groups.

Gene structure and conservative motifs were also important basis for studying gene evolution and gene duplication. We analyzed in detail the structure of the pomegranate *bZIP* gene and the number of introns and exons. Compared with other gene families, we found that the pomegranate *bZIP* gene structure was relatively simple, with the number of introns ranging from 0–11. In conclusion, most of the PgbZIP genes have a similar number of introns compared to other plant species (Liu et al., 2014; Wang et al., 2017; Fan et al., 2019; Li et al., 2020). In *Arabidopsis*, the subfamily-specific and conserved motifs may play important roles in the functional differentiation of AtbZIPs subfamilies. For example, most members of group A participate in the ABA biological pathway and regulate plant responses to abiotic stress (Choi et al., 2000; Finkelstein et al., 2000; Lopez-Molina et al., 2001). Therefore, the PgbZIP genes in group A could have similar functions. The significant feature of group C members was the extension of the leucine zipper region, which can be up to nine repeats. In addition, potential target sites for protein modification, such as phosphorylation sites that regulated nuclear translocation and DNA binding, were also preserved (Jakoby et al., 2002). Group D genes could participate in plant defense against pathogens (Jakoby et al., 2002). Group G gene and their homologues were mainly involved in the signal transduction of blue-violet light (Kircher et al., 1998).

Group H contained two genes that could be directly combined with light-induced gene promoters to regulate plant cell elongation, chloroplast synthesis, hormone synthesis, and anthocyanin biosynthesis (Chattopadhyay et al., 1998). Group S had the most members, but the number of well-researched genes was less. Members of this group not only play an important role in the sucrose metabolism pathway, but also could be activated and transcribed under the antibiotic stress (Strathmann et al., 2001). In our study, transcriptome data indicated that PgbZIPs were highly homologous to *Arabidopsis*, demonstrating similar roles in specific biological processes. It seems that the evolution events in the *bZIP* gene family members have happened before species divergence, which affected their number and function (Abdullah et al., 2021; Musavizadeh et al., 2021).

The expression intensity of pomegranate PgbZIPs in different tissues was further analyzed. The results showed that the genes were expressed in leaves, roots, flowers, seed coat, and envelope, except for *PgbZIP3*, *PgbZIP12*, *PgbZIP39*, *PgbZIP55*, and *PgbZIP59* which were hardly expressed in the tissues. This indicated that the *bZIP* gene family play an essential role in the growth and development of pomegranate. At the same time, the expression of *PgbZIP46* was found to be higher in all tissues than other genes, especially in roots, leaves, and flowers. This may be that this gene is closely related to growth, development, and stress in pomegranate. In addition, we speculated that several genes that were barely expressed in tissues may not be involved in the regulation of pomegranate development or stress.

Anthocyanins are water-soluble pigments involved in pathways of plant secondary metabolism. Anthocyanins are mainly found in flowers, leaves, seed coat, and fruits of plants in form of glycoside. Anthocyanins not only provide brilliant colors to plants, but also protect plants from ultraviolet radiation and pathogens (Leon et al., 2020). The current research on *HY5* and *HYH* of *Arabidopsis*, tomato and apple is more in-depth. For example, *HY5* and *HYH* in *Arabidopsis* are phytochrome receptors of the light signal pathway downstream. Not only the expression of *EBGs* and *LBGs* directly activate, *HY5* can positively regulate the transcriptional activation of *AtPAP1* (Strathmann et al., 2001; Jungeun et al., 2007; Zhang et al., 2011). In apple, *bZIP* transcription factor gene *MdHY5* could directly promote the expression of *MdMYB10* and *MdMYB1* genes, and positively regulate anthocyanin accumulation by enhancing the interaction with its downstream target genes (Yoshida et al., 2010; Wu et al., 2018; An et al., 2017; An et al., 2018). In addition, *SlHY5* gene silencing downregulated the accumulation of anthocyanins in tomato (Liu et al., 2018). The phylogenetic analyses demonstrate that *PgbZIP16* and *PgbZIP34* shared higher homology with *AtHY5* and *AtHYH*. Therefore, we speculated that *PgbZIP16* and *PgbZIP34* genes in pomegranate also had similar functions to *Arabidopsis*, tomato and apple.

Pomegranate, as an ancient fruit is a widely consumed fresh fruit; it is an economically important fruit tree crop in China. Preliminary research has been conducted on the coloring mechanism of the pomegranate peel, but the mechanism of flower color formation has not been studied in depth. In this study, we performed gene cloning, subcellular localization, and functional verification of *PgbZIP16* and

PgbZIP34 in the flower color formation mechanism of three kinds of ornamental pomegranate. The results suggested that the patterns of expression of both genes in red were significantly higher than those in white and pink; this outcome was consistent with the results of grape hyacinth and red pear (Cao et al., 2019; Wang et al., 2020). To investigate the functions of *PgbZIP16* and *PgbZIP34*, we constructed pBI121-*PgbZIP16* and pBI121-*PgbZIP34* overexpression vectors and transformed tobacco leaves. The conclusions showed that there was a consistency between anthocyanin content and gene expression, which both increased and then decreased. Both *PgbZIP16* and *PgbZIP34* promoted anthocyanin accumulation in tobacco leaves.

Compared with *PgbZIP34*, *PgbZIP16* played a more important role in anthocyanin accumulation. To determine the genetic relationship between *PgbZIP16* and structural genes, we used the dipstick method to transfer the constructed overexpression vector into *Arabidopsis* and obtained *PgbZIP16* overexpression strains. Histochemical staining for GUS activity showed that seedlings overexpressing *PgbZIP16* were specifically expressed in stem segments. Such an expression pattern suggested that *PgbZIP16* may be involved in the accumulation of anthocyanins at the early stages of *Arabidopsis* development. Furthermore, studies have shown that the overexpression of *PgbZIP16* significantly promoted the anthocyanin accumulation in the transgenic strain. Meanwhile, most genes on the anthocyanin synthesis pathway (*FLS*, *4CL*, *CHI*, *CHS*, *F3H*, *F3'H*, *DFR*, *UF3GT*, *UGT1*, *UGT2*, and *ANS*) and *PgbZIP16* gene expression were consistent. The expression of *PAL* and *C4H* was lower in *PgbZIP16* overexpressing plants than in the control, which may have resulted from the fact that *PAL* and *C4H*, as structural genes of the mangiferin synthesis pathway, were not directly involved in anthocyanin synthesis (Shin et al., 2013). Based on the present experimental study, further investigation of the relationship between *PgbZIP16* and other transcription factors and their role in the process of flower color formation is the focus of future work.

Conclusions

In this study, a total of 65 PgbZIP genes were identified in pomegranate using bioinformatics methods, and their *bZIP* structural domain was determined. We constructed a phylogenetic tree of pomegranate and *Arabidopsis*, and divided the *PgbZIP* genes into 13 groups. Due to the high conservation of *bZIP* genes, proteins with similar functions were clustered into one group, which provided a reliable basis for studying the functions of related genes in gene families in plants. Additionally, we identified two candidate genes in the anthocyanin biosynthesis using transcriptome data analysis and performed their gene cloning, subcellular localization, quantitative fluorescence analysis, transient expression, and Arabidopsis transformation. Our results indicated that *PgbZIP16* and *PgbZIP34* had similar regulatory mechanisms in anthocyanin accumulation. It is believed that future studies will elucidate the exact molecular mechanisms by which *PgbZIP16* interacts with other transcription factors to promote anthocyanin accumulation.

Supplementary Materials: The following are available online at https://doi.org/10.1186/s12870-022-03560-6, Table S1: Primers for the gene cloning, subcellular localization, and qRT-PCR.

References

Abdullah, Faraji S., Mehmood, F., Malik, H.M.T., Ahmed, I., Heidari, P., and Poczai, P. 2021. The GASA Gene Family in Cacao (Theobroma cacao, Malvaceae): Genome Wide Identification and Expression Analysis. *Agronomy*, *11*(7): 1425.

An, J., Qu, F., Yao, J., Wang, X., You, C., Wang, X., and Hao, Y. 2017. The *bZIP* transcription factor *MdHY5* regulates anthocyanin accumulation and nitrate assimilation in apple. *Hortic. Res.*, *4*: 17023.

An, J., Yao, J., Xu, R., You, C., Wang, X., and Hao, Y. 2018. Apple *bZIP* transcription factor MdbZIP44 regulates abscisic acid-promoted anthocyanin accumulation. *Plant Cell Environ.*, *41*(11): 2678–92.

Balakrishnan, S., Gao, S., Lercher, M., Hu, S., and Chen, W. 2019. Evolview v3: A webserver for visualization, annotation, and management of phylogenetic trees. *Nucleic Acids Research*, *47*: W270–W275.

Ben, L., Kim, K., Quah, C., Kim, W., and Shahimi, M. 2017. Anti-inflammatory potential of ellagic acid, gallic acid, and punicalagin A&B isolated from *Punica granatum*. *BMC Complement. Altern. Med.*, *17*(1): 47–57.

Bi, C., Yu, Y., Dong, C., Yang, Y., Zhai, Y., Du, F., Xia, C., Ni, Z., Kong, X., and Zhang, L. 2021. The *bZIP* transcription factor *TabZIP15* improves salt stress tolerance in wheat. *Plant Biotechnol. Journal*, *19*(2): 209–11.

Cao, S., and Liu, Y. 2019. Cloning and expression analysis of *MaHY5* transcription factor from grape hyacinth. *Acta Botanica Boreali-Occidentalia Sinica*, *39*(07): 1188–94.

Cassidy, A. 2018. Berry anthocyanin intake and cardiovascular health. *Molecular Aspects of Medicine*, *61*: 76–82.

Catalá, R., Medina, J., and Salinas, J. 2011. Integration of low temperature and light signaling during cold acclimation response in Arabidopsis. *Proc. Natl. Acad. Sci. USA.*, *108*: 16475–80.

Chater, J., Merhaut, D., Jia, Z., Mauk, P., and Preece, J. 2018. Fruit quality traits of ten California-grown pomegranate cultivars harvested over three months. *Scientia Horticulturae*, *237*(1): 11–19.

Chattopadhyay, S., Ang, L.H., Puente, P., Deng, X.W., and Wei, N. 1998. Arabidopsis *bZIP* protein HY5 directly interacts with light-responsive promoters in mediating light control of gene expression. *Plant Cell*, *10*(5): 673–83.

Chen, C., Chen, H., Zhang, Y., Thomas, H., Frank, M., He, Y., and Xia, R. 2020. TBtools: An integrative toolkit developed for interactive analyses of big biological data. *Molecular Plant*, *13*(8): 1194–1202.

Chen, S., Zhou, Y., Chen, Y., and Gu, J. 2018. fastp: An ultra-fast all-in-one FASTQ preprocessor. *Bioinformatics*, *34*(17): i884–i890.

Choi, H., Hong, J., Ha, J., Kang, J., and Kim, S. 2000. ABFs: A family of ABA responsive element binding factors. *Journal of Biolgical Chemistry*, *275*(3): 1723–30.

Clough, S., and Bent, A. 1998. Floral dip: A simplified method for Agrobacterium-mediated transformation of *Arabidopsis thaliana*. *Plant J.*, *16*(6): 735–43.

Crooks, G., Hon, G., Chandonia, J., and Brenner, S. 2004. WebLogo: A sequence logo generator. *Genome Research*, *14*(6): 88–90.

de Pascual-Teresa, S. 2014. Molecular mechanisms involved in the cardiovascular and neuroprotective effects of anthocyanins. *Archives of Biochemistry and Biophysics*, *559*: 68–74.

Dröge-Laser, W., Snoek, B., Snel, B., and Weiste, C. 2018. The Arabidopsis *bZIP* transcription factor family: An update. *Current Opinion in Plant Biology*, *45*(A): 36–49.

Ellenberger, T., Brandl, C., Struhl, K., and Harrison, S. 1992. The *GCN4* basic region leucine zipper binds DNA as a dimer of uninterrupted a. helices: Crystal structure of the protein-DNA complex. *Cell*, *71*(7):1223–37.

El-Gebali, S., Mistry, J., Bateman, A., Eddy, S., Luciani, A., Potter, S., Qureshi, M., Richardson, L., Salazar, G., and Smart, A. 2019. The Pfam protein families' database in 2019. *Nucleic Acids Res.*, *47*(D1): D427–D432.

Fan, L., Xu, L., Wang, Y., Tang, M., and Liu, L. 2019. Genome- and Transcriptome-Wide Characterization of *bZIP* Gene Family Identifies Potential Members Involved in Abiotic Stress Response and Anthocyanin Biosynthesis in Radish (*Raphanus sativus* L.). *International Journal of Molecular Sciences*, *20*(24): 6334.

Finkelstein, R., and Lynch, T. 2000. The Arabidopsis abscisic acid response gene ABI5 encodes a basic leucine zipper transcription factor. *Plant Cell*, *12*(4): 599–610.

Finn, R., Clements J., and Eddy, S. 2011. HMMER web server: Interactive sequence similarity searching. *Nucleic Acids Research*, *39*: 29–37.

Gai, W., Ma, X., Qiao, Y., Shi, B., Ul Haq, S., Li, Q., Wei, A., Liu, K., and Gong, Z. 2020. Characterization of the *bZIP* Transcription Factor Family in Pepper (*Capsicum annuum* L.): CabZIP25 Positively Modulates the Salt Tolerance. *Front Plant Science*, *11*: 139.

Gangappa, S., Botto, J. 2016. The Multifaceted Roles of HY5 in Plant Growth and Development. *Mol. Plant*, *9*(10): 1353–65.

Gibalová, A., Steinbachová, L., Hafidh, S., Bláhová, V., Gadiou, Z., Michailidis, C., Műller, K., Pleskot, R., Dupľáková, N., and Honys, D. 2017. Characterization of pollen-expressed *bZIP* protein interactions and the role of *ATbZIP18* in the male gametophyte. *Plant Reprod.*, *30*(1): 1–17.

Guan, Y., Ren, H., Xie, H., Ma, Z., and Chen, F. 2009. Identification and characterization of *bZIP*-type transcription factors involved in carrot (*Daucus carota* L.) somatic embryogenesis. *Plant Journal*, *60*(2): 207–17.

Jakoby, M., Weisshaar, B., Droge-Laser, W., Vicente-Carbajosa, J., Tiedemann, J., Kroj, T., and Parcy, F. 2002. *bZIP* transcription factors in Arabidopsis. *Trends Plant Science*, *7*(3): 106–11.

Jefferson, R.A., Kavanagh, T.A., Bevan, M.W. 1987. GUS fusions beta-glucuronidase as a sensitive and versatile gene fusion marker in higher plants. *EMBO J.*, *6*: 3901–3907.

Jungeun, L., Kun, H., Viktor, S., Horim, L., Pablo, F., Ying, G., Waraporn, T., Hongyu, Z., Ilha, L., and Xing, W.D. 2007. Analysis of Transcription Factor HY5 Genomic Binding Sites Revealed Its Hierarchical Role in Light Regulation of Development. *The Plant Cell*, *19*(3): 731–49.

Kelly, E., Vyas, P., and Weber, J.T. 2017. Biochemical properties and neuroprotective effects of compounds in various species of berries. *Molecules*, *23*(1): 26.

Kircher, S., Ledger, L., Hayashi, H., Weisshaar, B., and Frohnmeyer, H. 1998. CPRF4, a novel plant *bZIP* protein of the CPRF family : Comparative analysis of light dependent expression, post-transcriptional regulation, nuclear import, and heterodimerization. *Molecular and General Genetics*, *257*(6): 595–605.

Kristine, M., Rune, S., Unni, S., Cato, B., Trond, L., Peter, R., Michel, V., and Cathrine, L. 2009. Temperature and nitrogen effects on regulators and products of the flavonoid pathway: Experimental and kinetic model studies. *Plant, Cell, and Environment*, *3*(32): 286–99.

Kumar, S., Stecher, G., and Tamura, K. 2016. MEGA7: Molecular evolutionary genetics analysis version 7.0 for bigger datasets. *Mol. Biol. Evol.*, *33*(7): 1870–74.

Landi, M., Tattini, M., Kevin, Gould S. 2015. Multiple functional roles of anthocyanins in plant-environment interactions, *Environmental and Experimental Botany*, *119*: 4–17.

Landschulz, W., Johnson, P., and McKnight, S. 1998. The leucine zipper: A hypothetical structure common to a new class of DNA binding proteins. *Science*, *240*(4860): 1759–64.

Leon, R., Lightbourn, L., Melina, L., and Amarillas L. 2020. Differential Gene Expression of Anthocyanin Biosynthetic Genes under Low Temperature and Ultraviolet-B Radiation in Bell Pepper (*Capsicum annuum*). *International Journal of Agriculture and Biology*, *23*(3): 531–38.

Lefort, V., Longueville, J-E., and Gascuel, O. 2017. SMS: Smart model selection in PhyML. *Molecular Biology and Evolution*, *34*(9): 2422–24.

Letunic, I., and Peer, B. 2018. 20 years of the SMART protein domain annotation resource. *Nucleic Acids Research*, *46*(D1): D493–D496.

Li, X., Zhang, Y., Yuan, Y., Sun, Y., Qin, Z., and Deng, H. 2016. Protective effects of selenium, vitamin E, and purple carrot anthocyanins on D-Galactose-induced oxidative damage in blood, liver, heart, and kidney rats. *Biological Trace Element Research*, *173*(2): 433–42.

Li, D., Fu, F., Zhang, H., and Song, F. 2016. Genome-wide systematic characterization of the *bZIP* transcriptional factor family in tomato (*Solanum lycopersicum* L.). *BMC Genomic*, *16* (771): 60–78.

Li, Y., Meng, D., Li, M., and Cheng, L. 2016. Genome-wide identification and expression analysis of the *bZIP* gene family in apple (*Malus domestica*). *Tree Genetics and Genomes*, *12*(82): 1–17.

Li, H., Li, L., ShangGuan, G., Jia, C., Deng, S., Noman, M., Liu, Y., Guo, Y., Han, L., Zhang, X., Dong, Y., Ahmad, N., Du, L., Li, H., and Yang, J. 2020. Genome-wide identification and expression analysis of –gene family in *Carthamus tinctorius* L. *Scientific Reports*, *10*(1): 15521.

Liu, J., Chen, N., Chen, F., Cai, B., Dal Santo, S., Tornielli, G., Pezzotti, M., and Cheng, Z. 2014. Genome-wide analysis and expression profile of the *bZIP* transcription factor gene family in grapevine (*Vitis vinifera*). *BMC Genomics*, *15*(281): 1–18.

Liu, C., Chi, C., Jin, L., Zhu, J., Yu, J., and Zhou, Y. 2018. The *bZIP* transcription factor HY5 mediates CRY1a-induced anthocyanin biosynthesis in tomato. *Plant Cell Environ*, *41*(8): 1762–75.

Livak, K., and Schmittgen, T. 2001. Analysis of relative gene expression data using real-time quantitative PCR and the $2^{-\Delta\Delta CT}$ method. *Methods*, *25*(4): 402–408.

Lopez-Molina, L., Mongrand, S., and Chua, N. 2001. A post-germination developmental arrest checkpoint is mediated by abscisic acid and requires the *ABI5* transcription factor in Arabidopsis. *PNAS*, *98*(8): 4782–87.

Luo, X., Li, H., Wu, Z., Yao, W., Zhao, P., Cao, D., Yu, H., Li, K., Poudel, K., Zhao, D., Zhang, F., Xia, X., Chen, L., Wang, Q., Jing, D., and Cao, S. 2019. The pomegranate (*Punica granatum* L.) draft genome dissects genetic divergence between soft- and hard-seeded cultivars. *Plant Biotechnology Journal*, *18*(4): 955–68.

Luo, Q., Liu, R., Zeng, L., Wu, Y., Jiang, Y., Yang, Q., and Nie, Q. 2020. Isolation and molecular characterization of *NtMYB4a*, a putative transcription activation factor involved in anthocyanin synthesis in tobacco. *Gene*, *760*: 144990.

Ma, H., Liu, C., Li, Z., Ran, Q., Xie, G., Wang, B., Fang, S., Chu, J., and Zhang, J. 2018. *ZmbZIP4* Contributes to Stress Resistance in Maize by Regulating ABA Synthesis and Root Development. *Plant Physiol.*, *178*(2): 753–70.

Magali, L., Patrice, D., Gert, T., Kathleen, M., Yves, M., Yves, V., Pierre, R., and Stephane, R. 2020. PlantCARE: A database of plant cis-acting regulatory elements and a portal to tools for *in silico* analysis of promoter sequences. *Nucleic Acids Research*, *30*(1): 325–27.

Marchler-Bauer, A., Bo, Y., Han, L., He, J., Lanczycki, C., Lu, S., Chitsaz, F., Derbyshire, M., Geer, R., and Gonzales, N. 2017. CDD/SPARCLE: Functional classification of proteins via subfamily domain architectures. *Nucleic Acids Research*, *45*(D1): D200–D203.

Musavizadeh, Z., Najafi-Zarrini, H., Kazemitabar, S.K., Hashemi, S.H., Faraji, S., Barcaccia, G., and Heidari, P. 2021. Genome-Wide Analysis of Potassium Channel Genes in Rice: Expression of the OsAKT and OsKAT Genes under Salt Stress. *Genes*, *12*(5): 784.

Naing, A., and Kim, C. 2021. Abiotic stress-induced anthocyanins in plants: Their role in tolerance to abiotic stresses. *Physiologia Plantarum*, *6*: 1–13.

Nguyen, N.H., Jeong, C.Y., Kang, G.H., Yoo, S.D., Hong, S.W., and Lee, H. 2015. MYBD employed by HY5 increases anthocyanin accumulation via repression of *MYBL2* in Arabidopsis. *Plant Journal*, *84*(6): 1192–1205.

Nijhawan, A., Jain, M., Tyagi, A., and Khurana, J. 2008. Genomic survey and gene expression analysis of the basic leucine zipper transcription factor family in rice. *Plant Physiol.*, *146*(2): 333–50.

Perez-Rodriguez, P., Riano-Pachon, D., Correa, L., Rensing, S., Kersten, B., Mueller-Roeber, B. 2010. PlnTFDB: Updated content and new features of the plant transcription factor database. *Nucleic Acids Research*, *38*: 822–27.

Qin, G., Xu, C., Ming, R., Tang, H., Guyot, R., Kramer, E., Hu, Y., Yi, X., Qi, Y., Xu, X., Gao, Z., Pan, H., Jian, J., Tian, Y., Yue, Z., and Xu, Y. 2017. The pomegranate (*Punica granatum* L.) genome and the genomics of punicalagin biosynthesis. *Plant Journal*, *91*(6): 1108–28.

Schepens, I., Duek, P., and Fankhauser, C. 2004. Phytochrome mediated light signaling in Arabidopsis. *Curr. Opin. Plant Biol.*, *7*(5): 564–69.

Shen, H., and Chou, K. 2010. Cell-PLoc 2.0: An improved package of web-servers for predicting subcellular localization of proteins in various organisms. *Natural Science*, *2*(10): 1090–1103.

Shin, D., Choi, M., Kim, K., Bang, G., Cho, M., Choi, S., Choi, G., and Park, Y. 2013. HY5 regulates anthocyanin biosynthesis by inducing the transcriptional activation of the *MYB75/PAP1* transcription factor in Arabidopsis. *FEBS Letters*, *587*(10): 1543–47.

Shin, J., Park, E., and Choi, G. 2007. PIF3 regulates anthocyanin biosynthesis in an HY5-dependent manner with both factors directly binding anthocyanin biosynthetic gene promoters in Arabidopsis. *The Plant Journal for Cell and Molecular Biology*, *49*(6): 981–94.

Strathmann, A., Kuhlmann, M., Heinekamp, T., and Droge-Laser, W. 2001. BZI-1 specifically heterodimerises with the tobacco *bZIP* transcription factors BZI-2, BZI-3/TBZF, and BZI-4, and is functionally involved in flower development. *Plant Journal*, *28*(4): 397–408.

Stracke, R., Favory, J., Gruber, H., Bartelniewoehner, L., Bartels, S., Binkert, M., Funk, M., Weisshaar, B., and Roman, U. 2010. The Arabidopsis *bZIP* transcription factor *HY5* regulates expression of the *PFG1/MYB12* gene in response to light and ultraviolet-B radiation. *Plant Cell and Environment*, *33*(1): 88–103.

Talanian, R., Mcknight, C., and Kim, P. 1990. Sequence-specific DNA-binding by a short peptide dimer. *Science*, *249*(4970): 769–71.

Thompson, J., Gibson, T., Plewniak, F., Jeanmougin, F., and Higgins, D. 1997. The CLUSTAL_X windows interface: Flexible strategies for multiple sequence alignment aided by quality analysis tools. *Nucleic Acids Research*, *25*(24): 4876–82.

Toh, S., McCourt, P., and Tsuchiya, Y. 2012. HY5 is involved in strigolactone-dependent seed germination in Arabidopsis. *Plant Signal Behav.*, *7*(5): 556–58.

Vidmantas Bendokas, Kristina Skemiene, Sonata Trumbeckaite, Vidmantas Stanys, Sabina Passamonti, Vilmante Borutaite, and Julius Liobikas. 2020. Anthocyanins: From plant pigments to health benefits at mitochondrial level. *Critical Reviews in Food Science and Nutrition*, *60*(19): 3352–65.

Wang, Y., Zhang, X., Zhao, Y., Yang, J., He, Y., Li, G., Ma, W., Huang, X., and Su, J. 2020. Transcription factor PyHY5 binds to the promoters of *PyWD40* and *PyMYB10* and regulates its expression in red pear "Yunhongli No. 1". *Plant Physiology Biochemistry*, *154*: 665–74.

Wang, X., Chen, X., Yang, T., Cheng, Q., and Cheng, Z. 2017. Genome-Wide Identification of *bZIP* Family Genes Involved in Drought and Heat Stresses in Strawberry (*Fragaria vesca*). *International Journal of Genomics*, *2017*: 3981031.

Wei, K., Chen, J., Wang, Y., Chen, Y., Chen, S., Lin, Y., Pan, S., Zhong, X., and Xie, D. 2012. Genome-wide analysis of *bZIP*-encoding genes in maize. *DNA Research*, *19*(6): 463–76.

Weiste, C., Pedrotti, L., Selvanayagam, J., Muralidhara, P., Fröschel, C., Novák, O., Ljung, K., Hanson, J., and Dröge-Laser, W. 2017. The Arabidopsis *bZIP11* transcription factor links low-energy signaling to auxin-mediated control of primary root growth. *PLoS Genet.*, *13*(2): e1006607.

Wu, S., Zhu, P., Jia, B., Yang, J., Shen, Y., Cai, X., Sun, X., Zhu, Y., and Sun, M. 2018. A Glycine soja group S2 *bZIP* transcription factor *GsbZIP67* conferred bicarbonate alkaline tolerance in Medicago sativa. *BMC Plant Biology*, *18*(1): 234.

Xu, Y., Zhu, X., Gong, Y., Xu, L., Wang, Y., and Liu, L. 2012. Evaluation of reference genes for gene expression studies in radish (*Raphanus sativus* L.) using quantitative real-time PCR. *Biochemical and Biophysical Research Communications*, *424*(3): 398–403.

Yoshida, T., Fujita, Y., Sayama, H., Kidokoro, S., Maruyama, K., Mizoi, J., Shinozaki, K., and Yamaguchi-Shinozaki, K. 2010. *AREB1*, *AREB2*, and *ABF3* are master transcription factors that cooperatively regulate ABRE-dependent ABA signaling involved in drought stress tolerance and require ABA for full activation. *Plant Journal*, *61*(4): 672–85.

Yuan, Z., Fang, Y., Zhang, T., Fei, Z., Han, F., Liu, C., Liu, M., Xiao, W., Zhang, W., Wu, S., Zhang, M., Ju, Y., Xu, H., Dai, H., Liu, Y., Chen, Y., Wang, L., Zhou, J., Guan, D., Yan, M., Xia, Y., Huang, X., Liu, D., Wei, H., and Zheng, H. 2018. The pomegranate (*Punica granatum* L.) genome provides insights into fruit quality and ovule developmental biology. *Plant Biotechnology Journal*, *16*(7): 1363–74.

Zhang, Y., Gao, W., Li, H., Wang, Y., Li, D., Xue, C., Liu, Z., Liu, M., and Zhao, J. 2020. Genome-wide analysis of the *bZIP* gene family in Chinese jujube (*Ziziphus jujuba* Mill.). *BMC Genomics*, *21*(1): 483.

Zhang, H., He, H., Wang, X., Wang, X., Yang, X., Li, L., and Deng, X. 2011. Genome-wide mapping of the *HY5*-mediated gene networks in Arabidopsis that involve both transcriptional and post-transcriptional regulation. *Plant J.*, *65*(3): 346–58.

Zhang, H., Koes, R., Shang, H., Fu, Z., Wang, L., Dong, X., Zhang, J., Passeri, V., Li, Y., Jiang, H., Gao, J., Li, Y., and Wang, H. 2019. Identification and functional analysis of three new anthocyanin R2R3-MYB genes in Petunia. *Plant Direct*, *3*(1): e00114.

Zhao, K., Chen, S., Yao, W., Cheng, Z., Zhou, B., and Jiang, T. 2021. Genome-wide analysis and expression profile of the *bZIP* gene family in poplar. *BMC Plant Biology*, *21*: Article No.122.

Index